T0296522

LONDON MATHEMATICAL SOCIETY LECTURE NOTE SERIES

Managing Editor: Professor N.J. Hitchin, Mathematical Institute,
University of Oxford, 24–29 St Giles, Oxford OX1 3LB, United Kingdom

The titles below are available from booksellers, or from Cambridge University Press at www.cambridge.org

46   *p*-adic Analysis: a short course on recent work, N. KOBLITZ
59   Applicable differential geometry, M. CRAMPIN & F.A.E. PIRANI
86   Topological topics, I.M. JAMES (ed)
88   FPF ring theory, C. FAITH & S. PAGE
90   Polytopes and symmetry, S.A. ROBERTSON
96   Diophantine equations over function fields, R.C. MASON
97   Varieties of constructive mathematics, D.S. BRIDGES & F. RICHMAN
99   Methods of differential geometry in algebraic topology, M. KAROUBI & C. LERUSTE
100 Stopping time techniques for analysts and probabilists, L. EGGHE
105 A local spectral theory for closed operators, I. ERDELYI & WANG SHENGWANG
107 Compactification of Siegel moduli schemes, C.-L. CHAI
109 Diophantine analysis, J. LOXTON & A. VAN DER POORTEN (eds)
113 Lectures on the asymptotic theory of ideals, D. REES
116 Representations of algebras, P.J. WEBB (ed)
119 Triangulated categories in the representation theory of finite-dimensional algebras, D. HAPPEL
121 Proceedings of *Groups - St Andrews 1985*, E. ROBERTSON & C. CAMPBELL (eds)
128 Descriptive set theory and the structure of sets of uniqueness, A.S. KECHRIS & A. LOUVEAU
130 Model theory and modules, M. PREST
131 Algebraic, extremal & metric combinatorics, M.-M. DEZA, P. FRANKL & I.G. ROSENBERG (eds)
140 Geometric aspects of Banach spaces, E.M. PEINADOR & A. RODES (eds)
141 Surveys in combinatorics 1989, J. SIEMONS (ed)
144 Introduction to uniform spaces, I.M. JAMES
146 Cohen-Macaulay modules over Cohen-Macaulay rings, Y. YOSHINO
148 Helices and vector bundles, A.N. RUDAKOV *et al*
149 Solitons, nonlinear evolution equations and inverse scattering, M. ABLOWITZ & P. CLARKSON
150 Geometry of low-dimensional manifolds 1, S. DONALDSON & C.B. THOMAS (eds)
151 Geometry of low-dimensional manifolds 2, S. DONALDSON & C.B. THOMAS (eds)
152 Oligomorphic permutation groups, P. CAMERON
153 L-functions and arithmetic, J. COATES & M.J. TAYLOR (eds)
155 Classification theories of polarized varieties, TAKAO FUJITA
158 Geometry of Banach spaces, P.F.X. MÜLLER & W. SCHACHERMAYER (eds)
159 Groups St Andrews 1989 volume 1, C.M. CAMPBELL & E.F. ROBERTSON (eds)
160 Groups St Andrews 1989 volume 2, C.M. CAMPBELL & E.F. ROBERTSON (eds)
161 Lectures on block theory, BURKHARD KÜLSHAMMER
163 Topics in varieties of group representations, S.M. VOVSI
164 Quasi-symmetric designs, M.S. SHRIKANDE & S.S. SANE
166 Surveys in combinatorics, 1991, A.D. KEEDWELL (ed)
168 Representations of algebras, H. TACHIKAWA & S. BRENNER (eds)
169 Boolean function complexity, M.S. PATERSON (ed)
170 Manifolds with singularities and the Adams-Novikov spectral sequence, B. BOTVINNIK
171 Squares, A.R. RAJWADE
172 Algebraic varieties, GEORGE R. KEMPF
173 Discrete groups and geometry, W.J. HARVEY & C. MACLACHLAN (eds)
174 Lectures on mechanics, J.E. MARSDEN
175 Adams memorial symposium on algebraic topology 1, N. RAY & G. WALKER (eds)
176 Adams memorial symposium on algebraic topology 2, N. RAY & G. WALKER (eds)
177 Applications of categories in computer science, M. FOURMAN, P. JOHNSTONE & A. PITTS (eds)
178 Lower K- and L-theory, A. RANICKI
179 Complex projective geometry, G. ELLINGSRUD *et al*
180 Lectures on ergodic theory and Pesin theory on compact manifolds, M. POLLICOTT
181 Geometric group theory I, G.A. NIBLO & M.A. ROLLER (eds)
182 Geometric group theory II, G.A. NIBLO & M.A. ROLLER (eds)
183 Shintani zeta functions, A. YUKIE
184 Arithmetical functions, W. SCHWARZ & J. SPILKER
185 Representations of solvable groups, O. MANZ & T.R. WOLF
186 Complexity: knots, colourings and counting, D.J.A. WELSH
187 Surveys in combinatorics, 1993, K. WALKER (ed)
188 Local analysis for the odd order theorem, H. BENDER & G. GLAUBERMAN
189 Locally presentable and accessible categories, J. ADAMEK & J. ROSICKY
190 Polynomial invariants of finite groups, D.J. BENSON
191 Finite geometry and combinatorics, F. DE CLERCK *et al*
192 Symplectic geometry, D. SALAMON (ed)
194 Independent random variables and rearrangement invariant spaces, M. BRAVERMAN
195 Arithmetic of blowup algebras, WOLMER VASCONCELOS
196 Microlocal analysis for differential operators, A. GRIGIS & J. SJÖSTRAND
197 Two-dimensional homotopy and combinatorial group theory, C. HOG-ANGELONI *et al*
198 The algebraic characterization of geometric 4-manifolds, J.A. HILLMAN
199 Invariant potential theory in the unit ball of $\mathbf{C}^n$, MANFRED STOLL
200 The Grothendieck theory of dessins d'enfant, L. SCHNEPS (ed)
201 Singularities, JEAN-PAUL BRASSELET (ed)
202 The technique of pseudodifferential operators, H.O. CORDES
203 Hochschild cohomology of von Neumann algebras, A. SINCLAIR & R. SMITH
204 Combinatorial and geometric group theory, A.J. DUNCAN, N.D. GILBERT & J. HOWIE (eds)
205 Ergodic theory and its connections with harmonic analysis, K. PETERSEN & I. SALAMA (eds)
207 Groups of Lie type and their geometries, W.M. KANTOR & L. DI MARTINO (eds)

208   Vector bundles in algebraic geometry, N.J. HITCHIN, P. NEWSTEAD & W.M. OXBURY (eds)
209   Arithmetic of diagonal hypersurfaces over finite fields, F.Q. GOUVÊA & N. YUI
210   Hilbert C*-modules, E.C. LANCE
211   Groups 93 Galway / St Andrews I, C.M. CAMPBELL et al (eds)
212   Groups 93 Galway / St Andrews II, C.M. CAMPBELL et al (eds)
214   Generalised Euler-Jacobi inversion formula and asymptotics beyond all orders, V. KOWALENKO et al
215   Number theory 1992–93, S. DAVID (ed)
216   Stochastic partial differential equations, A. ETHERIDGE (ed)
217   Quadratic forms with applications to algebraic geometry and topology, A. PFISTER
218   Surveys in combinatorics, 1995, PETER ROWLINSON (ed)
220   Algebraic set theory, A. JOYAL & I. MOERDIJK
221   Harmonic approximation, S.J. GARDINER
222   Advances in linear logic, J.-Y. GIRARD, Y. LAFONT & L. REGNIER (eds)
223   Analytic semigroups and semilinear initial boundary value problems, KAZUAKI TAIRA
224   Computability, enumerability, unsolvability, S.B. COOPER, T.A. SLAMAN & S.S. WAINER (eds)
225   A mathematical introduction to string theory, S. ALBEVERIO et al
226   Novikov conjectures, index theorems and rigidity I, S. FERRY, A. RANICKI & J. ROSENBERG (eds)
227   Novikov conjectures, index theorems and rigidity II, S. FERRY, A. RANICKI & J. ROSENBERG (eds)
228   Ergodic theory of Z^d actions, M. POLLICOTT & K. SCHMIDT (eds)
229   Ergodicity for infinite dimensional systems, G. DA PRATO & J. ZABCZYK
230   Prolegomena to a middlebrow arithmetic of curves of genus 2, J.W.S. CASSELS & E.V. FLYNN
231   Semigroup theory and its applications, K.H. HOFMANN & M.W. MISLOVE (eds)
232   The descriptive set theory of Polish group actions, H. BECKER & A.S. KECHRIS
233   Finite fields and applications, S. COHEN & H. NIEDERREITER (eds)
234   Introduction to subfactors, V. JONES & V.S. SUNDER
235   Number theory 1993–94, S. DAVID (ed)
236   The James forest, H. FETTER & B. GAMBOA DE BUEN
237   Sieve methods, exponential sums, and their applications in number theory, G.R.H. GREAVES et al
238   Representation theory and algebraic geometry, A. MARTSINKOVSKY & G. TODOROV (eds)
240   Stable groups, FRANK O. WAGNER
241   Surveys in combinatorics, 1997, R.A. BAILEY (ed)
242   Geometric Galois actions I, L. SCHNEPS & P. LOCHAK (eds)
243   Geometric Galois actions II, L. SCHNEPS & P. LOCHAK (eds)
244   Model theory of groups and automorphism groups, D. EVANS (ed)
245   Geometry, combinatorial designs and related structures, J.W.P. HIRSCHFELD et al
246   p-Automorphisms of finite p-groups, E.I. KHUKHRO
247   Analytic number theory, Y. MOTOHASHI (ed)
248   Tame topology and o-minimal structures, LOU VAN DEN DRIES
249   The atlas of finite groups: ten years on, ROBERT CURTIS & ROBERT WILSON (eds)
250   Characters and blocks of finite groups, G. NAVARRO
251   Gröbner bases and applications, B. BUCHBERGER & F. WINKLER (eds)
252   Geometry and cohomology in group theory, P. KROPHOLLER, G. NIBLO, R. STÖHR (eds)
253   The q-Schur algebra, S. DONKIN
254   Galois representations in arithmetic algebraic geometry, A.J. SCHOLL & R.L. TAYLOR (eds)
255   Symmetries and integrability of difference equations, P.A. CLARKSON & F.W. NIJHOFF (eds)
256   Aspects of Galois theory, HELMUT VÖLKLEIN et al
257   An introduction to noncommutative differential geometry and its physical applications 2ed, J. MADORE
258   Sets and proofs, S.B. COOPER & J. TRUSS (eds)
259   Models and computability, S.B. COOPER & J. TRUSS (eds)
260   Groups St Andrews 1997 in Bath, I, C.M. CAMPBELL et al
261   Groups St Andrews 1997 in Bath, II, C.M. CAMPBELL et al
262   Analysis and logic, C.W. HENSON, J. IOVINO, A.S. KECHRIS & E. ODELL
263   Singularity theory, BILL BRUCE & DAVID MOND
264   New trends in algebraic geometry, K. HULEK, F. CATANESE, C. PETERS & M. REID (eds)
265   Elliptic curves in cryptography, I. BLAKE, G. SEROUSSI & N. SMART
267   Surveys in combinatorics, 1999, J.D. LAMB & D.A. PREECE (eds)
268   Spectral asymptotics in the semi-classical limit, M. DIMASSI & J. SJÖSTRAND
269   Ergodic theory and topological dynamics, M.B. BEKKA & M. MAYER
270   Analysis on Lie groups, N.T. VAROPOULOS & S. MUSTAPHA
271   Singular perturbations of differential operators, S. ALBEVERIO & P. KURASOV
272   Character theory for the odd order theorem, T. PETERFALVI
273   Spectral theory and geometry, E.B. DAVIES & Y. SAFAROV (eds)
274   The Mandlebrot set, theme and variations, TAN LEI (ed)
275   Descriptive set theory and dynamical systems, M. FOREMAN et al
276   Singularities of plane curves, E. CASAS-ALVERO
277   Computational and geometric aspects of modern algebra, M. D. ATKINSON et al
278   Global attractors in abstract parabolic problems, J.W. CHOLEWA & T. DLOTKO
279   Topics in symbolic dynamics and applications, F. BLANCHARD, A. MAASS & A. NOGUEIRA (eds)
280   Characters and automorphism groups of compact Riemann surfaces, THOMAS BREUER
281   Explicit birational geometry of 3-folds, ALESSIO CORTI & MILES REID (eds)
282   Auslander-Buchweitz approximations of equivariant modules, M. HASHIMOTO
283   Nonlinear elasticity, Y. FU & R. W. OGDEN (eds)
284   Foundations of computational mathematics, R. DEVORE, A. ISERLES & E. SÜLI (eds)
285   Rational points on curves over finite fields, H. NIEDERREITER & C. XING
286   Clifford algebras and spinors 2ed, P. LOUNESTO
287   Topics on Riemann surfaces and Fuchsian groups, E. BUJALANCE, A.F. COSTA & E. MARTÌNEZ (eds)
288   Surveys in combinatorics, 2001, J. HIRSCHFELD (ed)
289   Aspects of Sobolev-type inequalities, L. SALOFF-COSTE
290   Tits buildings and the model theory of groups, K. TENT (ed)
291   A quantum groups primer, S. MAJID

London Mathematical Society Lecture Note Series. 262

Analysis and Logic

Edited by

Catherine Finet & Christian Michaux
University of Mons-Hainaut

Authors:

C. Ward Henson
University of Illinois at Urbana-Champaign

José Iovino
The University of Texas at San Antonio

Alexander S. Kechris
California Institute of Technology

Edward Odell
The University of Texas at Austin

 CAMBRIDGE
UNIVERSITY PRESS

CAMBRIDGE
UNIVERSITY PRESS

32 Avenue of the Americas, New York NY 10013-2473, USA

Cambridge University Press is part of the University of Cambridge.

It furthers the University's mission by disseminating knowledge in the pursuit of education, learning and research at the highest international levels of excellence.

www.cambridge.org
Information on this title: www.cambridge.org/9780521648615

© Cambridge University Press 2002

First published 2002

A catalogue record for this publication is available from the British Library

ISBN 978-0-521-64861-5 Paperback

Dedicated to Maurice Boffa (1939 – 2001)

Contents

Preface   *page* ix
Introduction   xi

Ultraproducts in Analysis
by C. Ward Henson and José Iovino   1
1   Introduction   3
2   Normed Space Structures   10
3   Signatures   13
4   Ultrapowers of Normed Space Structures   17
5   Positive Bounded Formulas   21
6   Basic Model Theory I   31
7   Quantifier-Free Formulas   34
8   Ultraproducts of Normed Space Structures   38
9   Basic Model Theory II   43
10 Isomorphic Ultrapowers   55
11 Alternative Formulations of the Theory   72
12 Homogeneous Structures   75
13 More Model Theory   79
14 Types   96
References   107
Index of Notation   111
Index   112

Actions of Polish Groups and Classification Problems
by Alexander S. Kechris   115
1   Introduction   117
2   The General Glimm-Effros Dichotomy   120
3   Actions of Polish Groups   123

4   Actions of Countable Groups   126
5   Actions of Locally Compact Groups   131
6   Actions of the Infinite Symmetric Group   133
7   Turbulence I: Overview   137
8   Turbulence II: Basic Facts   143
9   Turbulence III: Induced Actions   146
10 Turbulence IV: Some Examples   150
11 Turbulence V: Calmness   152
12 Turbulence VI: The First Main Theorem   157
13 Turbulence VII: The Second Main Theorem   163
References   183
Index   186

**On Subspaces, Asymptotic Structure, and Distortion
of Banach Spaces; Connections with Logic
by Edward Odell**

                                                               189
1   Introduction   191
2   Background material: The 60's and 70's   194
3   The unconditional basic sequence problem
    and connections with distortion   208
4   Gowers' dichotomy: A block Ramsey Theorem   217
5   Distortion   223
6   Asymptotic structure   233
7   Ordinal Indices   244
8   The homogeneous Banach space problem   252
9   Concluding remarks   254
References   256
Index   265

Preface

In 1997 the Analysis and the Mathematical Logic teams of the University of Mons-Hainaut, and the Analysis team of the University of Paris 6 organized an international conference entitled "Analyse & Logique". It took place at the University of Mons-Hainaut, Mons, Belgium from 25 to 29 August 1997. The scientific committee consisted of Maurice Boffa (Mons), Gilles Godefroy (Paris 6; Columbia, Missouri), Tim Gowers (Cambridge), Boris S. Kashin (Moscow), Angus Macintyre (Oxford), Olek Pelczynski (Warsaw), Françoise Point (Mons), Alexander A. Razborov (Moscow), Stanimir Troyanski (Sofia), and Lior Tzafriri (Jerusalem). Members of the organizing committee were Robert Deville (Bordeaux), Catherine Finet (Mons), Jean-Pierre Gossez (Bruxelles), C. Ward Henson (Urbana), Chris Impens (Gent), John Jayne (London), Alain Louveau (Paris 6), Christian Michaux (Mons), André Pétry (Liège), Gilles Pisier (Paris 6; Texas A & M), Jean Schmets (Liège), and Jan Van Casteren (Antwerpen).

This conference was the third of a cycle of conferences initiated by Catherine Finet at Mons; the previous ones were held in 1987 and 1992.

The meeting was a true success; more than one hundred and twenty mathematicians from all over the world participated. The main topics discussed at the meeting were the numerous connections between Analysis and Logic.

The purpose of this volume is to report the content of the three mini-courses given by C. Ward Henson (Urbana), Alexander S. Kechris (Caltech) and Edward Odell (Austin), respectively.

Plenary lectures were presented by Nigel J. Cutland (Hull), Gabriel Debs (Le Havre, Paris 6), Lou van den Dries (Urbana), Richard Haydon (Oxford), Joram Lindenstrauss (Jerusalem), Donald A. Martin (Los Angeles), Amos Nevo (Haïfa), David Preiss (London) and Nicole Tomczak-Jaegermann (Edmonton).

There were twenty-five contributed papers presented in parallel sessions.

A special issue of the journal *Annals of Pure and Applied Logic* was devoted to the meeting. It appeared in July 2001 as Volume 111/1–2 and contains papers by some of the plenary speakers and contributors.

The conference was supported financially by the Association Stefan Banach, the Centre National de la Recherche en Belgique, the Equipe d'Analyse de l'Université de Paris 6, the Fonds National de la Recherche Scientifique, the Fonds voor Wetenschappelijk Onderzoek, the Ministère de l'Education de la Communauté Française de Belgique, the International Association for the Promotion of Cooperation with Scientists from the Independent States of the Former Soviet Union (INTAS), The United Nations Educational, Scientific and Cultural Organization (UNESCO), the University of Mons-Hainaut, and the town of Mons. We add special mention of the Belgian National Bank, which supported the project Analysis and Logic for several years through its program for fundamental research.

Last but not least, our thanks to the departmental Secretaries Lyane Bouchez and Anne-Marie Saucez. They provided calm reason in moments of panic and kept the conference on an even keel throughout. Thanks also to all of those who helped us, especially to the juniors of the department, to Jacques Lion, and to Michèle Boffa.

The Editors
Catherine Finet, *Mons*
Christian Michaux, *Mons*

December 2001

Introduction

The articles in this book had their origins in three mini-courses offered at the conference "Analyse & Logique" held August 25–29, 1997, at the University of Mons-Hainaut in Mons, Belgium. For a long time there have been rich connections between analysis and logic; these articles bear witness that this relationship is still very active, and continues to be important for both areas.

Here we briefly describe these three articles; each one has a more detailed Introduction as its first chapter.

Part One:
Ultraproducts in Analysis
by C. Ward Henson and José Iovino

Applications of model theory in functional analysis have been pursued since the mid 1960s, beginning with the introduction of Banach space ultraproducts by Bretagnolle, Dacunha-Castelle, and Krivine, and of nonstandard hulls by Luxemburg. These constructions have been widely and successfully used in many parts of analysis. This paper presents the basic aspects of a systematic model theoretic framework within which these tools are naturally situated.

The logic developed here has its origin in the question: "what does a normed space structure have in common with its Banach space ultrapowers?" To give a precise answer it is necessary to introduce a suitable formal language of *positive bounded formulas* together with a semantics of *approximate satisfaction*. The resulting theory is developed here for a very general class of structures based on normed spaces; our treatment has the most general context possible within functional analysis, in which the Banach space ultraproduct and nonstandard hull constructions apply.

The model theory for normed space structures that is described here has turned out to have many unexpected features and to be a close analogue of ordinary first order model theory. A large part of the basic aspects of this theory is outlined here. This model theory has been developed since the mid 1970s and has been used from time to time in applications of nonstandard analysis (especially of the nonstandard hull construction). By now it has shown itself to be the "correct" analogue of first order logic for the normed structures that arise in functional analysis.

In this article are discussed such basic model theoretic topics as elementary equivalence and elementary extension, logical compactness, Löwenheim-Skolem theorems, unions of chains, saturated and strongly homogeneous models, axiomatizability of classes of structures using positive bounded sentences, quantifier elimination, and spaces of model-theoretic types.

Many results in the literature regarding the structure of Banach space ultraproducts resonate immediately with the model theoretic ideas presented here. Some suggestive examples of this are given in various remarks throughout the paper.

Part Two:
Actions of Polish Groups and
Classification Problems
by Alexander S. Kechris

This survey concerns the theory of definable actions of Polish groups, the structure and classification of their orbit spaces, and the closely related study of definable equivalence relations. This work is motivated by basic foundational questions, like understanding the nature of complete classification of mathematical objects by invariants, up to some notion of equivalence, and creating a mathematical framework for measuring the complexity of such classification problems. This theory, which has been growing rapidly over the last few years, is developed within the context of descriptive set theory, which provides the basic underlying concepts and methods. There are natural interactions of it with other areas of mathematics, such as the theory of topological groups, topological dynamics, ergodic theory and its relationships with the theory of operator algebras, model theory, and recursion theory.

Classically, in various branches of dynamics one studies actions of special groups, such as the additive groups of integers or reals, Lie groups, or, more generally, (second countable) locally compact groups. One of the goals of the theory is to expand the scope of this subject by considering the more comprehensive class of *Polish groups* (separable completely metrizable topological groups). One of the main problems concerning a given definable action of a Polish group G on a Polish space X is to give a complete classification by invariants of members of X up to orbit equivalence. (*Orbit equivalence* being the equivalence relation induced by the orbits of the action.) This is a special case of the more general problem of completely classifying elements of a given Polish space X up to some definable equivalence relation E on that space. This means finding a set of invariants I and a map $c\colon X \to I$ such that $xEy \Leftrightarrow c(x) = c(y)$; for this to have any significance, both I, c must be "explicit" or "definable" too. Typical examples of this kind of problem are: the classification of countable models of a theory up to isomorphism, the classification of the irreducible unitary representations of a locally compact group up to unitary equivalence, the classification of measure preserving transformations up to conjugacy, etc.

Chapters 2–7 of this paper give a survey of certain aspects of the program discussed in this introduction, but with very few technical details. Chapters 8–13 give a somewhat detailed technical exposition of Hjorth's recent theory of turbulence.

Part Three:
On Subspaces, Asymptotic Structure, and Distortion of Banach Spaces; Connections with Logic
by Edward Odell

A number of key problems in the theory of infinite dimensional Banach spaces, those that were central to the study of the general structure of a Banach space, finally yielded their secrets in the 1990's. This survey gives an extensive discussion of these problems and their solutions, as well as a presentation of important problems that remain open. At the foundation of this work are certain deep connections between Banach space theory and logic (especially set theory).

One of these connections involves Ramsey theory. This played a direct role in such results as Rosenthal's ℓ_1-theorem. More recently, *approximate* Ramsey theorems have emerged, especially as developed by Gowers leading to the proof of his spectacular Dichotomy Theorem. Generally speaking, Ramsey theorems assert that a function into a finite set can be restricted to some sort of substructure on which it is constant. In an approximate Ramsey theorem one seeks a substructure where the function is nearly constant in some appropriate sense; this represents a joining of analysis and Ramsey theory.

Among the topics discussed in this survey are: the unconditional basic sequence problem and its connections with distortion; Gowers' block Ramsey theorem for Banach spaces; asymptotic structure, involving a type of logical connection between the finite dimensional (or local) structure of a space X and the infinite dimensional structure of X; certain ordinal indices constructed (e.g., by Szlenk and Bourgain) to study the structure of infinite dimensional Banach spaces; and the homogeneous Banach space problem. (i.e., if X is isomorphic to all of its infinite dimensional subspaces, must X be isomorphic to Hilbert space?)

The authors of this book are grateful to Catherine Finet and Christian Michaux and their colleagues in Mons, for having organized such an interesting meeting at the interface between analysis and logic, one of a series of such meetings. The authors hope that this book will give continuing recognition of the success of these meetings, and especially of their unusual, interdisciplinary character.

The Authors

C. Ward Henson, *Urbana*

José Iovino, *San Antonio*

Alexander S. Kechris, *Pasadena*

Edward Odell, *Austin*

December 2001

Ultraproducts in Analysis

C. Ward Henson and José Iovino

Department of Mathematics
University of Illinois
Urbana, Illinois 61801 USA

henson@math.uiuc.edu
http://www.math.uiuc.edu/~henson/

Department of Mathematics
University of Texas at San Antonio
San Antonio, Texas 78249 USA

iovino@math.utsa.edu
http://www.math.utsa.edu/~iovino/

1
Introduction

The ideas and methods of model theory are being applied today in nearly all parts of mathematics. Here we concentrate on a framework for applications in functional analysis. Model theory has already provided several tools for the research analyst, of which the most important are: (a) the Banach space *ultraproduct* and *nonstandard hull* constructions in functional analysis; (b) spaces of (model-theoretic) *types* as used in the geometry of Banach spaces; and (c) *Loeb measure spaces* in stochastic analysis and its applications. In this paper we explain a systematic model theoretic framework within which these tools (especially (a) and (b)) are naturally situated. Our main intended audience consists of analysts who are familiar with the ultraproduct construction, and this perspective has strongly influenced our presentation of the material. We expect that many model theorists will also find something of interest in this subject. In particular, we indicate the initial steps of a program for introducing the key ideas and methods of model theory into functional analysis in a systematic and comprehensive way.

Applications of model theory in functional analysis have been pursued since the mid 1960s, beginning with the introduction of Banach space ultraproducts by Bretagnolle, Dacunha-Castelle, and Krivine [BDCK66] [Kri67] [DCK72] and nonstandard hulls by Luxemburg [Lux69b]. These two constructions were first used at about the same time and they are essentially the same; however, these initial steps led to largely independent lines of research that have still not been fully integrated.

In nonstandard analysis, two important publications in the late 1960s were Abraham Robinson's book [Rob66] and the conference proceedings [Lux69a]. These books showed a wide range of areas within analysis in

Acknowledgment. Research for and preparation of this paper were partially supported by NSF Grants DMS 96-26628 and DMS 99-70009.

which one could expect fruitful applications of model theory, and they are still of value.

Robinson's initial goal was to provide a rigorous foundation for using infinitesimals in analysis. Banach space ultraproducts were introduced to investigate $L_p(\mu)$-spaces and related spaces. These methods have evolved considerably and they are now being used across the entire spectrum of mathematical analysis.

An up to date presentation of many applications to which Robinson's ideas have led, many of which are in functional analysis, can be found in [ACH97]. That book is intended to be accessible to all mathematicians, and includes introductory material on nonstandard analysis. Surveys of how ultraproducts and nonstandard analysis have been applied in functional analysis include [Hei80], [Sim82], and [HM83b].

There is a definite thought process that leads from (Banach space) ultraproducts to the model theory that is presented here, as we now indicate. Noting that ultraproducts are widely and effectively used in functional analysis, it becomes natural to analyze, as precisely as possible, what a normed space structure has in common with its ultrapowers. This problem can be formulated in an abstract way as is done in the next paragraphs. (For convenience we use some of the terminology that is introduced later in the paper.)

Given two normed space structures $\mathcal{M}$ and $\mathcal{N}$ with the same signature, let us consider them to be *ultrapower equivalent* if there exist ultrafilters D and E for which the ultrapowers $(\mathcal{M})_D$ and $(\mathcal{N})_E$ are isomorphic. (Since norms are included as distinguished functions in the structures we consider, every isomorphism of normed space structures is necessarily isometric.) It is easy to see that $\mathcal{M}$ is ultrapower equivalent to any of its ultrapowers.

It is clear that the relation of ultrapower equivalence is reflexive and symmetric; eventually we prove that it is transitive, but this is far from clear at the beginning. Therefore we introduce the transitive closure: define $\mathcal{M} \sim \mathcal{N}$ to mean that there is a finite sequence of normed space structures $\mathcal{M}_1, \ldots, \mathcal{M}_n$ such that $\mathcal{M}_1 = \mathcal{M}$, $\mathcal{M}_n = \mathcal{N}$, and for each $k = 1, \ldots, n-1$ we have that $\mathcal{M}_k$ and $\mathcal{M}_{k+1}$ are ultrapower equivalent. In other words, $\sim$ is the coarsest equivalence relation on normed space structures under which (a) $\mathcal{M} \sim (\mathcal{M})_D$ for every structure $\mathcal{M}$ and every ultrafilter D, and (b) $\mathcal{M} \sim \mathcal{N}$ whenever $\mathcal{M}$ and $\mathcal{N}$ are isomorphic.

Understanding what it is that normed space structures have in common with their ultrapowers is the same as understanding the equivalence relation $\sim$. Our analysis shows that $\sim$ is identical to a relation of *elementary equivalence* with respect to a certain logic for normed space structures. The formal sentences of this logic are the *positive bounded sentences* of a certain first order language. Elementary equivalence is taken with respect to a semantics of *approximate satisfaction*. (All of this is explained in Chapter 5.) Given two normed space structures $\mathcal{M}$ and $\mathcal{N}$ with the same signature, we write $\mathcal{M} \equiv_A \mathcal{N}$ if, for every positive bounded sentence σ in their common signature, σ is approximately satisfied in $\mathcal{M}$ if and only if σ is approximately satisfied in $\mathcal{N}$. When this condition holds we say that $\mathcal{M}$ and $\mathcal{N}$ are *approximately elementarily equivalent*.

Our solution to the problem of understanding the equivalence relation $\sim$ is to show that it is identical to $\equiv_A$. Indeed, it turns out that these relations are the same as the relation of ultrapower equivalence itself. The easy direction is Corollary 9.4: if $\mathcal{M}$ and $\mathcal{N}$ have isomorphic ultrapowers, then $\mathcal{M} \equiv_A \mathcal{N}$. The converse, which has a difficult proof using combinatorial ideas going back to a related argument of Shelah [She71], is Theorem 10.7.

The model theory for normed space structures that is described here has turned out to have many unexpected features and to be a close analogue of ordinary first order model theory. A large part of the basic aspects of this theory is outlined here. This model theory has been developed since the mid 1970s (see especially [Hen75] and [Hen76]) and has been used from time to time in applications of nonstandard analysis (especially of the nonstandard hull construction) in functional analysis. By now it has shown itself to be the "correct" analogue of first order logic for the normed structures that arise in functional analysis. (In addition to the results presented here, see the Lindström-type maximality theorem for this logic proved by the second author [Iov01].)

Perhaps the most novel aspect of this theory is the relation $\models_A$ of approximate satisfaction. (See Definition 5.9.) It is natural to ask how necessary or intrinsic to this setting the relation $\models_A$ may be. One indication that $\models_A$ is an essential ingredient of any model theory that is adequate for handling ultrapowers of normed space structures is given by Proposition 9.26. Suppose $\mathcal{M}$ is a normed space structure, σ is a positive bounded sentence in the signature of $\mathcal{M}$, and D is any countably incomplete ultrafilter. Then σ is approximately satisfied in $\mathcal{M}$ if and only if σ is satisfied (in the usual semantics of ordinary model theory) in the ultrapower $(\mathcal{M})_D$. (Corollary 9.3 shows that this is also equivalent to the statement that σ is *approximately* satisfied in $(\mathcal{M})_D$.) Note that this yields an alternative "soft" way to define approximate satisfaction for positive bounded sentences, in which approximations do not appear at all. The same equivalences are true for positive bounded formulas applied to elements of $\mathcal{M}$.

In order to cover as many aspects of this model theory as possible in a limited space, we have omitted many proofs. Usually the proofs being left out are reasonably routine, at least for mathematicians who have had some exposure to the basic parts of formal logic (first order logic). Often these proofs have already appeared somewhere in the literature, if usually in a somewhat more restricted framework than the one found here, and we give pointers to that literature.

We conclude this introduction with a summary of each chapter.

2. Normed Space Structures

The general class of structures to which our model theory applies is introduced. These are many-sorted structures, in which each sort is a normed space. The rest of the structure consists of a family of functions between various sorts; these functions must be uniformly continuous on each bounded subset of their domains. Most of the standard normed structures of functional analysis are of this type, including normed spaces, normed vector lattices, normed algebras, and the like.

3. Signatures

In order to treat normed space structures in a general setting, it is convenient to introduce *signatures* for them; these provide a basis for indexing the sorts and functions that make up each structure. These are also helpful in giving exact definitions of such concepts as *embedding* and *isomorphism* for normed space structures.

4. Ultrapowers of Normed Space Structures

The Banach space ultrapower construction is well known and widely used for normed spaces, and it extends in a completely natural way to the normed space structures that are considered here. Indeed, the requirement that all functions in a normed space structure are uniformly continuous (and hence bounded) on bounded sets is exactly what is needed to permit the ultrapower construction to be applied to such a structure.

5. Positive Bounded Formulas

Here we introduce the formal language of *positive bounded formulas* in the signature of a given normed space structure, as well as the semantics of *approximate satisfaction*. These are the fundamental concepts on which the rest of our model theory is built.

A key aspect of approximate satisfaction is its well-behaved topological nature. This is illustrated by the Perturbation Lemma (Proposition 5.15) which expresses a sense in which satisfaction of positive bounded formulas is itself uniformly continuous on bounded sets. There are many important consequences of this, including the fact that the completion of any normed space structure is an approximate elementary extension of it.

6. Basic Model Theory I

The key model theoretic concepts of *approximate elementary equivalence* and *approximate elementary extension* are defined and their most basic properties are proved.

7. Quantifier-Free Formulas

The expressive power of quantifier-free positive bounded formulas in the pure language of normed spaces (with no additional structure) is explored.

8. Ultraproducts of Normed Space Structures

The ultraproduct of a family of normed space structures $(\mathcal{M}_\xi \mid \xi \in \Lambda)$ with respect to any ultrafilter D on Λ is defined. Note that this cannot be defined in complete generality; even in the simple case where each $\mathcal{M}_\xi$ consists of a normed space X (not depending on the index ξ) together with a distinguished element x_ξ of X, the ultraproduct can be naturally defined for every D if and only if there is a real number r such that the bound $\|x_\xi\| \leq r$ holds for all $\xi \in \Lambda$. Similar considerations apply to boundedness and uniform continuity of the distinguished functions in the structures $\mathcal{M}_\xi$. We define what it means for a family $(\mathcal{M}_\xi \mid \xi \in \Lambda)$ to be *uniform* and note that this is precisely what is needed to be able to define the ultraproduct of the family. The most basic classes of normed space structures have the property that they are *uniform*, in the sense that any family taken from the class is uniform. This is true of normed spaces, normed lattices, and normed algebras, as is familiar. For expansions of such structures the requirement of uniformity is generally natural and easy to verify.

9. Basic Model Theory II

The fundamental theorem of ultraproducts is proved for the ultraproduct construction and the model theory that are being considered here. This is the analogue of the well known Theorem of Łoś from ordinary model theory. In this context it says the following: suppose $(\mathcal{M}_\xi \mid \xi \in \Lambda)$ is a uniform family of normed space structures of the same signature, and let $\mathcal{N}$ be the Banach space ultraproduct of this family with respect to an ultrafilter D on Λ. Let σ be any positive bounded sentence in this signature. Then σ is approximately satisfied by $\mathcal{N}$ if and only if for each approximation σ' of σ, the set $\{\xi \in \Lambda \mid \sigma'$ is satisfied by $\mathcal{M}_\xi\}$ is an element of the ultrafilter D. The same equivalences are true for positive bounded formulas applied to elements of $\mathcal{N}$.

An immediate consequence of this theorem about ultraproducts is the Compactness Theorem for the model theory presented in this paper. Löwenheim-Skolem theorems and saturated structures are also considered in this chapter.

10. Isomorphic Ultrapowers

This chapter treats the analogue for this model theory of the Keisler-Shelah Theorem [Kei61] [She71] on isomorphism of ultrapowers and ultraproducts. Among other things, this yields the existence of ultrapowers of normed space structures that are κ-saturated, for any cardinal number κ.

11. Alternative Formulations of the Theory

The isomorphism results in the previous chapter are used to show that the expressive power of this model theory is not affected by various natural changes in how the sort of the real numbers is handled or in how the syntax of positive bounded formulas is defined.

12. Homogeneous Structures

The existence of ultrapowers that are κ-saturated and strongly κ-homogeneous is proved, using the theorems about isomorphic ultraproducts from an earlier chapter. This provides a family of richly endowed approximate elementary extensions of a given normed space structure, and these structures play an important role in applications of model theory.

13. More Model Theory

Let $\mathcal{D}$ be a uniform class of structures with a common signature, and consider the class $\mathcal{C}$ of structures for that signature that can be embedded in some ultraproduct of members of $\mathcal{D}$. It is shown that $\mathcal{C}$ consists of precisely the normed space structures for that signature that satisfy a certain set of positive bounded sentences in which only universal quantifiers appear.

A model theoretic characterization of axiomatizability for classes of normed space structures with a common signature is also discussed here. Let $\mathfrak{D}$ be a uniform class of structures with a common signature, and suppose that $\mathfrak{D}$ is closed under isomorphisms and ultraproducts. Let $\mathfrak{C}$ be contained in $\mathfrak{D}$. We show that $\mathfrak{C}$ is axiomatizable within $\mathfrak{D}$ by positive bounded sentences if and only if $\mathfrak{C}$ is closed under isomorphisms and ultraproducts and $\mathfrak{D} \setminus \mathfrak{C}$ is closed under ultrapowers (*i.e.* $\mathfrak{C}$ is closed under "ultraroots" within $\mathfrak{D}$).

This chapter also introduces a natural concept of *quantifier elimination* for normed space structures, and gives a model theoretic criterion for it. This concept often provides an effective basis for understanding what is definable by positive bounded formulas in a given structure or class of structures.

A more general result concerning the approximation of arbitrary positive bounded formulas by simpler formulas is also given here.

For many classes of structures from functional analysis, the structure of ultraproducts has been extensively investigated. When this work is combined with the results presented in this chapter, it is often possible to prove axiomatizability and even sometimes to prove quantifier elimination for interesting classes of such structures. In several cases this perspective raises interesting new questions about classes of normed space structures that have been previously studied; especially, it emphasizes the importance of understanding whether these classes are closed under "ultraroots" and whether they satisfy the mapping extension criterion for quantifier elimination.

14. Types

Spaces of types are a characteristic feature of modern applications of model theory in all areas of mathematics. Quantifier-free types have been used significantly in the geometry of Banach spaces, especially in connection with the class of stable spaces introduced (with direct motivation from work in model theory) by Krivine and Maurey [KM81]. Some fundamental aspects of this topic are developed in this chapter.

2

Normed Space Structures

2.1 Definition. A *normed space structure* $\mathfrak{M}$ consists of the following items:

(i) A family $(M^{(s)} \mid s \in S)$ of normed spaces.
(ii) A collection of functions of the form

$$F \colon M^{(s_1)} \times \cdots \times M^{(s_m)} \to M^{(s_0)},$$

each of which is uniformly continuous (and hence bounded) on every bounded subset of its domain.

The normed spaces $M^{(s)}$ are called the *sorts* of $\mathfrak{M}$, and we say that $\mathfrak{M}$ is *based on* $(M^{(s)} \mid s \in S)$. If every sort of $\mathfrak{M}$ is a Banach space (*i.e.*, is complete in its norm), we say that $\mathfrak{M}$ is a *Banach space structure*.

Some of the functions of $\mathfrak{M}$ may have arity 0; they correspond to distinguished elements of the sorts of $\mathfrak{M}$. These elements are called the *constants* of $\mathfrak{M}$. The other functions may be referred to as the *operations* of $\mathfrak{M}$.

We require that every normed space structure include a special sort and certain special functions; we will refer to these functions as the *required operations* of the structure. Namely, if $\mathfrak{M}$ is a normed space structure based on $(M^{(s)} \mid s \in S)$, then:

· The field $\mathbb{R}$ must occur as a sort of $\mathfrak{M}$; *i.e.*, the sort index set S contains a distinguished element $s = s_{\mathbb{R}}$ for which $M^{(s)}$ is $\mathbb{R}$.
· For every $s \in S$, the vector space operations and the norm of $M^{(s)}$ must occur as functions of $\mathfrak{M}$, and the additive identity $0_{M^{(s)}}$ of $M^{(s)}$ must occur as a constant of $\mathfrak{M}$. (By *vector space operations* we mean $+$ and $-$ as binary functions from $M^{(s)} \times M^{(s)}$ into $M^{(s)}$ as well as scalar multiplication as a binary function from $\mathbb{R} \times M^{(s)}$ into $M^{(s)}$.)

10

· The distinguished operations on $\mathbb{R} = M^{(s_\mathbb{R})}$ consist of the usual field operations $+$, $-$, and $\times$, as well as the absolute value function (which is the norm on this sort). In addition to the additive identity 0 we require that each rational number be a distinguished constant in this sort.

If $\mathfrak{M}$ is based on $(M^{(s)} \mid s \in S)$ and $(F_i \mid i \in I)$ is a list of the functions of $\mathfrak{M}$, we may write

$$\mathfrak{M} = (M^{(s)}, F_i \mid s \in S, i \in I).$$

The special sort $\mathbb{R}$ as well as the functions and constants which are required to be present in every normed space structure (see above) need not be listed explicitly in the preceding notation.

If $\mathfrak{M}$ is based on $(M^{(s)} \mid s \in S)$ and $a \in M^{(s)}$ for some $s \in S$, we refer to a as an *element* of $\mathfrak{M}$. Note that we make no assumptions about how the sorts of $\mathfrak{M}$ are related to each other as sets. In particular $a \in M^{(s)}$ may be true for several different values of s. This means that when dealing with an element a of a normed space structure $\mathfrak{M}$, it is necessary to specify the sort of which a is taken to be an element.

2.2 Examples. We give a number of examples of normed space structures to indicate the wide range of possibilities.

(i) Normed spaces X over $\mathbb{R}$: the sorts are X and $\mathbb{R}$, and the functions are the vector space operations, the additive identity 0_X and the norm of X, as well as the field operations, the additive identity 0 and the absolute value function on $\mathbb{R}$.

(ii) Normed spaces X over $\mathbb{C}$: these can be regarded as normed space structures in several ways. For example we may add $\mathbb{C}$ as a sort together with its field structure and absolute value, and the scalar multiplication operation as a map from $\mathbb{C} \times X$ into X, as well as the inclusion map from $\mathbb{R}$ into $\mathbb{C}$. Alternatively, we may simply include a unary function from X into itself, corresponding to scalar multiplication by $\sqrt{-1}$, in addition to the usual operations that come from regarding X as a normed space over $\mathbb{R}$.

(iii) Normed vector lattices $(X, \vee, \wedge)$: this is the result of expanding the normed space structure corresponding to X (see above) by adding the lattice operations $\vee$ and $\wedge$ on X and the functions max and min on $\mathbb{R}$.

(iv) Normed algebras: multiplication is included as an operation; if the algebra has a multiplicative identity, it may be included as a constant.

(v) C^*-algebras: multiplication and the *-map are included as operations.

(vi) Hilbert spaces with inner product, in which the pairing is included as a function.

(vii) Dual pairs (X, X'), where X is a Banach space, X' is the dual of X, and the pairing between X and X' is included as a function.

(viii) Triples (X, X', X''), where X' and X'' are the dual and the double dual of X and the pairing between X and X', the pairing between X' and X'', and the embedding $X \to X''$ are included as functions.

(ix) Operator spaces, which include for each $n \geq 1$ a real-valued function of n^2 arguments mapping each $n \times n$ matrix (a_{ij}) of elements of the underlying Banach space to its operator norm.

(x) If $\mathcal{M}$ is a normed space structure and a is an element of $\mathcal{M}$, then the expansion $(\mathcal{M}, a)$ is a normed space structure, in which a is a constant.

(xi) If $\mathcal{M}$ is a normed space structure, and T is a bounded linear operator between sorts of $\mathcal{M}$, then the expansion $(\mathcal{M}, T)$ is a normed space structure, in which T is a distinguished function.

(xii) If $\mathcal{M}$ is a normed space structure, $M^{(s)}$ is a sort of $\mathcal{M}$, and A is a given subset of $M^{(s)}$, then $\mathcal{M}$ can be expanded by adding the real-valued function $x \mapsto \text{dist}(x, A)$, where x ranges over $M^{(s)}$ and dist denotes the distance function with respect to the norm on $M^{(s)}$. The same can be done with subsets of finite cartesian products of sorts.

3

Signatures

In order to treat normed space structures $\mathfrak{M}$ in general, it is convenient to introduce formally a way of indexing the sorts and functions of $\mathfrak{M}$; this is the *signature* of $\mathfrak{M}$.

3.1 Definition. Let $\mathfrak{M}$ be a normed space structure based on $(M^{(s)} \mid s \in S)$. A *signature* L for $\mathfrak{M}$ consists of the following items:

- A sort index set S.
- An element $s_{\mathbb{R}} \in S$ such that $M^{(s_{\mathbb{R}})} = \mathbb{R}$.
- For each designated function

$$F \colon M^{(s_1)} \times \cdots \times M^{(s_m)} \to M^{(s_0)}$$

of $\mathfrak{M}$, a triple of the form

$$\big(f, (s_1, \ldots, s_m), s_0\big),$$

where f is a syntactic symbol called a *function symbol* for F. In this case, we write $F = f^{\mathfrak{M}}$ and call F the *interpretation* of f in $\mathfrak{M}$. We call $s_1 \times \cdots \times s_m$ the *domain* of f and s_0 the *range* of f, and express this by writing (purely formally)

$$f \colon s_1 \times \cdots \times s_m \to s_0.$$

The number m is called the *arity* of f. If $m = 0$, we call f a *constant symbol* and write $f : s_0$. In this case $f^{\mathfrak{M}}$ is an element of $M^{(s_0)}$.

Each of the required functions of a normed space structure has a specific designated function symbol, and these are taken to be the same for each signature. For example, the additive identity of a sort $M^{(s)}$ is associated to a constant symbol designated $0 : s$, the vector addition and subtraction on this sort correspond to function symbols whose associated triples are $(+, (s, s), s)$ and $(-, (s, s), s)$, and so forth.

We express the fact that L is a signature for $\mathcal{M}$ by saying that $\mathcal{M}$ is an L-*structure*.

3.2 Definition. If L, L' are signatures, we write $L \subseteq L'$ if the following two conditions hold:

· The sort index set of L is a subset of the sort index set of L'.
· Every triple of the form $\bigl(f, (s_1, \ldots, s_m), s_0\bigr)$ which is in L is also in L'.

In this case we say L' is an *extension* of L and that L is a *subsignature* of L'.

Informally, $L \subseteq L'$ means that every sort and function symbol of L is also in L', and that they are used in precisely the same way in L' as they are in L.

If L, L' are signatures and $L \subseteq L'$, we call L' an *extension by constants* of L if L and L' have the same sort index set and if every function symbol of L' that is not in L is a constant symbol. In this situation we write $L' = L(C)$ where C is the set of constant symbols that have been added to L to get L'.

3.3 Definition. If L is a signature, the *cardinality of* L is the cardinal number $\operatorname{card}(L)$ defined to be equal to

$$\operatorname{card}(S) + \operatorname{card}(\{\, f \mid f \text{ a function symbol of } L \,\}) + \aleph_0.$$

In particular, we say L is *countable* if it has a finite or countable number of sorts and distinguished function symbols.

3.4 Definition. Let L be a signature and suppose that $\mathcal{M}$ and $\mathcal{N}$ are normed space L-structures based on $(M^{(s)} \mid s \in S)$ and $(N^{(s)} \mid s \in S)$, respectively. We say that $\mathcal{M}$ is a *substructure* of $\mathcal{N}$ (or that $\mathcal{N}$ is an *extension* of $\mathcal{M}$) if $M^{(s)} \subseteq N^{(s)}$ for every $s \in S$ and $f^{\mathcal{N}}$ extends $f^{\mathcal{M}}$, for every function symbol f of L. We write $\mathcal{M} \subseteq \mathcal{N}$ to indicate that $\mathcal{M}$ is a substructure of $\mathcal{N}$.

Let $\mathcal{N}$ be as above and let $(A^{(s)} \mid s \in S)$ be a family of sets such that $A^{(s)} \subseteq N^{(s)}$ for every $s \in S$. The substructure of $\mathcal{N}$ *generated by* $(A^{(s)} \mid s \in S)$ is the L-substructure $\mathcal{M}$ of $\mathcal{N}$ that results from closing $(A^{(s)} \mid s \in S)$ and $\mathbb{R} = M^{(s_{\mathbb{R}})}$ under all the functions of $\mathcal{M}$. We say that $\mathcal{N}$ is *finitely generated* if $\mathcal{N}$ is generated by a family $(A^{(s)} \mid s \in S)$ such that $A^{(s)}$ is finite for every $s \in S$ and $A^{(s)} = \emptyset$ for all but finitely many $s \in S$.

Informally we may say that $\mathcal{M} \subseteq \mathcal{N}$ holds if and only if $M^{(s)}$ is a normed subspace of $N^{(s)}$ for every $s \in S$ and $f^{\mathcal{N}}$ maps $\mathcal{M}$ into itself, for every function symbol f of L.

3.5 Definition. Let L be a signature and suppose that $\mathcal{M}$ and $\mathcal{N}$ are normed space L-structures based on $(M^{(s)} \mid s \in S)$ and $(N^{(s)} \mid s \in S)$, respectively. An *isomorphism* between $\mathcal{M}$ and $\mathcal{N}$ consists of a collection of surjective isometries $T^{(s)} \colon M^{(s)} \to N^{(s)}$, for $s \in S$, which commute with the interpretations of the function symbols of L in the following sense: whenever $f \colon s_1 \times \cdots \times s_m \to s_0$ is a function symbol of L and $a_k \in M^{(s_k)}$ for $k = 1, \ldots, m$, we have

$$f^{\mathcal{N}}(T^{(s_1)}(a_1), \ldots, T^{(s_m)}(a_m)) = T^{(s_0)}(f^{\mathcal{M}}(a_1, \ldots, a_m)).$$

We say that $\mathcal{M}$ and $\mathcal{N}$ are *isomorphic*, and write $\mathcal{M} \cong \mathcal{N}$, if there exists an isomorphism between $\mathcal{M}$ and $\mathcal{N}$. (Sometimes we say *isometric isomorphism* to emphasize that isomorphisms must be norm preserving; since the norm on each sort is included as a distinguished function in every normed space structure, it must be preserved by every isomorphism.)

An *embedding* (equivalently, *isometric embedding*) of $\mathcal{M}$ into $\mathcal{N}$ is an isomorphism between $\mathcal{M}$ and a substructure of $\mathcal{N}$.

An *automorphism* (equivalently, *isometric automorphism*) of $\mathcal{M}$ is an isomorphism between $\mathcal{M}$ and $\mathcal{M}$.

Note that if $\mathcal{M}$ and $\mathcal{N}$ are normed space L-structures and $(T^{(s)} \mid s \in S)$ is an embedding of $\mathcal{M}$ into $\mathcal{N}$, then $T^{(s_{\mathbb{R}})}$ must be the identity map of $M^{(s_{\mathbb{R}})}$ to $N^{(s_{\mathbb{R}})}$.

3.6 Definition. Let L be a signature and suppose that $\mathcal{M}$ is an L-structure based on $(M^{(s)} \mid s \in S)$.

(i) The *completion* of $\mathcal{M}$ is the Banach space L-structure $\overline{\mathcal{M}}$ defined as follows. For each $s \in S$ let $\overline{M^{(s)}}$ denote the norm completion of $M^{(s)}$. Then $\overline{\mathcal{M}}$ is based on $(\overline{M^{(s)}} \mid s \in S)$, and for every function symbol f of L, $f^{\overline{\mathcal{M}}}$ is defined to be the unique continuous extension of $f^{\mathcal{M}}$. (The domain D of $f^{\overline{\mathcal{M}}}$ is determined since $\overline{\mathcal{M}}$ is an L-structure based on a specific family of sorts, and D has the domain of $f^{\mathcal{M}}$ as a dense subspace.)

(ii) If $\mathcal{M} \subseteq \mathcal{N}$, the *closure* of $\mathcal{M}$ in $\mathcal{N}$ is the substructure $\mathcal{M}'$ of $\mathcal{N}$ defined as follows. For each $s \in S$ let $(M^{(s)})'$ denote the closure of $M^{(s)}$ in $\mathcal{N}$. Then $\mathcal{M}'$ is the substructure of $\mathcal{N}$ that is based on $((M^{(s)})' \mid s \in S)$. (Continuity of the functions of $\mathcal{N}$ ensures that there is such a substructure.)

Evidently each normed space structure is a substructure of its completion and of its closure in any extension.

3.7 Definition. Let L, L' be signatures with $L \subseteq L'$ and suppose that $\mathcal{N}$ is an L'-structure. The *reduct* of $\mathcal{N}$ to L is the L-structure that results by removing from $\mathcal{N}$ the sorts and functions that are indexed by L' but not by L. We say that a normed space $\mathcal{N}$ is an *expansion* of a structure $\mathcal{M}$ if $\mathcal{M}$ is a reduct of $\mathcal{N}$.

We say that a normed space L'-structure $\mathcal{N}$ is an *expansion by constants* of an L-structure $\mathcal{M}$ if L' is an extension by constants of L and $\mathcal{N}$ is an expansion of $\mathcal{M}$.

Suppose $L(C)$ is an extension by constants of L, $\mathcal{M}$ is a normed space L-structure based on $(M^{(s)} \mid s \in S)$, and the $L(C)$-structure $\mathcal{N}$ is an expansion of $\mathcal{M}$. In this situation we denote $\mathcal{N}$ by $(\mathcal{M}, \mathbf{A})$, where $\mathbf{A}$ is the function defined on C by setting $\mathbf{A}(c) = c^{\mathcal{N}}$ for each $c \in C$. Suppose $\mathcal{M}$ is based on $(M^{(s)} \mid s \in S)$. Note that $\mathbf{A}$ may be any function from C to elements of $\mathcal{M}$ which satisfies the requirement that for each $c \in C$, if the range of c is s, then $\mathbf{A}(c) \in M^{(s)}$. We refer to the function $\mathbf{A}$ as an *interpretation of C in $\mathcal{M}$*.

4

Ultrapowers of Normed Space Structures

Ultrafilter Limits

Let X be a topological space. Further, let $(x_\xi)_{\xi \in \Lambda}$ be an indexed family in X. If D is an ultrafilter on Λ and $x \in X$, we write

$$\lim_{\xi,D} x_\xi = x$$

and say that x is the D-*limit* of $(x_\xi)_{\xi \in \Lambda}$ if for every neighborhood U of x, the set $\{\, \xi \in \Lambda \mid x_\xi \in U \,\}$ is in the ultrafilter D. A basic fact from general topology is that X is a compact Hausdorff space if and only if for every indexed family $(x_\xi)_\xi \in \Lambda$ in X and every ultrafilter D on Λ the D-limit of $(x_\xi)_{\xi \in \Lambda}$ exists and is unique.

Ultraproducts of Normed Spaces

Let $(X_\xi \mid \xi \in \Lambda)$ be a family of normed spaces. Define

$$\ell_\infty\big(\Lambda, (X_\xi \mid \xi \in \Lambda)\big) = \{\, (x_\xi)_{\xi \in \Lambda} \in \prod_{\xi \in \Lambda} X_\xi \mid \sup_{\xi \in \Lambda} \|x_\xi\| < \infty \,\}.$$

One sees that $\ell_\infty(\Lambda, (X_\xi \mid \xi \in \Lambda))$ is naturally a normed space with respect to the norm defined by

$$\|(x_\xi)_{\xi \in \Lambda}\|_\infty = \sup_{\xi \in \Lambda} \|x_\xi\|.$$

An ultrafilter D on Λ induces a continuous seminorm on $\ell_\infty(\Lambda, (X_\xi \mid \xi \in \Lambda))$ by defining

$$\|(x_\xi)_{\xi \in \Lambda}\| = \lim_{\xi,D} \|x_\xi\|$$

for each $(x_\xi)_{\xi \in \Lambda} \in \ell_\infty(\Lambda, (X_\xi \mid \xi \in \Lambda))$. The set N_D of families $(x_\xi)_{\xi \in \Lambda}$ in $\ell_\infty(\Lambda, (X_\xi \mid \xi \in \Lambda))$ such that $\|(x_\xi)_{\xi \in \Lambda}\| = 0$ is a closed subspace of

17

$\ell_\infty(\Lambda, (X_\xi \mid \xi \in \Lambda))$. We define

$$\left(\prod_{\xi \in \Lambda} X_\xi\right)_D = \left(\ell_\infty(\Lambda, (X_\xi \mid \xi \in \Lambda))\right)\Big/ N_D.$$

The space $(\prod_{\xi \in \Lambda} X_\xi)_D$ is called the *D-ultraproduct of* $(X_\xi \mid \xi \in \Lambda)$. If $(x_\xi)_{\xi \in \Lambda} \in \ell_\infty(\Lambda, (X_\xi \mid \xi \in \Lambda))$ we denote the equivalence class of $(x_\xi)_{\xi \in \Lambda}$ in $(\prod_{\xi \in \Lambda} X_\xi)_D$ by $((x_\xi)_{\xi \in \Lambda})_D$.

If $X_\xi = X$ for every $\xi \in \Lambda$, the space $(\prod_{\xi \in \Lambda} X_\xi)_D$ is called the *D-ultrapower of* X and is denoted $(X)_D$.

If X is a normed space and D is an ultrafilter on Λ, then the map $T \colon X \to (X)_D$ defined by $T(x) = ((x_\xi)_{\xi \in \Lambda})_D$, where $x_\xi = x$ for every $\xi \in \Lambda$, is an isometric embedding. It is called the *diagonal embedding* of X into $(X)_D$. This embedding is generally not surjective; it is, however, when the ultrafilter D is principal or the space X is finite dimensional.

Ultrapowers of Functions

4.1 Notation. If X is a normed space and r is a nonnegative real number, we let $\mathcal{B}_r(X)$ denote the closed ball of radius r around 0 in X.

4.2 Definition. Suppose that $X^{(0)}, \ldots, X^{(m)}$ are normed spaces and that

$$F \colon X^{(1)} \times \cdots \times X^{(m)} \to X^{(0)}$$

is uniformly continuous on a subset B of $X^{(1)} \times \cdots \times X^{(m)}$. A *modulus of uniform continuity for* F *on* B is a map $\Delta \colon (0, \infty) \to (0, \infty)$ such that for any $(x_1, \ldots, x_m), (y_1, \ldots, y_m) \in B$ and any $\epsilon > 0$,

$$\max_k \|x_k - y_k\| < \Delta(\epsilon) \quad \text{implies} \quad \|F(x_1, \ldots, x_m) - F(y_1, \ldots, y_m)\| \leq \epsilon.$$

Suppose that $X^{(0)}, \ldots, X^{(m)}$ are normed spaces and that

$$F \colon X^{(1)} \times \cdots \times X^{(m)} \to X^{(0)}$$

is uniformly continuous on every bounded subset of its domain. Given an ultrafilter D on Λ, we define a function

$$(F)_D \colon (X^{(1)})_D \times \cdots \times (X^{(m)})_D \to (X^{(0)})_D$$

as follows: if $(x_\xi^k)_{\xi \in \Lambda} \in \ell_\infty(\Lambda, (X^{(k)})^\Lambda)$ for $k = 1, \ldots, m$, then

$$(F)_D\big(((x_\xi^1)_{\xi \in \Lambda})_D, \ldots, ((x_\xi^m)_{\xi \in \Lambda})_D\big) = \left((F(x_\xi^1, \ldots, x_\xi^m))_{\xi \in \Lambda}\right)_D.$$

The uniform continuity assumption on F ensures that $(F)_D$ is well-defined; in fact, we have the following:

4.3 Proposition. *Let F and $(F)_D$ be as above. Any modulus of uniform continuity for F on $\mathcal{B}_{r_1}(X^{(1)}) \times \cdots \times \mathcal{B}_{r_m}(X^{(m)})$ is also a modulus of uniform continuity for $(F)_D$ on $\mathcal{B}_{r_1}((X^{(1)})_D) \times \cdots \times \mathcal{B}_{r_m}((X^{(m)})_D)$.*

Proof. This is a consequence of the definition of D-limit. ☐

The fact that $(F)_D$ is well-defined depends in part on the observation that if B is a bounded convex subset of the domain of F and F is uniformly continuous on B, then F is uniformly bounded on B. Indeed, if the diameter of B is $\leq d$, x is any element of B, and Δ is any modulus of uniform continuity for F on B, then for any $y \in B$ we have

$$\|F(y)\| \leq \|F(x)\| + d \cdot \Delta(1)^{-1}.$$

Ultrapowers of Normed Space Structures

4.4 Definition. Let L be a signature and suppose that $\mathcal{M}$ is a normed space L-structure based on $(M^{(s)} \mid s \in S)$. If D is an ultrafilter, the *D-ultrapower* of $\mathcal{M}$ is the normed space L-structure defined as follows. The sorts of $(\mathcal{M})_D$ are the D-ultrapowers $(M^{(s)})_D$, for $s \in S$, and for every function symbol f of L, one sets $f^{(\mathcal{M})_D} = (f^{\mathcal{M}})_D$.

4.5 Remark. If $\mathcal{M}$ is a normed space structure based on $(M^{(s)} \mid s \in S)$ and $T^{(s)} \colon M^{(s)} \to (M^{(s)})_D$ is the diagonal embedding for each $s \in S$, then the family $(T^{(s)} \mid s \in S)$ is an embedding of $\mathcal{M}$ into $(\mathcal{M})_D$. We call this family the *diagonal embedding* of $\mathcal{M}$ into $(\mathcal{M})_D$.

Nonstandard Hulls

Let $\mathcal{M}$ be a normed space L-structure and let D be an ultrafilter on the index set Λ. The ultrapower $(\mathcal{M})_D$ is an example of a *nonstandard hull* of $\mathcal{M}$. This more general construction is widely used in applications of nonstandard analysis. See [HL85] [Lin88] [Hen97] for background on this construction and the general area of nonstandard analysis.

Given a set theoretic superstructure V of which $\mathcal{M}$ is an element, let $^*V$ be the D-ultrapower nonstandard extension of V constructed using the ultrafilter D. (See, for example, Section II.4 in [HL85] or the proof of Theorem 2.28 in [Hen97].) The *-mapping from V into $^*V$ takes $\mathcal{M}$ to the internal object $^*\mathcal{M}$; this object is represented by the constant family $(\mathcal{M} \mid \xi \in \Lambda)$.

In general, when $^*V$ is any nonstandard extension of V, the nonstandard hull is constructed from $^*\mathcal{M}$ by (a) keeping *finite* elements of

$^*\mathcal{M}$ (elements whose internal norm is a finite element of $^*\mathbb{R}$) and (b) identifying two finite elements x, y of the same sort of $^*\mathcal{M}$ if $x - y$ is *infinitesimal* (*i.e.* its internal norm is an infinitesimal element of $^*\mathbb{R}$). The uniform continuity requirement on the operations F of $\mathcal{M}$ ensures that $^*F$ maps finite arguments to finite values and preserves infinitesimal equivalence, so it gives rise to a well-defined quotient operation. The resulting structure is a normed space L-structure, denoted by $\widehat{\mathcal{M}}$.

Again consider the situation where $^*V$ is the D-ultrapower nonstandard extension of V. It is easy to check that finite elements of $^*\mathbb{R}$ have representatives which are bounded families of real numbers; likewise, infinitesimal elements of $^*\mathbb{R}$ are represented by families of real numbers whose D-limit is 0. It follows that the nonstandard hull $\widehat{\mathcal{M}}$ constructed using this $^*V$ is isomorphic, as a normed space L-structure, to $(\mathcal{M})_D$.

5

Positive Bounded Formulas

In this chapter we introduce the key ingredients of the logic for normed space structures that is described in this paper. These are the *positive bounded formulas* and the concept of *approximate satisfaction* of such formulas in normed space structures.

Let L be a signature for a normed space structure $\mathcal{M}$ based on $(M^{(s)} \mid s \in S)$. Recall that S has a distinguished element $s = s_{\mathbb{R}}$ for which $M^{(s)} = \mathbb{R}$ is the sort of real numbers.

We begin considering $\mathcal{M}$ from the model theoretic point of view, introducing a formal language based on L and a semantics according to which this language is interpreted in $\mathcal{M}$. In addition to the symbols of the signature L, we also need for each element s of the sort index set S, a countable set of symbols called the *variables of sort s*.

We begin defining the formal language by introducing the set of *terms of L*, or *L-terms*. Each term is a finite string of symbols, each of which may be a variable or a function symbol of L, or one of the symbols (or , which are used for punctuation. In this many-sorted context, each term is associated with a unique sort which indicates its range. The formal definition is recursive.

5.1 Definition. An *L-term with range of sort s* is a string which can be obtained by finitely many applications of the following rules of formation:

· If x is a variable of L of sort s, then x is a term with range of sort s.
· If f is a function symbol of L with $f \colon s_1 \times \cdots \times s_m \to s_0$ and t_k is a term with range of sort s_k for $k = 1, \ldots, m$, then $f(t_1, \ldots, t_m)$ is a term with range of sort s_0.

We say that t is a *term* if t is a term with range of sort s for some $s \in S$; further, t is a *real-valued term* if t is a term with range of sort $s_{\mathbb{R}}$.

5.2 Definition. Let L be a signature for normed space structures. The set of *positive bounded L-formulas* is defined recursively as follows.

· If t is a real-valued term and r is a fixed rational number, then the expressions

$$r \leq t, \qquad t \leq r$$

are positive bounded formulas.

· If φ_1 and φ_2 are positive bounded formulas, then the expressions

$$(\varphi_1 \wedge \varphi_2), \qquad (\varphi_1 \vee \varphi_2)$$

are positive bounded formulas.

· If φ is a positive bounded formula, x is a variable, and r is a positive rational number, then the expressions

$$\exists x(\|x\| \leq r \wedge \varphi),$$
$$\forall x(\|x\| \leq r \rightarrow \varphi)$$

are positive bounded formulas.

Thus, the positive bounded formulas are built from the *atomic formulas*

$$r \leq t, \qquad t \leq r$$

by finite application of the *positive connectives* $\wedge$, $\vee$ and the *bounded quantifiers*

$$\exists x(\|x\| \leq r \wedge \ldots),$$
$$\forall x(\|x\| \leq r \rightarrow \ldots)$$

If t is a term and r_1, r_2 are real numbers, we write $r_1 \leq t \leq r_2$ as an abbreviation of the positive bounded formula $(r_1 \leq t \wedge t \leq r_2)$. Similarly, we write $t = r$ as an abbreviation for $r \leq t \wedge t \leq r$. When the context allows it, we often omit the outer parentheses in formulas of the form $(\varphi_1 \wedge \varphi_2)$ or $(\varphi_1 \vee \varphi_2)$. Sometimes we also write $\bigwedge_{i=1}^{n} \varphi_i$ and $\bigvee_{i=1}^{n} \varphi_i$ as abbreviations of $\varphi_1 \wedge \cdots \wedge \varphi_n$ and $\varphi_1 \vee \cdots \vee \varphi_n$, respectively.

The bounded quantifiers may also be abbreviated. We write $\exists_r x\, \varphi$ and $\forall_r x\, \varphi$ as abbreviations of $\exists x(\|x\| \leq r \wedge \varphi)$ and $\forall x(\|x\| \leq r \rightarrow \varphi)$, respectively.

Let φ be a positive bounded formula. A *subformula* of φ is any string of consecutive symbols of φ which is a positive bounded formula in its own right. A variable x is said to be *free* in a positive bounded formula

φ if there is at least one occurrence of x in φ which is not within any subformula of φ of the form $\exists_r x\, \psi$ or $\forall_r x\, \psi$. A *positive bounded sentence* is a positive bounded formula without free variables.

If t is a term and $x_1, \ldots, x_n$ are distinct variables, we will write $t(x_1, \ldots, x_n)$ to indicate that all variables occurring in t are among $x_1, \ldots, x_n$. Similarly, if φ is a positive bounded formula, we write $\varphi(x_1, \ldots, x_n)$ to indicate that all variables occurring free in φ are among $x_1, \ldots, x_n$. Also, $\Gamma(x_1, \ldots, x_n)$ denotes a set of positive bounded formulas whose free variables are among $x_1, \ldots, x_n$.

Each term t defines a function $t^{\mathcal{M}}$ on $\mathcal{M}$, as specified in the following definition:

5.3 Definition. Suppose that $\mathcal{M}$ is a normed space L-structure based on $(M^{(s)} \mid s \in S)$. Let $t(x_1, \ldots, x_n)$ be an L-term, where x_k is a variable of sort s_k for $k = 1, \ldots, n$, and let $a_1, \ldots, a_n$ be such that $a_k \in M^{(s_k)}$ for $k = 1, \ldots, n$. We define the *evaluation of t in $\mathcal{M}$ at $a_1, \ldots a_n$*, denoted

$$t^{\mathcal{M}}[a_1, \ldots, a_n],$$

as follows. The definition is by induction on the complexity of t. It ensures that $t^{\mathcal{M}}[a_1, \ldots, a_n] \in M^{(s)}$ holds whenever t is a term with range of sort s.

· If t is x_k, then $t^{\mathcal{M}}[a_1, \ldots, a_n]$ is a_k;
· If t is $f(t_1, \ldots, t_m)$, where f is a function symbol of L and $t_1, \ldots, t_m$ are terms, then $t^{\mathcal{M}}[a_1, \ldots, a_n]$ is

$$f^{\mathcal{M}}(t_1^{\mathcal{M}}[a_1, \ldots, a_n], \ldots, t_m^{\mathcal{M}}[a_1, \ldots, a_n]).$$

5.4 Remark. Note that $t^{\mathcal{M}}[a_1, \ldots, a_n]$ depends not only on t, $\mathcal{M}$ and $a_1, \ldots, a_n$, but also on a given list of variables which is not explicit in the notation. The context will make it clear which list of variables is being used.

5.5 Definition. Suppose that $\mathcal{M}$ is a normed space L-structure based on $(M^{(s)} \mid s \in S)$. Let $\varphi(x_1, \ldots, x_n)$ be a positive bounded L-formula, where x_k is a variable of sort s_k for $k = 1, \ldots, n$, and let $a_1, \ldots, a_n$ be such that $a_k \in M^{(s_k)}$ for $k = 1, \ldots, n$. We define the relation

$$\mathcal{M} \models \varphi[a_1, \ldots, a_n].$$

When this holds, we say that $\mathcal{M}$ *satisfies* $\varphi(x_1, \ldots, x_n)$ at $a_1, \ldots, a_n$, or that $\varphi(x_1, \ldots, x_n)$ is *true in* $\mathcal{M}$ *at* $a_1, \ldots, a_n$. The definition of $\mathcal{M} \models \varphi[a_1, \ldots, a_n]$ is by induction on the complexity of φ.

· If φ is $t(x_1,\ldots,x_n) \leq r$, where t is a real-valued term and r is a rational number, then $\mathcal{M} \models \varphi[a_1,\ldots,a_n]$ if and only if

$$t^{\mathcal{M}}[a_1,\ldots,a_n] \leq r.$$

· If φ is $r \leq t(x_1,\ldots,x_n)$, where t is a real-valued term and r is a rational number, then $\mathcal{M} \models \varphi[a_1,\ldots,a_n]$ if and only if

$$r \leq t^{\mathcal{M}}[a_1,\ldots a_n].$$

· If φ is $(\psi_1 \wedge \psi_2)$, then $\mathcal{M} \models \varphi[a_1,\ldots,a_n]$ if and only if

$$\mathcal{M} \models \psi_1[a_1,\ldots,a_n] \quad \text{and} \quad \mathcal{M} \models \psi_2[a_1,\ldots,a_n].$$

· If φ is $(\psi_1 \vee \psi_2)$, then $\mathcal{M} \models \varphi[a_1,\ldots,a_n]$ if and only if

$$\mathcal{M} \models \psi_1[a_1,\ldots,a_n] \quad \text{or} \quad \mathcal{M} \models \psi_2[a_1,\ldots,a_n].$$

· If φ is $\exists_r x\, \psi(x, x_1,\ldots,x_n)$, where r is a positive rational number and x is a variable of sort s, then $\mathcal{M} \models \varphi[a_1,\ldots,a_n]$ if and only if

$$\mathcal{M} \models \psi[a, a_1,\ldots,a_n], \quad \text{for some } a \in M^{(s)} \text{ with } \|a\| \leq r.$$

· If φ is $\forall_r x\, \psi(x, x_1,\ldots,x_n)$, where r is a positive rational number and x is a variable of sort s, then $\mathcal{M} \models \varphi[a_1,\ldots,a_n]$ if and only if

$$\mathcal{M} \models \psi[a, a_1,\ldots,a_n], \quad \text{for every } a \in M^{(s)} \text{ with } \|a\| \leq r.$$

5.6 Definition. If $\mathcal{M}$ is a normed space L-structure and $\Gamma(x_1,\ldots,x_n)$ is a set of positive bounded L-formulas, we write

$$\mathcal{M} \models \Gamma[a_1,\ldots,a_n]$$

if $\mathcal{M} \models \varphi[a_1,\ldots,a_n]$ for every $\varphi \in \Gamma$. In this case we say that $\mathcal{M}$ *satisfies* $\Gamma(x_1,\ldots,x_n)$ *at* $a_1,\ldots,a_n$, or that $\Gamma(x_1,\ldots,x_n)$ is *true in* $\mathcal{M}$ *at* $a_1,\ldots,a_n$.

Approximations

We now define the concept of *approximation* of a positive bounded formula. Intuitively, if φ is a positive bounded formula, an approximation φ' of φ is a positive bounded formula that results from relaxing all the norm estimates in φ. We indicate that φ' is an approximation of φ by writing $\varphi' > \varphi$ or equivalently $\varphi < \varphi'$. The formal definition is by induction on the complexity of φ:

If φ is:	The approximations of φ are:
$r \leq t$	$r' \leq t$, where $r' < r$
$t \leq r$	$t \leq r'$, where $r' > r$
$(\psi_1 \wedge \psi_2)$	$(\psi_1' \wedge \psi_2')$, where $\psi_i' > \psi_i$, for $i = 1, 2$
$(\psi_1 \vee \psi_2)$	$(\psi_1' \vee \psi_2')$, where $\psi_i' > \psi_i$, for $i = 1, 2$
$\exists_r x\, \psi$	$\exists_{r'} x\, \psi'$, where $r' > r$ and $\psi' > \psi$
$\forall_r x\, \psi$	$\forall_{r'} x\, \psi'$, where $r' < r$ and $\psi' > \psi$

5.7 Remark. Suppose that $\mathcal{M}$ is an L-structure and $\varphi(x_1, \ldots, x_n)$ is a positive bounded formula. Then

$$\mathcal{M} \models \varphi[a_1, \ldots, a_n] \quad \text{implies} \quad \mathcal{M} \models \varphi'[a_1, \ldots, a_n], \quad \text{for every } \varphi' > \varphi.$$

5.8 Notation. If Γ is a set of positive bounded formulas, we let Γ^+ denote the set of approximations of formulas in Γ.

Approximate Satisfaction

5.9 Definition. Suppose that $\mathcal{M}$ is a normed space L-structure based on $(M^{(s)} \mid s \in S)$. Let $\varphi(x_1, \ldots, x_n)$ be an L-formula, where x_k is a variable of sort s_k, for $k = 1, \ldots, n$, and let $a_1, \ldots, a_n$ be such that $a_k \in M^{(s_k)}$, for each k. We say that $\mathcal{M}$ *approximately satisfies* $\varphi(x_1, \ldots, x_n)$ at $a_1, \ldots, a_n$, or that $\varphi(x_1, \ldots, x_n)$ is *approximately true* at $a_1, \ldots, a_n$ in $\mathcal{M}$, and we write

$$\mathcal{M} \models_A \varphi[a_1, \ldots, a_n],$$

if $\mathcal{M} \models \varphi'[a_1, \ldots, a_n]$ for every approximation φ' of φ.

5.10 Definition. If $\mathcal{M}$ is a normed space L-structure and $\Gamma(x_1, \ldots, x_n)$ is a set of positive bounded L-formulas, we say that $\mathcal{M}$ *approximately satisfies* $\Gamma(x_1, \ldots, x_n)$ at $a_1, \ldots, a_n$ (or that $\Gamma(x_1, \ldots, x_n)$ is *approximately true* in $\mathcal{M}$ at $a_1, \ldots, a_n$), and write $\mathcal{M} \models_A \Gamma[a_1, \ldots, a_n]$, if $\mathcal{M} \models_A \varphi[a_1, \ldots, a_n]$ for every $\varphi \in \Gamma$.

5.11 Remark. Let Γ be as in the preceding definition and define Γ^+ as in 5.8. Then $\mathcal{M} \models_A \Gamma[a_1, \ldots, a_n]$ if and only if $\mathcal{M} \models \Gamma^+[a_1, \ldots, a_n]$.

5.12 Notation. Suppose that $\mathcal{M}$ is a normed space L-structure based on $(M^{(s)} \mid s \in S)$ and let $\varphi(x_1, \ldots, x_n)$ be an L-formula, where x_k is a variable of sort s_k, for $k = 1, \ldots, n$. The relations $\mathcal{M} \models \varphi[a_1, \ldots, a_n]$

and $\mathcal{M} \models_A \varphi[a_1, \ldots, a_n]$ are defined only when $a_k \in M^{(s_k)}$ holds for $k = 1, \ldots, n$. Informally, we express this requirement by saying that $a_1, \ldots, a_n$ are elements *of suitable sorts of* $\mathcal{M}$.

Weak Negation

The negation connective is not allowed in positive bounded formulas, nor is the implication connective, except when it occurs as part of a bounded universal quantifier. However, for every positive bounded formula φ there is a positive bounded formula $\mathrm{neg}(\varphi)$ which in Banach space model theory plays a role analogous to that of the negation of φ. The operation neg is defined by induction on the complexity of formulas, as follows.

If φ is:	$\mathrm{neg}(\varphi)$ is:
$r \leq t$	$t \leq r$
$t \leq r$	$r \leq t$
$(\psi_1 \wedge \psi_2)$	$(\mathrm{neg}(\psi_1) \vee \mathrm{neg}(\psi_2))$
$(\psi_1 \vee \psi_2)$	$(\mathrm{neg}(\psi_1) \wedge \mathrm{neg}(\psi_2))$
$\exists_r x\, \psi$	$\forall_r x\, \mathrm{neg}(\psi)$
$\forall_r x\, \psi$	$\exists_r x\, \mathrm{neg}(\psi)$

5.13 Remarks. Suppose that $\varphi(x_1, \ldots, x_n), \varphi'(x_1, \ldots, x_n)$ are positive bounded L-formulas. Then the following conditions hold whenever $\mathcal{M}$ is a normed space L-structure and $a_1, \ldots, a_n$ are elements of suitable sorts of $\mathcal{M}$.

(i) $\mathrm{neg}(\mathrm{neg}(\varphi))$ is φ.
(ii) $\varphi < \varphi'$ if and only if $\mathrm{neg}(\varphi') < \mathrm{neg}(\varphi)$.
(iii) If $\mathcal{M} \not\models \varphi[a_1, \ldots, a_n]$, then $\mathcal{M} \models \mathrm{neg}(\varphi)[a_1, \ldots, a_n]$.
(iv) If $\varphi < \varphi'$ and $\mathcal{M} \models \mathrm{neg}(\varphi')[a_1, \ldots, a_n]$, then $\mathcal{M} \not\models \varphi[a_1, \ldots, a_n]$.

5.14 Proposition. *Suppose that* $\mathcal{M}$ *is a normed space L-structure. Let* $\varphi(x_1, \ldots, x_n)$ *be a positive bounded L-formula, and let $a_1, \ldots, a_n$ be elements of suitable sorts of* $\mathcal{M}$. *Then* $\mathcal{M} \not\models_A \varphi[a_1, \ldots, a_n]$ *if and only there exists $\varphi' > \varphi$ such that* $\mathcal{M} \models_A \mathrm{neg}(\varphi')[a_1, \ldots, a_n]$.

Proof. If $\mathcal{M} \not\models_A \varphi[a_1, \ldots, a_n]$, there exists $\varphi' > \psi$ such that $\mathcal{M} \not\models \varphi'[a_1, \ldots, a_n]$. Then $\mathcal{M} \models \mathrm{neg}(\varphi')[a_1, \ldots, a_n]$ and hence

$$\mathcal{M} \models_A \mathrm{neg}(\varphi')[a_1, \ldots, a_n]. \qquad (*)$$

Assume, conversely, that there exists $\varphi' > \varphi$ such that (*) holds and take formulas ψ, ψ' such that $\varphi < \psi < \psi' < \varphi'$. By Remark 5.13,

$$\mathcal{M} \models \mathrm{neg}(\psi')[a_1, \ldots, a_n]$$

and hence $\mathcal{M} \not\models \psi[a_1, \ldots, a_n]$, so $\mathcal{M} \not\models_A \varphi[a_1, \ldots, a_n]$. $\qquad\square$

Perturbation Lemma

If $\mathcal{M}$ is a normed space L-structure and t is an L-term, then $t^{\mathcal{M}}$ (see Definition 5.3) is a uniformly continuous function on any bounded subset of its domain. The next result shows that positive bounded formulas have a similar property.

This result first appeared, in the context of Banach spaces without additional structure, as [HM83a, Proposition 2.3]. It is the basis for many of the most important features of positive bounded formulas and approximate satisfaction.

5.15 Proposition (Perturbation Lemma). *Let a normed space L-structure $\mathcal{M}$ and a positive bounded L-formula $\varphi(x_1, \ldots, x_n)$ be given. Then for every $r > 0$ and every approximation φ' of φ there exists $\delta > 0$ such that whenever $a_1, \ldots, a_n$ and $b_1, \ldots, b_n$ are elements of suitable sorts of $\mathcal{M}$ we have that*

$$\mathcal{M} \models \bigwedge_{1 \le i \le n} \|a_i\| \le r \ \wedge \bigwedge_{1 \le i \le n} \|a_i - b_i\| \le \delta \ \wedge \ \varphi[a_1, \ldots, a_n]$$

implies

$$\mathcal{M} \models \varphi'[b_1, \ldots, b_n].$$

Proof. By induction on the complexity of φ. $\qquad\square$

5.16 Corollary. *Let $\varphi(x_1, \ldots, x_n)$ be a positive bounded L-formula and let $\mathcal{M}$ be a normed space L-structure. Then the set*

$$\varphi^{\mathcal{M}} = \{(a_1, \ldots, a_n) \mid \mathcal{M} \models_A \varphi[a_1, \ldots, a_n]\}$$

is norm-closed.

Proof. Suppose $a_1, \ldots, a_n$ are elements of suitable sorts of $\mathcal{M}$ such that $(a_1, \ldots, a_n) \notin \varphi^{\mathcal{M}}$. By definition, there exists an approximation ψ' of φ such that

$$\mathcal{M} \not\models \psi'[a_1, \ldots, a_n].$$

Choose another approximation ψ such that $\varphi < \psi < \psi'$. Let r be a real number such that $\|a_j\| + 1 \leq r$ for all $j = 1, \ldots, n$. From the Perturbation Lemma obtain δ with $0 < \delta \leq 1$ for the formulas ψ, ψ' and the norm bound r. Whenever $b_1, \ldots, b_n$ are elements of suitable sorts with $\|b_j - a_j\| < \delta$ for all $j = 1, \ldots, n$ it follows that

$$\mathcal{M} \not\models \psi[b_1, \ldots, b_n]$$

and hence

$$\mathcal{M} \not\models_A \varphi[b_1, \ldots, b_n],$$

so $(b_1, \ldots, b_n) \notin \varphi^{\mathcal{M}}$. This proves that $\varphi^{\mathcal{M}}$ is closed for the norm topology. $\qquad\Box$

Satisfaction and Approximate Satisfaction

Suppose L is a signature and $\mathcal{M}$ is a normed space L-structure. If $\varphi(x_1, \ldots, x_n)$ is a positive bounded L-formula without quantifiers, it is easy to show that for any elements $a_1, \ldots, a_n$ of suitable sorts of $\mathcal{M}$,

$$\mathcal{M} \models_A \varphi[a_1, \ldots, a_n] \quad \text{if and only if} \quad \mathcal{M} \models \varphi[a_1, \ldots, a_n].$$

It follows that the same equivalence holds when φ is a positive bounded L-formula in which every bounded quantifier is universal (*i.e.*, of the form $\forall_r x$).

However, for a general positive bounded formula φ, it is usually false that $\models_A$ and $\models$ have the same meaning. The most general thing one can say (see Remark 5.7) is that for any normed space L-structure $\mathcal{M}$ and any elements $a_1, \ldots, a_n$ of suitable sorts of $\mathcal{M}$,

$$\mathcal{M} \models \varphi[a_1, \ldots, a_n] \quad \text{implies} \quad \mathcal{M} \models_A \varphi[a_1, \ldots, a_n].$$

For example, consider the following simple positive bounded sentence σ, where x, y are variables of sort s:

$$\exists_1 x \exists_1 y (\|x\| = 1 \wedge \|y\| = 1 \wedge \|x + y\| = 1 \wedge \|x - y\| = 1).$$

It is easy to show that $\mathcal{M} \models \sigma$ is true if and only if the normed space $M^{(s)}$ contains a subspace linearly isometric to ℓ_∞^2. On the other hand, $\mathcal{M} \models_A \sigma$ holds if and only if ℓ_∞^2 is finitely representable in $M^{(s)}$ (*i.e.*, for each $\epsilon > 0$, the normed space $M^{(s)}$ has a subspace $(1 + \epsilon)$-isomorphic to ℓ_∞^2. Finally, it is easy to construct Banach spaces X such that ℓ_∞^2 is finitely representable in X while X contains no subspace linearly isometric to ℓ_∞^2. If X is such a space, $X \models_A \sigma$ while $X \not\models \sigma$.

Alternative Formulation of Quantifiers

There are two essentially equivalent ways to formulate bounded universal quantifiers. Consider a positive bounded formula σ of the form

$$\forall y(\|y\| \leq r \to \varphi).$$

As the term "positive" is normally used in formal logic, this is not a positive formula, since the atomic formula $\|y\| \leq r$ is implicitly negated due to its position in front of the implication connective. Some logicians may find this to be a bothersome abuse of terminology. However, when a positive bounded formula is interpreted using the semantics of approximate satisfaction, this syntactic issue disappears.

One could systematically replace conditions of the form $\|y\| \leq r$ in the bounded universal quantifier by $\|y\| < r$, thus making the formulas considered in this theory "positive" in the usual technical sense; let us write $\tilde{\psi}$ for the result of making this change in every bounded universal quantifier in a positive bounded formula ψ. Thus we can write $\tilde{\sigma}$ as

$$\forall y(\|y\| < r \to \tilde{\varphi}).$$

An easy argument by induction shows that if τ is any approximation of σ, then σ implies $\tilde{\sigma}$ which in turn implies τ, when they are interpreted in any normed space structure. Therefore we also have τ implies $\tilde{\tau}$.

Our entire theory could be based on the positive formulas $\tilde{\sigma}$; we would regard the positive formulas $\tilde{\tau}$ as the approximations of $\tilde{\sigma}$. The relationships in the previous paragraph yield that σ is approximately satisfied in a given normed space structure $\mathcal{M}$ if and only if $\tilde{\sigma}$ is approximately satisfied in $\mathcal{M}$ (in the sense that every approximation $\tilde{\tau}$ is true in $\mathcal{M}$). Hence there is no difference between σ and $\tilde{\sigma}$ when the approximate semantics is used.

We chose to formulate this theory so that the two bounded quantifiers $\exists_r y$ and $\forall_r y$ are relativized to the same condition, namely $\|y\| \leq r$, in the belief that this would seem more natural to a wider mathematical audience.

Earlier Approaches

In a slightly different form, positive bounded formulas and approximate satisfaction were introduced for Banach spaces (for the most part considered without additional structure) in [Hen76]; see also [HM83b] and [HH86]. Here we consider a somewhat larger collection of positive bounded formulas than was used in those papers, but the differences are slight and inessential. (See the discussion of alternative formulations in Chapter 11.)

Formal languages for structures included among the ones considered here were also introduced by Krivine [Kri72] [Kri74] and by Stern [Ste74] [Ste75] [Ste76] [Ste77] [Ste78]. The formulas considered in these papers had limited quantifier complexity, either universal, existential, or universal-existential, and the focus of the work was on subspace properties or on relationships such as local reflexivity. These papers did not present a fully developed model theory for Banach spaces nor were they adequate to completely characterize the properties shared by a Banach space and its ultrapowers. These papers do contain many interesting results giving axiomatizations of familiar classes of spaces, and these results can be adapted to yield axiomatizations by positive bounded sentences for the most part. See the remarks in Chapter 13.

6

Basic Model Theory I

6.1 Definition. Let L be a signature for normed space structures. A *positive bounded theory* in L is a set of positive bounded L-sentences (*i.e.*, formulas without free variables). If Γ is a positive bounded theory in L and $\mathcal{M}$ is a normed space L-structure, we say that $\mathcal{M}$ *is an approximate model* of Γ if $\mathcal{M} \models_A \Gamma$. This is equivalent to saying that every sentence in Γ^+ is true in $\mathcal{M}$.

If $\mathcal{M}$ is a normed space L-structure, the *positive bounded theory of* $\mathcal{M}$, denoted $\mathrm{Th}_A(\mathcal{M})$, is the set of positive bounded L-sentences which are approximately satisfied by $\mathcal{M}$.

6.2 Definition. Suppose that $\mathcal{M}$ and $\mathcal{N}$ are normed space L-structures based on $(M^{(s)} \mid s \in S)$ and $(N^{(s)} \mid s \in S)$, respectively.

(i) We say that $\mathcal{M}$ and $\mathcal{N}$ are *approximately elementarily equivalent*, and write

$$\mathcal{M} \equiv_A \mathcal{N},$$

if $\mathcal{M}$ and $\mathcal{N}$ approximately satisfy exactly the same positive bounded L-sentences *i.e.* if $\mathrm{Th}_A(\mathcal{M}) = \mathrm{Th}_A(\mathcal{N})$.

(ii) If $\mathcal{M} \subseteq \mathcal{N}$ we say that $\mathcal{M}$ is an *approximate elementary substructure of* $\mathcal{N}$, and write

$$\mathcal{M} \preceq_A \mathcal{N},$$

if the following condition holds: whenever $\varphi(x_1, \ldots, x_n)$ is a positive bounded L-formula and $a_1, \ldots, a_n$ are elements of suitable sorts of $\mathcal{M}$, we have

$$\mathcal{M} \models_A \varphi[a_1, \ldots, a_n] \quad \text{if and only if} \quad \mathcal{N} \models_A \varphi[a_1, \ldots, a_n].$$

31

In this case, we also say that $\mathcal{N}$ is an *approximate elementary extension* of $\mathcal{N}$.

(iii) An *approximate elementary embedding* of $\mathcal{M}$ into $\mathcal{N}$ consists of a family of maps $(T^{(s)} \mid s \in S)$ with $T^{(s)} \colon M^{(s)} \to N^{(s)}$ for $s \in S$ such that the following condition holds: whenever $\varphi(x_1, \ldots, x_n)$ is a positive bounded L-formula and $a_1, \ldots, a_n$ are elements of $\mathcal{M}$, where x_k is a variable of sort s_k and $a_k \in M^{(s_k)}$ for $k = 1, \ldots, n$, we have

$$\mathcal{M} \models_A \varphi[a_1, \ldots, a_n] \quad \text{iff} \quad \mathcal{N} \models_A \varphi[T^{(s_1)}(a_1), \ldots, T^{(s_n)}(a_n)].$$

6.3 Remarks.

(i) Every approximate elementary embedding of one normed space structure into another is an embedding.

(ii) The collection of approximate elementary embeddings is closed under composition.

(iii) Every isomorphism between normed space structures is an approximate elementary embedding.

6.4 Proposition. *Let $\mathcal{M}$ and $\mathcal{N}$ be normed space L-structures. The following conditions are equivalent:*

(i) $\mathcal{M} \equiv_A \mathcal{N}$;

(ii) *For every positive bounded L-sentence φ and every approximation φ' of φ,*

$$\mathcal{M} \models \varphi \quad \text{implies} \quad \mathcal{N} \models \varphi'.$$

Proof. The implication (i)$\Rightarrow$(ii) is trivial. For the converse, note that condition (ii) implies immediately that $\mathrm{Th}_A(\mathcal{M}) \subseteq \mathrm{Th}_A(\mathcal{N})$. To complete the proof suppose that (ii) is true but that there exists a positive bounded sentence ψ such that $\mathcal{N} \models_A \psi$ and $\mathcal{M} \not\models_A \psi$. Then there exists an approximation ψ' of ψ such that $\mathcal{M} \models \mathrm{neg}(\psi')$ by Proposition 5.14. Let σ, σ' be approximations of ψ such that $\psi < \sigma < \sigma' < \psi'$. By 5.13(iii) we have $\mathrm{neg}(\psi') < \mathrm{neg}(\sigma')$. Hence by the assumed condition (ii) we get $\mathcal{N} \models \mathrm{neg}(\sigma')$. By Remark 5.13(iv) we have $\mathcal{N} \not\models \sigma$ which contradicts the assumption that $\mathcal{N} \models_A \psi$. $\qquad\square$

6.5 Remark. Proposition 6.4 shows the usefulness of the weak negation operation. This result yields that if $\mathcal{M}$ and $\mathcal{N}$ are normed space L-structures with $\mathrm{Th}_A(\mathcal{M}) \subseteq \mathrm{Th}_A(\mathcal{N})$, then $\mathrm{Th}_A(\mathcal{M}) = \mathrm{Th}_A(\mathcal{N})$ (that is, $\mathcal{M} \equiv_A \mathcal{N}$.

6.6 Proposition (Tarski-Vaught Test for $\preceq_A$). *Suppose $\mathcal{M}, \mathcal{N}$ are normed space L-structures with $\mathcal{M} \subseteq \mathcal{N}$. The following conditions are equivalent:*

(i) $\mathcal{M} \preceq_A \mathcal{N}$;

(ii) *For every positive bounded L-formula $\varphi(x_1, \ldots, x_n, y)$ and every approximation φ' of φ the following condition holds: if $a_1, \ldots, a_n$ are elements of suitable sorts of $\mathcal{M}$ and b is in a suitable sort of $\mathcal{N}$, and if*

$$\mathcal{N} \models \varphi[a_1, \ldots, a_n, b],$$

then there exists c in a suitable sort of $\mathcal{M}$ such that

$$\mathcal{N} \models \varphi'[a_1, \ldots, a_n, c].$$

Proof. The implication from the first condition to the second is immediate from the definition of $\preceq_A$. For the converse direction, one proves the equivalence

$$\mathcal{M} \models_A \varphi[a_1, \ldots, a_n] \quad \text{if and only if} \quad \mathcal{N} \models_A \varphi[a_1, \ldots, a_n]$$

(for $a_1, \ldots, a_n$ in suitable sorts of $\mathcal{M}$) by induction on the complexity of φ, using the given condition to cover the case when φ begins with a quantifier. □

6.7 Corollary. *Let $\mathcal{M}$ and $\mathcal{N}$ be normed space L-structures with $\mathcal{M} \preceq_A \mathcal{N}$. Let $\overline{\mathcal{M}}$ be the completion of $\mathcal{M}$ and let $\mathcal{M}'$ be the closure of $\mathcal{M}$ in $\mathcal{N}$. Then*

(i) $\mathcal{M} \preceq_A \overline{\mathcal{M}}$ and

(ii) $\mathcal{M} \preceq_A \mathcal{M}' \preceq_A \mathcal{N}$.

Proof. Use the Perturbation Lemma (Proposition 5.15) to verify the second condition in Proposition 6.6. □

7

Quantifier-Free Formulas

In this chapter we consider normed spaces (over $\mathbb{R}$) without any additional structure, and we give some information about what can be expressed by quantifier-free positive bounded formulas.

Here L_0 is the signature for normed spaces. This has two sorts, one of which is for the scalar field $\mathbb{R}$, and function symbols corresponding to the fact that each sort is regarded as a normed space. If X is a normed space, we regard it as an L_0-structure in an obvious way, as follows: the distinguished functions in this L_0-structure include the field operations $+, -, \times$, the identities 0 and 1, and the absolute value function, all on $\mathbb{R}$; in addition there are the vector addition and subtraction functions (mapping $X \times X$ to X), the norm (mapping X to $\mathbb{R}$), the scalar multiplication (mapping $\mathbb{R} \times X$ to X), and the additive identity 0 in X. The signature also has constant symbols naming the rational numbers within the sort $\mathbb{R}$.

Let X and Y be normed spaces, and consider $\bar{a} = (a_1, \ldots, a_n) \in X^n$ and $\bar{b} = (b_1, \ldots, b_n) \in Y^n$. Let $\lambda \geq 1$. We say $\bar{a}$ and $\bar{b}$ are λ-*equivalent*, and write $\bar{a} \sim_\lambda \bar{b}$, if for every sequence of scalars $r_1, \ldots, r_n$ we have

$$\lambda^{-1} \left\| \sum_{j=1}^n r_j a_j \right\|_X \leq \left\| \sum_{j=1}^n r_j b_j \right\|_Y \leq \lambda \left\| \sum_{j=1}^n r_j a_j \right\|_X.$$

Equivalently, $\bar{a} \sim_\lambda \bar{b}$ holds iff there is a linear isomorphism T from the linear span of $\bar{a}$ onto the linear span of $\bar{b}$ satisfying $T(a_j) = b_j$ for all $j = 1, \ldots, n$ and with $\|T\| \leq \lambda$ and $\|T^{-1}\| \leq \lambda$. (We refer to a linear isomorphism satisfying these norm bounds as a λ-*isomorphism*.)

7.1 Theorem. *Let $\bar{a} = (a_1, \ldots, a_n)$ be a linearly independent sequence in a normed space X and let $1 \leq \lambda < \eta$. There exists a quantifier-free positive bounded L_0-formula φ and an approximation $\varphi' > \varphi$ such that*

34

for any normed space Y and any $\bar{b} = (b_1, \ldots, b_n) \in Y^n$ the following two conditions hold:

(i) *If $\bar{a} \sim_\lambda \bar{b}$ then $Y \models \varphi[\bar{b}]$;*
(ii) *if $Y \models \varphi'[\bar{b}]$ then $\bar{b} \sim_\eta \bar{a}$.*

Proof. The construction of φ and φ' is based on a familiar set of estimates and depends on the compactness of closed bounded sets in finite dimensional normed spaces.

Fix $\bar{a} = (a_1, \ldots, a_n)$ linearly independent in a normed space X. Let E be the linear span of $\bar{a}$. Fix r and $\epsilon > 0$ such that $\|a_i\| \le r$ for all $i = 1, \ldots, n$ and

$$\left\| \sum_{j=1}^{n} r_j a_j \right\| \ge \epsilon \cdot \max(|r_1|, \ldots, |r_n|)$$

for all $r_1, \ldots, r_n \in \mathbb{R}$. Fix δ close to 0 in the interval $0 < \delta < 1/2$, with its size to be determined later. We suppose that r, ϵ, δ are all in $\mathbb{Q}$.

Let F be a finite subset of $\mathbb{Q}^n$ such that for each $(r_1, \ldots, r_n) \in F$ we have

$$1 - \delta \le \left\| \sum_{j=1}^{n} r_j a_j \right\| \le 1 + \delta.$$

We take F large enough so that for any $z \in E$ with $\|z\| = 1$ there exists $(r_1, \ldots, r_n) \in F$ such that

$$\left\| z - \sum_{j=1}^{n} r_j a_j \right\| \le \delta.$$

We let $\varphi(v_1, \ldots, v_n)$ be the conjunction of the atomic formulas $\|v_j\| \le \lambda r$ for $j = 1, \ldots, n$ as well as the atomic formulas

$$(1 - \delta)\lambda^{-1} \le \left\| \sum_{j=1}^{n} r_j v_j \right\| \le (1 + \delta)\lambda$$

as $(r_1, \ldots, r_n)$ ranges over F. Note that since L_0 has a constant naming each rational number, this is a well-formed positive bounded quantifier-free formula in L_0.

We let φ' be formed in the same way as φ, except we replace each occurrence of δ by 2δ and we replace $r\lambda$ by $r\lambda + \delta$. It is clear that φ' is an approximation of φ.

The construction makes it clear that φ satisfies condition (i) in the Theorem. Suppose Y is any normed space and $\bar{b} = (b_1, \ldots, b_n)$ is any sequence in Y such that $Y \models \varphi'[b_1, \ldots, b_n]$. We must show that $\bar{b} \sim_\eta \bar{a}$. To do this it suffices to consider any element $z = \sum_{j=1}^n s_j a_j$ with $\|z\| = 1$ and to show that

$$\eta^{-1} \leq \left\| \sum_{j=1}^n s_j b_j \right\| \leq \eta.$$

Given such an element z, choose $(r_1, \ldots, r_n) \in F$ such that

$$\left\| z - \sum_{j=1}^n r_j a_j \right\| \leq \delta.$$

This sequence determines one of the atomic formulas that we included in φ', from which we conclude that

$$(1 - 2\delta)\lambda^{-1} \leq \left\| \sum_{j=1}^n r_j b_j \right\| \leq (1 + 2\delta)\lambda.$$

Our choice of ϵ ensures that

$$\delta \geq \left\| \sum_{j=1}^n r_j a_j - \sum_{j=1}^n s_j a_j \right\| \geq \epsilon \cdot \max(|r_j - s_j| \mid j = 1, \ldots, n).$$

Hence, using the triangle inequality we get

$$\left\| \sum_{j=1}^n r_j b_j - \sum_{j=1}^n s_j b_j \right\| \leq \frac{\delta}{\epsilon} n(r\lambda + \delta).$$

Let $\gamma = \frac{\delta}{\epsilon} n(r\lambda + \delta)$. Note that γ can be made arbitrarily small by choosing δ small enough. From these estimates we conclude that

$$(1 - 2\delta)\lambda^{-1} - \gamma \leq \left\| \sum_{j=1}^n s_j b_j \right\| \leq (1 + 2\delta)\lambda + \gamma.$$

This makes it clear that φ' satisfies (ii) as long as we choose δ small enough so that $(1 + 2\delta)\lambda + \gamma \leq \eta$ and $(1 - 2\delta)\lambda^{-1} - \gamma \geq \eta^{-1}$. $\qquad \square$

Note that the preceding Theorem cannot be true in precisely this form if $\bar{a} = (a_1, \ldots, a_n)$ is linearly dependent. Suppose otherwise, that formulas $\varphi' > \varphi$ exist satisfying the conditions of the Theorem. We may consider $\bar{a}$ embedded isometrically in an infinite dimensional normed space X. Let $\delta > 0$ be as in the Perturbation Lemma (Proposition 5.15) for the formulas $\varphi' > \varphi$ and for $r = \max(\|a_1\|, \ldots, \|a_n\|)$. There exists a perturbation $\bar{b}$ of $\bar{a}$ in X such that $\bar{b}$ is linearly independent and such that $\|b_j - a_j\| < \delta$ for all $j = 1, \ldots, n$. Since $X \models \varphi[\bar{a}]$ it follows from the choice of δ that $X \models \varphi'[\bar{b}]$. Therefore $\bar{a}$ is isomorphically equivalent to $\bar{b}$, from which it follows that $\bar{a}$ must be linearly independent.

However, the quantifier-free formulas shown to exist in the preceding Theorem are sufficient for most purposes. We will next use them to show that whenever $X \preceq_A Y$, the normed spaces X and Y have the same "local reflexivity" relationship that X has with its second dual. (However, there is a limit to this analogy; indeed, the second dual of X is rarely an approximate elementary extension of X.)

7.2 Corollary. *Let X and Y be normed spaces with $X \preceq_A Y$. Let E be a finite dimensional subspace of Y and $\eta > 1$. There exists an η-isomorphism T of E into X such that $T(x) = x$ for all $x \in E \cap X$. In particular, Y is finitely representable in X.*

Proof. Let X, Y, E, λ be as stated. Choose a basis $a_1, \ldots, a_m$ for E such that $a_1, \ldots, a_k$ is a basis for $E \cap X$, $k \leq m$. Taking $\lambda = 1$, let $\varphi < \varphi'$ be quantifier-free positive bounded formulas as in Theorem 7.1. Let $r = \max(\|a_{k+1}\|, \ldots, \|a_m\|)$ and $s > r$. By the choice of φ we have $Y \models \exists_r x_{k+1} \ldots \exists_r x_m \varphi[a_1, \ldots, a_k]$. Since $X \preceq_A Y$ this implies $X \models \exists_s x_{k+1} \ldots \exists_s x_m \varphi'[a_1, \ldots, a_k]$. Hence there exists $b_{k+1}, \ldots, b_m \in X$ such that $X \models \varphi'[a_1, \ldots, a_k, b_{k+1}, \ldots, b_m]$. Therefore

$$(a_1, \ldots, a_k, a_{k+1}, \ldots, a_m) \sim_\eta (a_1, \ldots, a_k, b_{k+1}, \ldots, b_m)$$

by the choice of φ'. That is, there is an η-isomorphism T such that for $j = 1, \ldots, m$ we have $T(a_j) = a_j$ and for $j = k+1, \ldots, m$ we have $T(a_j) = b_j$. Therefore T is the desired η-isomorphism of E into X. $\square$

In the case where Y is an ultrapower or a nonstandard hull of X, the preceding result was proved independently by Stern [Ste76] and by Henson and Moore [HM74a] [HM74b].

Note the following consequence of Corollary 7.2: suppose $\mathcal{M}$ and $\mathcal{N}$ are normed space L-structures and $\mathcal{M} \preceq_A \mathcal{N}$. If $s \in S$ and $M^{(s)}$ is finite dimensional, then $N^{(s)} = M^{(s)}$.

8

Ultraproducts of Normed Space Structures

Uniform Families of L-structures

8.1 Definition. Suppose that $(\mathcal{M}_\xi \mid \xi \in \Lambda)$ is a family of normed space L-structures and that $\mathcal{M}_\xi$ is based on $(M_\xi^{(s)} \mid s \in S)$ for each $\xi \in \Lambda$. Let $f \colon s_1 \times \cdots \times s_m \to s_0$ be a function symbol of L, so $f^{\mathcal{M}_\xi}$ denotes the interpretation of f in $\mathcal{M}_\xi$. We say that $(f^{\mathcal{M}_\xi} \mid \xi \in \Lambda)$ is a *uniform family of functions* if the following two conditions hold. (They express the requirement that the functions $f^{\mathcal{M}_\xi}$ are bounded and uniformly continuous on bounded subsets of their domains in a way that is *uniform* in the index $\xi \in \Lambda$.)

(i) (Uniform Boundedness Condition) For each sequence of positive real numbers $r_1, \ldots, r_m$ there exists a positive real number R such that for arbitrary $\xi \in \Lambda$ and

$$ x_\xi^k \in \mathcal{B}_{r_k}(M_\xi^{(s_k)}), \qquad k = 1, \ldots, m, $$

we have

$$ \|f^{\mathcal{M}_\xi}(x_\xi^1, \ldots, x_\xi^m)\| \le R. $$

When this holds, the number R is called a *uniform bound for* $(f_\xi^{\mathcal{M}} \mid \xi \in \Lambda)$ *on* $(\mathcal{B}_{r_1}(M_\xi^{(s_1)}) \times \cdots \times \mathcal{B}_{r_m}(M_\xi^{(s_m)}) \mid \xi \in \Lambda)$.

(ii) (Uniform Continuity Condition) For each sequence of positive real numbers $r_1, \ldots, r_m$ there exists a function $\Delta \colon (0, \infty) \to (0, \infty)$ such that for every $\xi \in \Lambda$, Δ is a modulus of uniform continuity for $f^{\mathcal{M}_\xi}$ on $\mathcal{B}_{r_1}(M_\xi^{(s_1)}) \times \cdots \times \mathcal{B}_{r_m}(M_\xi^{(s_m)})$ (see Definition 4.2). In this case, we say that Δ is *uniform modulus of uniform continuity for* $(f_\xi^{\mathcal{M}} \mid \xi \in \Lambda)$ *on* $(\mathcal{B}_{r_1}(M_\xi^{(s_1)}) \times \cdots \times \mathcal{B}_{r_m}(M_\xi^{(s_m)}) \mid \xi \in \Lambda)$.

We say that a family $(\mathcal{M}_\xi \mid \xi \in \Lambda)$ of normed space L-structures is *uniform* if for every function symbol f of L, the family of functions $(f^{\mathcal{M}_\xi} \mid \xi \in \Lambda)$ is uniform.

A class $\mathfrak{C}$ of normed space L-structures is called *uniform* if every family of structures in $\mathfrak{C}$ is uniform.

8.2 Remark. Suppose that $(\mathcal{M}_\xi \mid \xi \in \Lambda)$ is a family of normed space L-structures and that $\mathcal{M}_\xi$ is based on $(M_\xi^{(s)} \mid s \in S)$ for each $\xi \in \Lambda$. Let $f \colon s_1 \times \cdots \times s_m \to s_0$ be a function symbol of L. Suppose the uniform continuity condition holds for $(f_\xi^{\mathcal{M}} \mid \xi \in \Lambda)$ on $(\mathcal{B}_{r_1}(M_\xi^{(s_1)}) \times \cdots \times \mathcal{B}_{r_m}(M_\xi^{(s_m)}) \mid \xi \in \Lambda)$. In order for the uniform boundedness condition to hold for $(f_\xi^{\mathcal{M}} \mid \xi \in \Lambda)$ on $(\mathcal{B}_{r_1}(M_\xi^{(s_1)}) \times \cdots \times \mathcal{B}_{r_m}(M_\xi^{(s_m)}) \mid \xi \in \Lambda)$ it suffices for the family $(\|f^{\mathcal{M}_\xi}(0,\ldots,0)\| \mid \xi \in \Lambda)$ to be bounded in $\mathbb{R}$..

In particular, if f is a constant symbol, $(f^{\mathcal{M}_\xi} \mid \xi \in \Lambda)$ is uniform iff there is a uniform bound on $(\|f^{\mathcal{M}_\xi}\| \mid \xi \in \Lambda)$ (*i.e.*, the uniform continuity requirement is automatic).

8.3 Examples. The following are some examples of uniform classes of L-structures, for various signatures L:

(i) Normed spaces over $\mathbb{R}$.

(ii) Normed spaces over $\mathbb{C}$.

(iii) Normed vector lattices.

(iv) Normed algebras.

(v) C^*-algebras.

(vi) Hilbert spaces (with inner product included as a function).

(vii) Dual pairs (X, X'), where X is a Banach space, X' is the dual of X, and the pairing is included as a function.

(viii) Triples (X, X', X''), where X' and X'' are the dual and the double dual of X and the pairing between X and X', the pairing between X' and X'', and the embedding $X \to X''$ are included as functions.

(ix) Operator spaces (with the infinite family of matrix norms included as functions).

(x) Any uniform class $\mathfrak{C}$ of normed space structures can be expanded to a uniform class by adding to each structure in $\mathfrak{C}$ a list of bounded linear operators as functions, as long as for each additional function symbol f there is a uniform bound on the operator norms of $f^{\mathcal{M}}$ as $\mathcal{M}$ ranges over $\mathfrak{C}$.

8.4 Remarks. Let $\mathfrak{C}$ be a uniform class of normed space L-structures. There exists a set Γ of positive bounded sentences which expresses the specific conditions on $\mathfrak{C}$ that ensure it is uniform. That is, every structure in $\mathfrak{C}$ is an approximate model of Γ and the class of all approximate models of Γ is uniform itself. Moreover, the sentences in Γ can all be taken to be in prenex normal form with only $\forall$ quantifiers. The following two statement follow from this:

(i) If $\mathfrak{D}$ is the class of all L-structures isomorphic to a substructure of some structure in $\mathfrak{C}$, then $\mathfrak{D}$ is also a uniform class of L-structures.

(ii) If $\mathfrak{D}$ is the class of all L-structures that are approximately elementarily equivalent to some member of $\mathfrak{C}$, then $\mathfrak{D}$ is also a uniform class of L-structures.

Ultraproducts of Normed Space L-structures

Suppose that $(\mathfrak{M}_\xi \mid \xi \in \Lambda)$ is a family of normed space L-structures and that $\mathfrak{M}_\xi$ is based on $(M_\xi^{(s)} \mid s \in S)$ for each $\xi \in \Lambda$. Let $f\colon s_1 \times \cdots \times s_m \to s_0$ be a function symbol of L such that the family $(f^{\mathfrak{M}_\xi} \mid \xi \in \Lambda)$ of interpretations of f is uniform. Given an ultrafilter D on Λ, we define a function

$$\left(\prod_{\xi \in \Lambda} f^{M_\xi}\right)_D \colon \left(\prod_{\xi \in \Lambda} M_\xi^{(s_1)}\right)_D \times \cdots \times \left(\prod_{\xi \in \Lambda} M_\xi^{(s_m)}\right)_D \to \left(\prod_{\xi \in \Lambda} M_\xi^{(s_0)}\right)_D$$

as follows. If $(x_\xi^k)_{\xi \in \Lambda} \in \ell_\infty(\Lambda, (M_\xi^{(s_k)} \mid \xi \in \Lambda))$ for $k = 1, \ldots, m$, we let

$$\left(\prod_{\xi \in \Lambda} f^{M_\xi}\right)_D (((x_\xi^1)_{\xi \in \Lambda})_D, \ldots, ((x_\xi^m)_{\xi \in \Lambda})_D) =$$

$$((f^{M_\xi}(x_\xi^1, \ldots, x_\xi^m))_{\xi \in \Lambda})_D.$$

The fact that the family $(f^{\mathfrak{M}_\xi} \mid \xi \in \Lambda)$ is uniform ensures that the function $(\prod_{\xi \in \Lambda} f^{M_\xi})_D$ is well-defined. Note that any uniform modulus of uniform continuity for $(f^{\mathfrak{M}_\xi} \mid \xi \in \Lambda)$ on

$$(\, \mathcal{B}_{r_1}(M_\xi^{(s_1)}) \times \cdots \times \mathcal{B}_{r_m}(M_\xi^{(s_m)}) \mid \xi \in \Lambda \,)$$

is also a modulus of uniform continuity for $(\prod_{\xi \in \Lambda} f^{M_\xi})_D$ on

$$\mathcal{B}_{r_1}\left(\left(\prod_{\xi \in \Lambda} M_\xi^{(s_1)}\right)_D\right) \times \cdots \times \mathcal{B}_{r_m}\left(\left(\prod_{\xi \in \Lambda} M_\xi^{(s_m)}\right)_D\right).$$

Similarly, any uniform bound for $(f^{\mathfrak{M}_\xi} \mid \xi \in \Lambda)$ yields a bound for $(\prod_{\xi \in \Lambda} f^{M_\xi})_D$.

8.5 Definition. If $(\mathcal{M}_\xi \mid \xi \in \Lambda)$ is a uniform family of normed space L-structures such that $\mathcal{M}_\xi$ is based on $(M_\xi^{(s)} \mid s \in S)$ and D is an ultrafilter on Λ, the D-*ultraproduct* of $(\mathcal{M}_\xi \mid \xi \in \Lambda)$, denoted

$$\left(\prod_{\xi \in \Lambda} \mathcal{M}_\xi \right)_D$$

is the L-structure $\mathcal{M}$ defined as follows. The sorts of $\mathcal{M}$ are the normed space D-ultraproducts

$$\mathcal{M}^{(s)} = \left(\prod_{\xi \in \Lambda} M_\xi^{(s)} \right)_D$$

for each $s \in S$, and

$$f^{\mathcal{M}} = \left(\prod_{\xi \in \Lambda} f^{\mathcal{M}_\xi} \right)_D$$

for each function symbol $f \colon s_1 \times \cdots \times s_m \to s_0$ of L.

8.6 Remarks.

(i) The definition of uniform class of normed space structures is precisely what is needed to ensure that the normed space ultraproduct construction makes sense when applied to any family taken from the class. That is, if $(\mathcal{M}_\xi \mid \xi \in \Lambda)$ is a family of normed space L-structures, then the definition of $(\prod_{\xi \in \Lambda} \mathcal{M})_D$ makes sense for every ultrafilter D on Λ if and only if $(\mathcal{M}_\xi \mid \xi \in \Lambda)$ is uniform.

(ii) Many operations on normed space structures preserve uniformity. For example, if $\mathfrak{C}$ is a uniform class of structures, then the class $\mathfrak{D}$ that results from closing $\mathfrak{C}$ under ultraproducts is also uniform. See also Remark 8.4.

Nonstandard Hulls

Let D be an ultrafilter on the index set Λ and let $(\mathcal{M}_\xi \mid \xi \in \Lambda)$ be a uniform family of normed space L-structures. As indicated on page 19 for ultrapowers, the ultraproduct $(\prod_{\xi \in \Lambda} \mathcal{M})_D$ can be seen as a nonstandard hull, when the superstructure V and nonstandard extension $^*V$ are suitably chosen.

Namely, let V be any set theoretic superstructure which contains the set $S = \{\mathcal{M}_\xi \mid \xi \in \Lambda\}$ as an element. Let $^*V$ be the D-ultrapower nonstandard extension of V. The family $(\mathcal{M}_\xi \mid \xi \in \Lambda)$ corresponds to an internal element M of $^*S$; in particular, M is an internal normed space L-structure. For each function symbol f of L, the interpretation f^M of f in M is the internal function whose representative is the family $(f^{\mathcal{M}_\xi} \mid \xi \in \Lambda)$. The requirement that the family $(\mathcal{M}_\xi \mid \xi \in \Lambda)$ be uniform ensures that for each function symbol f of L, the internal function f^M takes finite arguments to finite values, and preserves the infinitesimal equivalence relation. Hence M gives rise to a natural quotient structure (keeping finite elements of M and identifying two of them if their difference is infinitesimal) which is a normed space L-structure, denoted by $\widehat{M}$. The nonstandard hull $\widehat{M}$ is isomorphic, as a normed space L-structure, to the ultraproduct $(\prod_{\xi \in \Lambda} \mathcal{M})_D$.

9
Basic Model Theory II

In this chapter we continue the development of basic model theory for normed space structures, including such fundamental results as the Compactness Theorem (Propositions 9.5 and 9.22).

Our first result gives a uniform version of the Perturbation Lemma (Proposition 5.15), strengthening a key property of positive bounded formulas and approximate satisfaction.

9.1 Proposition (Perturbation Lemma for Uniform Classes).
Suppose that $\mathfrak{C}$ is a uniform class of L-structures and $\varphi(x_1, \ldots, x_n)$ is a positive bounded L-formula. Then for every $r > 0$ and every approximation φ' of φ there exists $\delta > 0$ such that whenever $\mathcal{M}$ is a structure in $\mathfrak{C}$ and $a_1, \ldots, a_n, b_1, \ldots, b_n$ are elements of suitable sorts of $\mathcal{M}$, we have that

$$\mathcal{M} \models \bigwedge_{1 \leq i \leq n} \|a_i\| \leq r \ \wedge \bigwedge_{1 \leq i \leq n} \|a_i - b_i\| \leq \delta \ \wedge \ \varphi[a_1, \ldots, a_n]$$

implies

$$\mathcal{M} \models \varphi'[b_1, \ldots, b_n].$$

Proof. The proof of Proposition 5.15 depends only on being given specific moduli of uniform continuity and uniform bounds for each of the functions in the structure being considered. Therefore, if all structures considered come from a uniform class of L-structures, then the dependence of δ on r, φ, and φ' can be taken to be the same for all of them. $\square$

Ultraproducts and $\models_{\mathcal{A}}$

The next result is the analogue, in the model theory of normed space structures, of the result in first-order model theory that is known as

43

the *Fundamental Theorem on Ultraproducts* or as the *Łoś Theorem*, and which was proved by J. Łoś [Łoś55].

9.2 Proposition. *Let* $(\mathcal{M}_\xi \mid \xi \in \Lambda)$ *be a uniform family of normed space L-structures and let* $\varphi(x_1, \ldots, x_n)$ *be a positive bounded formula, where* x_k *is a variable of sort* s_k, *for* $k = 1, \ldots, n$. *If* $\mathcal{M}_\xi$ *is based on* $(M_\xi^{(s)} \mid s \in S)$ *and* $a_k \in \ell_\infty(\Lambda, (M_\xi^{(s_k)} \mid \xi \in \Lambda))$, *for* $k = 1, \ldots, n$, *then*

$$\left(\prod_{\xi \in \Lambda} \mathcal{M}_\xi \right)_D \models_A \varphi[\, (a_1)_D, \ldots, (a_n)_D \,]$$

if and only if for every approximation φ' *of* φ, *the set*

$$\{ \xi \in \Lambda \mid \mathcal{M}_\xi \models \varphi'[a_1(\xi), \ldots, a_n(\xi)] \}$$

is in the ultrafilter D.

Proof. The proof is by induction on the complexity of φ. Use Proposition 5.14.   □

9.3 Corollary. *If* $\mathcal{M}$ *is a normed space structure based on* $(M^{(s)} \mid s \in S)$ *and* $T^{(s)} \colon M^{(s)} \to (M^{(s)})_D$ *is the diagonal embedding for each* $s \in S$, *then the family* $(T^{(s)} \mid s \in S)$ *is an approximate elementary embedding of* $\mathcal{M}$ *into* $(\mathcal{M})_D$.

Proof. From Proposition 9.2.   □

9.4 Corollary. *If* $\mathcal{M}$ *and* $\mathcal{N}$ *are normed space L-structures and they have isomorphic ultrapowers, then* $\mathcal{M} \equiv_A \mathcal{N}$.

The converse of the preceding result is also true, in a strong form: in particular, if $\mathcal{M}$ and $\mathcal{N}$ are normed space L-structures and $\mathcal{M} \equiv_A \mathcal{N}$, then there exists an ultrafilter D such that $(\mathcal{M})_D$ is isomorphic to $(\mathcal{N})_D$. Chapter 10 below is devoted to proving a series of results of this kind.

In the rest of this chapter we regularly use notation established in Definition 3.7, which we now recall. Let L be a signature and $L(C)$ an extension of L by constants, with C being the set of new constants that has been added to L. For any L-structure $\mathcal{M}$, an arbitrary expansion of $\mathcal{M}$ to an $L(C)$-structure is denoted $(\mathcal{M}, \mathbf{A})$; $\mathbf{A}(c)$ is taken to be the interpretation of the constant symbol c in this expansion, for each $c \in C$. The function $\mathbf{A}$ is referred to as an *interpretation of C in $\mathcal{M}$*. Note that each constant c must be assigned a unique sort from L in order to specify the signature $L(C)$, and the function $\mathbf{A}$ must respect these sort assignments.

The Compactness Theorem

9.5 Proposition (Compactness Theorem). *Let Γ be a set of positive bounded L-sentences. Suppose that $\mathfrak{C}$ is a uniform class of normed space L-structures such that for each finite subset Σ of Γ^+ there exists a structure in $\mathfrak{C}$ which satisfies Σ. Then there exists an ultraproduct of structures from $\mathfrak{C}$ which approximately satisfies the entire set Γ.*

Proof. Let Λ be the set of finite subsets of Γ^+ and for each $\xi \in \Lambda$ let $\mathcal{M}_\xi$ be a normed space L-structure in $\mathfrak{C}$ which satisfies every sentence in ξ. For every finite subset Δ of Γ^+ let F_Δ be the set of all $\xi \in \Lambda$ such that $\Delta \subseteq \xi$. Note that F_Δ is contained in $\{\xi \in \Lambda \mid \mathcal{M}_\xi \models \Delta\}$. The family $\mathcal{F}$ of sets of the form F_Δ is closed under finite intersections. Then

$$\left(\prod_{\xi \in \Lambda} \mathcal{M}_\xi \right)_D \models_A \Gamma$$

for any ultrafilter D on Λ extending $\mathcal{F}$, by Proposition 9.2. $\qquad\square$

9.6 Remark. We will see in Proposition 9.22 that the ultraproduct constructed in the proof of Proposition 9.5 actually *satisfies* (not just approximately satisfies) every sentence in Γ.

9.7 Corollary. *Let $\mathfrak{C}$ be a uniform class of normed space L-structures which is closed under ultraproducts. The following conditions are equivalent for a positive bounded theory Γ in L:*

(i) *There exists a structure $\mathcal{M}$ in $\mathfrak{C}$ such that $\Gamma = \mathrm{Th}_A(\mathcal{M})$;*

(ii) *Every finite subset of Γ^+ is satisfied by some structure in $\mathfrak{C}$ and for every positive bounded L-sentence φ and every $\varphi' > \varphi$ either $\varphi \in \Gamma$ or $\mathrm{neg}(\varphi') \in \Gamma$.*

Proof. The implication (i)$\Rightarrow$(ii) follows from Proposition 5.14. To prove (ii)$\Rightarrow$(i), use Proposition 9.5 to obtain a structure $\mathcal{M}$ in $\mathfrak{C}$ such that $\mathcal{M} \models_A \Gamma$. If (i) fails to hold, then there is a positive bounded sentence φ such that $\mathcal{M} \models_A \varphi$ but $\varphi \notin \Gamma$. Let φ', φ'' be approximations of φ with $\varphi < \varphi' < \varphi''$. By (ii) we have $\mathrm{neg}(\varphi'') \in \Gamma$ and hence $\mathcal{M} \models_A \mathrm{neg}(\varphi'')$. By 5.13(iv) we conclude $\mathcal{M} \not\models \varphi'$ which contradicts our assumption that $\mathcal{M} \models_A \varphi$. $\qquad\square$

9.8 Definition. A set Γ of positive bounded L-sentences that satisfies $\Gamma = \mathrm{Th}_A(\mathcal{M})$ for some L-structure $\mathcal{M}$ is called a *complete* positive bounded theory in L.

9.9 Proposition. *Let* $\mathcal{M}$ *and* $\mathcal{N}$ *be normed space L-structures based on* $(M^{(s)} \mid s \in S)$ *and* $(N^{(s)} \mid s \in S)$, *respectively, such that* $\mathcal{M} \subseteq \mathcal{N}$. *The following conditions are equivalent:*

(i) $\mathcal{M} \preceq_A \mathcal{N}$;

(ii) *There exists an ultrapower* $(\mathcal{M})_D$ *of* $\mathcal{M}$ *and an elementary embedding* $(T^{(s)} \mid s \in S)$ *of* $\mathcal{N}$ *into* $(\mathcal{M})_D$ *such that* $T^{(s)}$ *is the identity on* $M^{(s)}$ *for each* $s \in S$.

Proof. The proof of (ii)⇒(i) is routine. We give the proof of (i)⇒(ii).

To begin, we add to L a set C of new constants of suitable sorts, with the intention of naming the elements of $\mathcal{N}$. Let $\mathbf{A}$ be an interpretation of C in $\mathcal{N}$ such that every element of $\mathcal{N}$ is of the form $\mathbf{A}(c)$ for some $c \in C$. We consider the $L(C)$-structure $(\mathcal{N}, \mathbf{A})$ and let $\Gamma = \mathrm{Th}_A(\mathcal{N}, \mathbf{A})$.

For each $c \in C$ let r_c be a real number satisfying $\|\mathbf{A}(c)\| < r_c$. Let $\mathfrak{C}$ be the class of all $L(C)$-structures of the form $(\mathcal{M}, \mathbf{B})$ such that for all $c \in C$ we have $\|\mathbf{B}(c)\| < r_c$ as well as having $\mathbf{B}(c) = \mathbf{A}(c)$ in every case where $\mathbf{A}(c)$ is an element of $\mathcal{M}$. Note that $\mathfrak{C}$ is a uniform class of $L(C)$-structures.

The hypothesis $\mathcal{M} \preceq_A \mathcal{N}$ ensures that each finite subset of Γ^+ is satisfied by some member of $\mathfrak{C}$. Using the Compactness Theorem (Proposition 9.5) we obtain a family of structures from $\mathfrak{C}$ and an ultrafilter D such that the D-ultraproduct of the family approximately satisfies the entire set Γ. This ultraproduct is of the form $((\mathcal{M})_D, \mathbf{B})$ for some function $\mathbf{B}$ from C to elements of $(\mathcal{M})_D$. Moreover, if $c \in C$ and $a = \mathbf{A}(c)$ is an element of $\mathcal{M}$, then $B(c)$ must be the image of the diagonal embedding (of $\mathcal{M}$ into $(\mathcal{M})_D$) applied to a. The desired elementary embedding of $\mathcal{N}$ into $(\mathcal{M})_D$ is obtained by mapping $\mathbf{A}(c)$ to $\mathbf{B}(c)$ for each $c \in C$.   □

Unions of Chains

If Λ is a linearly ordered set, a Λ-*chain* of normed space L-structures is a family $(\mathcal{M}_\xi \mid \xi \in \Lambda)$ such that $\mathcal{M}_\xi \subseteq \mathcal{M}_\eta$ for $\xi < \eta$. If $(\mathcal{M}_\xi \mid \xi \in \Lambda)$ is a chain of L-structures such that the family $(\mathcal{M}_\xi \mid \xi \in \Lambda)$ is uniform, we can define the *union* $\bigcup_{\xi \in \Lambda} \mathcal{M}_\xi$ as an L-structure as follows. If $\mathcal{M}_\xi$ is based on $(M_\xi^{(s)} \mid s \in S)$ for each $\xi \in \Lambda$, then $\bigcup_{\xi \in \Lambda} \mathcal{M}_\xi$ is based on $(\bigcup_{\xi \in \Lambda} M_\xi^{(s)} \mid s \in S)$, and if $f : s_1 \times \cdots \times s_m \to s_0$ is a function symbol of L, then the interpretation of f in $\bigcup_{\xi \in \Lambda} \mathcal{M}_\xi$ is the unique function

$$F: \bigcup_{\xi \in \Lambda} M_\xi^{(s_1)} \times \cdots \times \bigcup_{\xi \in \Lambda} M_\xi^{(s_m)} \to \bigcup_{\xi \in \Lambda} M_\xi^{(s_0)}$$

such that for $\xi \in \Lambda$ and $(x_1, \ldots, x_m) \in M_\xi^{(s_1)} \times \cdots \times M_\xi^{(s_m)}$ one has $F(x_1, \ldots, x_m) = f^{\mathfrak{M}_\xi}(x_1, \ldots, x_m)$.

The assumption that the chain $(\mathfrak{M}_\xi \mid \xi \in \Lambda)$ is a uniform family of L-structures is needed to ensure that the functions defined above are uniformly continuous on bounded subsets of their domains.

9.10 Definition. A chain of structures $(\mathfrak{M}_\xi \mid \xi \in \Lambda)$ is called an *approximate elementary chain* if $\mathfrak{M}_\xi \preceq_A \mathfrak{M}_\eta$ for all $\xi < \eta$.

9.11 Remark. If $(\mathfrak{M}_\xi \mid \xi \in \Lambda)$ is an approximate elementary chain, then the family $(\mathfrak{M}_\xi \mid \xi \in \Lambda)$ is uniform, so the union $\bigcup_{\xi \in \Lambda} \mathfrak{M}_\xi$ is well-defined.

9.12 Corollary. *If $(\mathfrak{M}_\xi \mid \xi \in \Lambda)$ is an approximate elementary chain, then $\mathfrak{M}_\xi \preceq_A \bigcup_{\xi \in \Lambda} \mathfrak{M}_\xi$ for every $\xi \in \Lambda$.*

Proof. Use Proposition 6.6. □

Löwenheim-Skolem Theorems

Recall that the *density character* of a topological space is the smallest cardinality of a dense subset of the space. For example, a space is separable if and only if its density character is $\aleph_0$. If A is a topological space, we denote its density character by density(A).

9.13 Proposition (Downward Löwenheim-Skolem Theorem).
Let $\mathfrak{M}$ be a normed space L-structure based on $(M^{(s)} \mid s \in S)$. Suppose that we are given sets $A^{(s)} \subseteq M^{(s)}$ for every $s \in S$ and a cardinal number κ such that density$(A^{(s)}) \le \kappa$ for every $s \in S$ and card$(L) \le \kappa$. Then there exists a closed substructure $\mathfrak{N}$ of $\mathfrak{M}$ based on $(N^{(s)} \mid s \in S)$ such that

(i) $\mathfrak{N} \preceq_A \mathfrak{M}$;
(ii) $A^{(s)} \subseteq N^{(s)} \subseteq M^{(s)}$, *for every $s \in S$;*
(iii) density$(N^{(s)}) \le \kappa$, *for every $s \in S$.*

Proof. For each $s \in S$ let $A_0^{(s)}$ be a dense subset of $A^{(s)}$ of cardinality at most κ. Without loss of generality we may assume that $A_0^{(s_\mathbb{R})}$ contains the set of rational numbers. By suitably enlarging these sets, we may obtain a family of sets $(N_0^{(s)} \mid s \in S)$ such that $A_0^{(s)} \subseteq N_0^{(s)} \subseteq M^{(s)}$ and $\operatorname{card}(N_0^{(s)}) \leq \kappa$ for every $s \in S$ and such that the following closure condition also holds: for every positive bounded formula $\varphi(x_1, \ldots, x_n, x_{n+1})$ and every approximation φ' of φ, with x_k a variable of sort s_k for $k = 1, \ldots, n+1$, if $\mathcal{M} \models \varphi[a_1, \ldots, a_n, a_{n+1}]$ with $a_k \in N_0^{(s_k)}$ for $k = 1, \ldots, n$ and $a_{n+1} \in M^{(s_{n+1})}$, then there exists $b \in N_0^{(s_{n+1})}$ such that $\mathcal{M} \models \varphi'[a_1, \ldots, a_n, b]$. It is possible to do this while maintaining the claimed cardinality bounds because L has at most κ many positive bounded formulas.

For each $s \in S$ we now let $N^{(s)}$ be the closure of $N_0^{(s)}$ in $M^{(s)}$. By considering atomic formulas in the closure condition above, one shows that there is a closed substructure $\mathcal{N}$ of $\mathcal{M}$ that is based on $(N^{(s)} \mid s \in S)$. Using the Perturbation Lemma (Proposition 5.15) one uses the closure condition above to show that $\mathcal{N} \subseteq \mathcal{M}$ satisfy the Tarski-Vaught test (Proposition 6.6) and hence that $\mathcal{N} \preceq_A \mathcal{M}$. $\qquad\square$

9.14 Proposition (Upward Löwenheim-Skolem Theorem). *Let $\mathcal{M}$ be a normed space L-structure based on $(M^{(s)} \mid s \in S)$. Suppose that κ is a cardinal number such that $\kappa \geq \operatorname{card}(L)$ and $\kappa \geq \operatorname{density}(M^{(s)})$ for every $s \in S$. Then there exists an approximate elementary extension $\mathcal{N}$ of $\mathcal{M}$ such that every infinite dimensional sort of $\mathcal{N}$ has density character equal to κ; $\mathcal{N}$ may be taken to be a Banach space structure.*

Proof. Let S_1 be the set of $s \in S$ such that $M^{(s)}$ is infinite dimensional. For each $s \in S$, let $M_0^{(s)}$ be a dense subset of $M^{(s)}$ of cardinality at most κ. Add to L a set C of new constants of suitable sorts, with the intention of naming the elements of the sets $M_0^{(s)}$; we require $\operatorname{card}(C) \leq \kappa$. Let $\mathbf{A}$ be an interpretation of C in $\mathcal{M}$ such that for each $s \in S$, every element of $M_0^{(s)}$ is of the form $\mathbf{A}(c)$ for some $c \in C$ of sort s. Extend the signature further by adding for each $s \in S_1$ a set $\{c_i^s \mid i < \kappa\}$ of new constant symbols of sort s. Let $L(C')$ be the resulting extension by constants of L.

We consider the following positive bounded theory in $L(C')$:

$$
\begin{aligned}
\Gamma = \operatorname{Th}_A(\mathcal{M}, \mathbf{A}) \\
\cup \{ \, \|c_i^s\| = 1 \mid i < \kappa, \ s \in S_1 \, \} \\
\cup \{ \, \|c_i^s - c_j^s\| = 1 \mid i < j < \kappa, \ s \in S_1 \, \}.
\end{aligned}
$$

Every finite subset of Γ^+ is satisfied by some expansion of $(\mathcal{M}, \mathbf{A})$ to $L(C')$. Moreover we can require that the interpretations of the constant symbols c_i^s are all of norm at most 1, so the set of expansions needed can be taken to be uniform. The Compactness Theorem (Proposition 9.5) yields an $L(C')$-structure $\mathcal{N}$ such that $\mathcal{N} \models_A \Gamma$; Replacing $\mathcal{N}$ by its completion and using Corollary 6.7 we may take $\mathcal{N}$ to be a Banach space structure. By renaming the elements of $\mathcal{N}$ if necessary, we may assume that $c^{\mathcal{N}} = \mathbf{A}(c)$ for all $c \in C$ and that $\mathcal{M}$ is in fact a substructure of $\mathcal{N}$. Since $\mathrm{Th}_A(\mathcal{M}, \mathbf{A}) \subseteq \Gamma$ it follows that $\mathcal{M} \preceq_A \mathcal{N}$.

It remains to consider the density character of the infinite dimensional sorts of $\mathcal{N}$, which may well be larger than κ. Suppose $\mathcal{N}$ is based on $(N^{(s)} \mid s \in S)$. Clearly, density$(N^{(s)}) \geq \kappa$, for every $s \in S_1$. Applying Proposition 9.13 we may reduce the density character of the infinite dimensional sorts of $\mathcal{N}$ as needed. □

9.15 Remark. Suppose $\mathcal{M}$ and $\mathcal{N}$ are based on $(M^{(s)} \mid s \in S)$ and $(N^{(s)} \mid s \in S)$, respectively, and $\mathcal{M} \equiv_A \mathcal{N}$. Then, for every nonnegative integer n, $M^{(s)}$ is n-dimensional if and only if $N^{(s)}$ is n-dimensional. If $\mathcal{M} \subseteq \mathcal{N}$ or vice versa, this forces $M^{(s)}$ and $N^{(s)}$ to be equal. (See Corollary 7.2.) Therefore, in passing from $\mathcal{M}$ to $\mathcal{N}$ in the Löwenheim-Skolem theorems, the finite dimensional sorts necessarily remain unchanged.

Saturated Structures

9.16 Definition. Let $\Gamma(x_1, \ldots, x_n)$ be a set of L-formulas and let $\mathcal{M}$ be a normed space L-structure. We say that $\Gamma(x_1, \ldots, x_n)$ is *satisfiable in* $\mathcal{M}$ if there exist elements $a_1, \ldots, a_n$ of suitable sorts of $\mathcal{M}$ such that $\mathcal{M} \models \Gamma[a_1, \ldots, a_n]$. We say that $\Gamma(x_1, \ldots, x_n)$ is *approximately satisfiable in* $\mathcal{M}$ if there exist elements $a_1, \ldots, a_n$ of suitable sorts of $\mathcal{M}$ such that $\mathcal{M} \models_A \Gamma[a_1, \ldots, a_n]$.

Note that Γ is approximately satisfiable in $\mathcal{M}$ if and only if Γ^+ is satisfiable in $\mathcal{M}$.

9.17 Definition. Let $\mathcal{M}$ be a normed space L-structure and let κ be an infinite cardinal. We say that $\mathcal{M}$ is κ-*saturated* if the following condition holds: whenever C is a set of constants such that card$(C) < \kappa$, $\mathbf{A}$ is an interpretation of C in $\mathcal{M}$, and $\Gamma(x_1, \ldots, x_n)$ is a set of positive bounded $L(C)$-formulas containing the formulas $\|x_k\| \leq r_k$ for $k = 1, \ldots, n$ (where $r_1, \ldots, r_n$ are nonnegative rational numbers), if every finite subset of Γ^+ is satisfiable in $(\mathcal{M}, \mathbf{A})$, then the entire set Γ is approximately satisfiable in $(\mathcal{M}, \mathbf{A})$.

It is easy to see that without the norm bound formulas it would not generally be possible for saturated structures to exist. (Consider the set of formulas $\Gamma(x) = \{n \leq \|x\| \mid n \in \mathbb{N}\}$; every finite subset of Γ is satisfiable in any nontrivial normed space, but Γ itself is certainly not approximately satisfiable in any normed space.)

Given any normed space L-structure $\mathcal{M}$ and any infinite cardinal number κ, there exists a κ-saturated $\mathcal{N}$ such that $\mathcal{M} \preceq_A \mathcal{N}$. We prove this fact at the end of the chapter by a unions-of-chains argument. Also, in Chapter 10 we prove that $\mathcal{M}$ has a κ-saturated ultrapower; the proof is a byproduct of proving a difficult theorem about isomorphic ultraproducts.

It is easier to prove the existence of κ-saturated normed space L-structures in the setting where L is countable and $\kappa = \aleph_1$, and we will do that next. Moreover, for many purposes this is the only degree of saturation that one needs for applications.

Recall that an ultrafilter D is said to be *countably incomplete* if D is not closed under countable intersections. It is equivalent to require the existence of elements J_n of D for each $n \in \mathbb{N}$ such that the intersection $\bigcap_{n \in \mathbb{N}} J_n$ is the empty set. Evidently any non-principal ultrafilter on a countable set is countably incomplete.

9.18 Proposition. *Let L be a countable signature and let D be a countably incomplete ultrafilter on a set Λ. Then for every uniform family $(\mathcal{M}_\xi \mid \xi \in \Lambda)$, the ultraproduct $(\prod_{\xi \in \Lambda} \mathcal{M}_\xi)_D$ is $\aleph_1$-saturated.*

Proof. In order to simplify the notation, we verify that $(\prod_{\xi \in \Lambda} \mathcal{M}_\xi)_D$ satisfies the condition of Definition 9.17 for $n = 1$ and $r_1 = 1$.

We have to prove the following statement: if C is a countable set of constant symbols, $\mathbf{A}$ is an interpretation of C in $(\prod_{\xi \in \Lambda} \mathcal{M}_\xi)_D$ and $\Gamma(x)$ is a set of positive bounded $L(C)$-formulas such that every finite subset of

$$\{\|x\| \leq 1\} \cup \Gamma(x)$$

is approximately satisfiable in $((\prod_{\xi \in \Lambda} \mathcal{M}_\xi)_D, \mathbf{A})$, then the entire set $\{\|x\| \leq 1\} \cup \Gamma(x)$ is approximately satisfied in $((\prod_{\xi \in \Lambda} \mathcal{M}_\xi)_D, \mathbf{A})$. For each $\xi \in \Lambda$ let $\mathbf{A}_\xi$ be an interpretation of C in $\mathcal{M}_\xi$ such that

$$\mathbf{A}(c) = ((\mathbf{A}_\xi(c))_{\xi \in \Lambda})_D, \quad \text{for every } c \in C.$$

Note that

$$\left((\prod_{\xi \in \Lambda} \mathcal{M}_\xi)_D, \mathbf{A} \right) = \left(\prod_{\xi \in \Lambda} (\mathcal{M}_\xi, \mathbf{A}_\xi) \right)_D.$$

Thus, since L is an arbitrary countable signature and C is also countable, it suffices to prove the following simpler statement:

If $\Gamma(x)$ is a set of positive bounded L-formulas and every finite subset of $\{ \|x\| \leq 1 \} \cup \Gamma(x)$ is approximately satisfiable in $(\prod_{\xi \in \Lambda} \mathcal{M}_\xi)_D$, then $\{ \|x\| \leq 1 \} \cup \Gamma(x)$ is approximately satisfiable in $(\prod_{\xi \in \Lambda} \mathcal{M}_\xi)_D$.

Suppose, then, that every finite subset of $\{ \|x\| \leq 1 \} \cup \Gamma(x)$ is approximately satisfiable in $(\prod_{\xi \in \Lambda} \mathcal{M}_\xi)_D$. Since L is countable, we may write

$$\Gamma(x) = \{ \varphi_n(x) \mid n \in \mathbb{N} \}.$$

Now for every $n \in \mathbb{N}$ let $\{ \varphi_n^k \mid n \in \mathbb{N} \}$ be a set of approximations of φ_n such that

· $\varphi_n' < \varphi_n^l < \varphi_n^k$, if $k < l$;
· For every approximation ψ of φ_n there exists $k \in \mathbb{N}$ such that $\varphi_n < \varphi_n^k < \psi$.

Since D is countably incomplete, we can fix a descending chain of elements of D

$$\Lambda = \Lambda_0 \supseteq \Lambda_1 \supseteq \ldots$$

such that $\bigcap_{k \in \mathbb{N}} \Lambda_k = \emptyset$.

Let $(r_k \mid k \geq 1)$ be a decreasing sequence of real numbers whose limit is 1. Let $X_0 = \Lambda$ and for each positive integer k define

$$X_k = \Lambda_k \cap \{ \xi \in \Lambda \mid \mathcal{M}_\xi \models \exists_{r_k} x \, (\varphi_1^k(x) \wedge \ldots \wedge \varphi_k^k(x)) \}.$$

Then $X_k \in D$ by Proposition 9.2. We have $X_k \supseteq X_{k+1}$ for every $k \in \mathbb{N}$ and $\bigcap_{k \in \mathbb{N}} X_k = \emptyset$, so for each $\xi \in \Lambda$ there exists a largest positive integer $k(\xi)$ such that $\xi \in X_{k(\xi)}$.

Suppose that x is a variable of sort s. We now define an element $a \in \ell_\infty(\Lambda, (M_\xi^{(s)} \mid \xi \in \Lambda))$ such that

$$\left(\prod_{\xi \in \Lambda} \mathcal{M}_\xi \right)_D \models_A \|(a)_D\| \leq 1, \qquad \left(\prod_{\xi \in \Lambda} \mathcal{M}_\xi \right)_D \models_A \Gamma[(a)_D].$$

For $\xi \in \Lambda$, if $k(\xi) = 0$, let $a(\xi) = 0_{M_\xi^{(s)}}$; otherwise, let $a(\xi)$ be such that

$$\mathcal{M}_\xi \models \|a(\xi)\| \leq 1 \, \wedge \, (\varphi_1^{k(\xi)} \wedge \cdots \wedge \varphi_{k(\xi)}^{k(\xi)})[a(\xi)].$$

If $k \in \mathbb{N}$ and $\xi \in X_k$, we have $k \leq k(\xi)$, so $\mathcal{M}_\xi \models \varphi_k^{k(\xi)}[a(\xi)]$. It follows from Proposition 9.2 that $(\prod_{\xi \in \Lambda} \mathcal{M}_\xi)_D \models_A \Gamma[(a)_D]$. $\square$

9.19 Remark. Note that the ultrafilters constructed in the proofs of Propositions 9.5 and 9.9 are both countably incomplete. Hence, by

Proposition 9.18, the ultrapowers constructed in those proofs are both $\aleph_1$-saturated.

Now we give some applications of the existence of saturated normed space structures.

9.20 Proposition. *Let* $\mathcal{M}$ *be an* L-*structure and* $\varphi(x_1, \ldots, x_n)$ *a positive bounded* L-*formula. If* $\mathcal{M}$ *is* $\aleph_1$-*saturated, then for any elements* $a_1, \ldots, a_n$ *of suitable sorts of* $\mathcal{M}$, *we have*

$$\mathcal{M} \models_{\mathcal{A}} \varphi[a_1, \ldots, a_n] \qquad \text{if and only if} \qquad \mathcal{M} \models \varphi[a_1, \ldots, a_n].$$

Proof. By induction on the complexity of φ. The $\aleph_1$-saturation is used to handle the case when φ begins with a bounded quantifier. $\square$

9.21 Proposition. *Every* $\aleph_1$-*saturated normed space structure is complete.*

Proof. Suppose that $\mathcal{M}$ is an $\aleph_1$-saturated L-structure and that (a_n) is a Cauchy sequence in a sort $M^{(s)}$ of $\mathcal{M}$. Without loss of generality, assume $\|a_n\| \leq 1$ for every index n. For each positive ϵ choose a positive integer $N = N(\epsilon)$ such that

$$m, n \geq N \quad \text{implies} \quad \|a_m - a_n\| \leq \epsilon.$$

Let $C = \{ c_n \mid n \in \mathbb{N} \}$ be a set of new constant symbols of sort s and define a set $\Gamma(x)$ of $L(C)$-formulas as follows.

$$\Gamma(x) = \{ \|x\| \leq 1 \} \cup \{ \|x - c_n\| \leq \epsilon \mid n \geq N(\epsilon), \epsilon > 0 \}.$$

Define an interpretation $\mathbf{A}$ of C in $\mathcal{M}$ by letting $\mathbf{A}(c_n) = a_n$. Every finite subset of Γ is approximately satisfiable in $(\mathcal{M}, \mathbf{A})$, so by the $\aleph_1$-saturation of $\mathcal{M}$, the set Γ is approximately satisfiable in $(\mathcal{M}, \mathbf{A})$. If a is the element of $M^{(s)}$ that approximately satisfies $\Gamma(x)$, it is clear that (a_n) converges to a. $\square$

9.22 Proposition (Compactness Theorem, Second Version).
Let $\mathfrak{C}$ *be a uniform class of normed space* L-*structures, and suppose that* Γ *is a set of positive bounded* L-*sentences such that for each finite subset* Σ *of* Γ^+ *there exists a structure in* $\mathfrak{C}$ *which satisfies* Σ. *Then there exists an ultraproduct of structures from* $\mathfrak{C}$ *which satisfies (not just approximately satisfies) the entire set* Γ.

Proof. The ultrafilter D used in the proof of Proposition 9.5 is countably incomplete; use Propositions 9.18 and 9.20. $\square$

Existence of Saturated Structures

In the following definition we borrow a term from nonstandard analysis. (See [HL85].)

9.23 Definition. Let $\mathcal{M}, \mathcal{N}$ be normed space L-structures with $\mathcal{M} \preceq_A \mathcal{N}$. We say that $\mathcal{N}$ is an *enlargement of* $\mathcal{M}$ if the following condition holds: whenever C is a set of constants and $\mathbf{A}$ is an interpretation of C in $\mathcal{M}$, and $\Gamma(x_1, \ldots, x_n)$ is a set of positive bounded $L(C)$-formulas containing the formulas $\|x_k\| \leq r_k$ for $k = 1, \ldots, n$ (where $r_1, \ldots, r_n$ are nonnegative rational numbers), if every finite subset of Γ^+ is satisfiable in $(\mathcal{M}, \mathbf{A})$, then the entire set Γ is approximately satisfiable in $(\mathcal{N}, \mathbf{A})$.

Suppose κ is a cardinal such that $\operatorname{card}(L) \leq \kappa$ and $\operatorname{card}(M^{(s)}) \leq \kappa$ for every sort s of L. Note that in Definition 9.23 it suffices only to consider sets of constants C with $\operatorname{card}(C) \leq \kappa$.

It is easy to construct ultrapower enlargements of a given normed space structure, as the following argument shows.

9.24 Lemma. *Let $\mathcal{M}$ be a normed space L-structure, and let κ be an infinite cardinal such that $\operatorname{card}(L) \leq \kappa$ and $\operatorname{card}(M^{(s)}) \leq \kappa$ for every sort s of L. Let I be the collection of finite subsets of κ. There is a countably incomplete ultrafilter D on I such that $(\mathcal{M})_D$ is an enlargement of $\mathcal{M}$.*

Proof. For each ordinal $\alpha < \kappa$, let $S_\alpha = \{i \in I \mid \alpha \in i\}$. Note that the family $\{S_\alpha \mid \alpha < \kappa\}$ has the finite intersection property, so it is contained in an ultrafilter D on I. (Every such D is countably incomplete.) We will show that $(\mathcal{M})_D$ is an enlargement of $\mathcal{M}$.

Let C be a set of new constants with $\operatorname{card}(C) \leq \kappa$, and suppose $\mathbf{A}$ is an interpretation of C in $\mathcal{M}$. Further, let $\Gamma(x_1, \ldots, x_n)$ be a set of positive bounded $L(C)$-formulas containing the formulas $\|x_k\| \leq r_k$ for $k = 1, \ldots, n$ (where $r_1, \ldots, r_n$ are nonnegative rational numbers) such that every finite subset of Γ^+ is satisfiable in $(\mathcal{M}, \mathbf{A})$. We must show that Γ is approximately satisfiable in $((\mathcal{M})_D, \mathbf{A})$. Our assumptions on L, C, κ ensure that there is a surjection from κ onto Γ^+; let the value of this surjection at $\alpha < \kappa$ be φ_α. We define functions $a_1, \ldots, a_n$ from I into suitable sorts of $\mathcal{M}$ such that for each $i \in I$ the n-tuple $(a_1(i), \ldots, a_n(i))$ satisfies the formulas $\varphi_\alpha(x_1, \ldots, x_n)$ in $(\mathcal{M}, \mathbf{A})$ for each $\alpha \in i$ and satisfies the norm bounds $\|a_j(i)\| \leq r_j + 1$ for $j = 1, \ldots, n$. Using Proposition 9.2 it follows that $((a_1)_D, \ldots, (a_n)_D)$ satisfies $\Gamma^+(x_1, \ldots, x_n)$ in $((\mathcal{M})_D, \mathbf{A})$. $\qquad\square$

9.25 Proposition. *Let $\mathcal{M}$ be a normed space L-structure. For every* *infinite cardinal κ, $\mathcal{M}$ has a κ-saturated approximate elementary extension.*

Proof. By increasing κ if necessary (for example, replacing κ by κ^+) we may assume κ is regular. By induction we construct an approximate elementary chain $(\mathcal{M}_\alpha \mid \alpha < \kappa)$ such that $\mathcal{M}_0 = \mathcal{M}$ and for each $\alpha < \kappa$, $\mathcal{M}_{\alpha+1}$ is an enlargement of $\mathcal{M}_\alpha$. (At limit ordinals we take unions.) Let $\mathcal{N}$ be the union of the chain $(\mathcal{M}_\alpha \mid \alpha < \kappa)$. By Corollary 9.12, $\mathcal{M} \preceq_A \mathcal{N}$. We claim that $\mathcal{N}$ is κ-saturated. Let C be a set of constants such that $\mathrm{card}(C) < \kappa$ and let $\mathbf{A}$ be any interpretation of C in $\mathcal{N}$. Since κ is regular, there exists $\alpha < \kappa$ such that all the values of $\mathbf{A}$ are elements of $\mathcal{M}_\alpha$. The elements of $\mathcal{N}$ needed to verify Definition 9.17 for $(\mathcal{N}, \mathbf{A})$ can be found in $\mathcal{M}_{\alpha+1}$. $\qquad\square$

Satisfaction and Approximate Satisfaction

The results about $\aleph_1$-saturated structures given earlier in this chapter can be used as the basis for an alternative point of view toward defining $\models_A$ between normed space structures and positive bounded formulas. Take any normed space L-structure $\mathcal{M}$, and consider any ultrapower $(\mathcal{M})_D$ of $\mathcal{M}$ with D being a countably incomplete ultrafilter. For example, take D to be any non-principal ultrafilter on a countable index set. The next result shows that approximate satisfaction of positive bounded formulas in $\mathcal{M}$ could be defined using ordinary model theoretic satisfaction of those formulas in $(\mathcal{M})_D$. (In particular, this turns out to be independent of D, as long as D is countably incomplete.)

9.26 Proposition. *Let D be any countably incomplete ultrafilter. If $\mathcal{M}$ is a normed space L-structure, $\varphi(x_1,\ldots,x_n)$ is a positive bounded formula, and $a_1,\ldots,a_n$ are elements of suitable sorts of $\mathcal{M}$, the following conditions are equivalent:*

(i) $\mathcal{M} \models_A \varphi[a_1,\ldots,a_n]$;
(ii) $(\mathcal{M})_D \models \varphi[a_1,\ldots,a_n]$.

Proof. The implication (ii)$\Rightarrow$(i) is given by Corollary 9.3, and (i)$\Rightarrow$(ii) by Propositions 9.18 and 9.20 (and Corollary 9.3). $\qquad\square$

10

Isomorphic Ultrapowers

In this chapter we prove isomorphism theorems for ultrapowers and ultraproducts of normed space structures. These results show that there is a very tight connection between (a) properties that are preserved under the ultraproduct construction and (b) properties that are expressible using the logic for normed space structures that is described in this paper.

Let L be a signature and let $\mathcal{M}$ and $\mathcal{N}$ be two normed space L-structures. If $\mathcal{M}$ and $\mathcal{N}$ have isomorphic ultrapowers, by Corollary 9.4 they must be approximately elementarily equivalent. Theorem 10.7 below gives the converse (in a strong form). Together these results show that ultrapower equivalence of $\mathcal{M}$ and $\mathcal{N}$ is the same as approximate elementary equivalence. (See the discussion of this issue in the Introduction.)

Theorem 10.8 below is a similar result for ultraproducts. Among other things, it shows that the ultrafilter guaranteed by Theorem 10.7 can be chosen in a highly uniform way and that the ultrapowers in question can be taken to be highly saturated. The uniformity will be exploited in Chapter 12 to prove the existence of ultrapowers that are highly homogeneous (in addition to being highly saturated).

The results in this chapter are analogous to the Keisler-Shelah Theorem in ordinary model theory. (See [She71] and Chapter 6 in [CK90].) Moreover, our proof follows a similar line of argument, with adjustments appropriate to the handling of positive bounded formulas and their approximations.

Theorem 10.7 is a special case of Theorem 10.8. We have included both arguments because the proof of Theorem 10.8 is a complicated elaboration of the proof of Theorem 10.7, and because the ultrapower isomorphism result in Theorem 10.7 will be of interest in its own right to many readers.

Combinatorial Lemmas

Throughout this section, λ, τ and κ will denote infinite cardinals with $\lambda = 2^\kappa$ and μ will denote the least cardinal α such that $\lambda^\alpha > \lambda$. Using basic cardinal arithmetic, one can verify that $\kappa < \mu \leq \lambda$ and μ is a regular cardinal.

10.1 Definition. Let F be a set of functions from λ into μ and let D be a proper filter on λ. We say that the pair (F, D) is τ-*consistent* if the following two conditions hold:

(1) D is generated by a subset of cardinality at most τ (*i.e.*, there exists a subset E of D of cardinality at most τ such that E is closed under finite intersections and every element of D is a superset of some element of E).

(2) Whenever we are given

    · a cardinal $\beta < \mu$,
    · a family $(f_i \mid i < \beta)$ of distinct elements of F, and
    · a family $(\sigma_i \mid i < \beta)$ of ordinals less than μ,

    then D together with

$$\{\xi < \lambda \mid f_i(\xi) = \sigma_i, \text{ for every } i < \beta\}$$

    generate a proper filter on λ (*i.e.*, when this set is added to D, the resulting family still has the finite intersection property).

10.2 Remarks.

(1) If (F, D) is τ-consistent, $\tau \leq \tau'$, and $F' \subseteq F$, then (F', D) is τ'-consistent.

(2) Suppose that $(\tau_i \mid i < \delta)$ is an increasing family of infinite cardinals such that $\sup_i \tau_i = \tau$. Suppose also that (F_i, D_i) is τ_i-consistent for all $i < \delta$ and that $F_j \supseteq F_i$ and $D_j \subseteq D_i$ for $j < i < \delta$. Then the pair $\left(\bigcap_i F_i, \bigcup_i D_i \right)$ is τ-consistent.

10.3 Remark. The pair (F, D) is τ-consistent in the sense of Definition 10.1 if and only if the triple $(F, \emptyset, D)$ is τ-consistent as defined in [She71]; see also page 394 in [CK90].

10.4 Lemma. *There is a set F of functions from λ into μ such that* $\mathrm{card}(F) = 2^\lambda$ *and* $(F, \{\lambda\})$ *is μ-consistent.*

Proof. See Lemma 1 of [She71] (or Lemma 6.1.10 in [CK90]).   □

10.5 Lemma. *Suppose that* $\tau \geq \mu$, (F, D) *is* τ-*consistent and* $(X_i \mid i < \tau)$ *is a family of subsets of* λ. *Then there exist* $F' \subseteq F$ *and* $D' \supseteq D$ *such that* $\mathrm{card}(F \setminus F') \leq \tau$, (F', D') *is* τ-*consistent, and for every* $i < \tau$, *either* $X_i \in D'$ *or* $\lambda \setminus X_i \in D'$.

Proof. See Lemma 4(B) of [She71] (or Lemma 6.1.13(ii) of [CK90]). $\square$

10.6 Lemma. *Suppose that the following conditions hold:*

· L *is a signature and* x *is a variable in* L *of sort* s,
· $\mathcal{M}$ *is a normed space* L-*structure,*
· M *is a dense subset of the sort of* $\mathcal{M}$ *indexed by* s *and* $\mathrm{card}(M) < \mu$,
· (F, D) *is* τ-*consistent,*
· $(\varphi_i(x, y_{i,1}, \ldots, y_{i,n(i)}) \mid i < \tau)$ *is a family of positive bounded formulas that is closed under conjunctions, and*
· *for each* $i < \tau$, $\bar{a}_i$ *is a function on* λ *such that for all* $\xi < \lambda$, $\bar{a}_i(\xi) = (a_{i,1}(\xi), \ldots, a_{i,n(i)}(\xi))$, *where* $a_{i,1}(\xi), \ldots, a_{i,n(i)}(\xi)$ *are elements of suitable sorts of* $\mathcal{M}$, *and*

$$\{ \xi < \lambda \mid \mathcal{M} \models \exists x \varphi_i(x, y_{i,1}, \ldots, y_{i,n(i)})[a_{i,1}(\xi), \ldots, a_{i,n(i)}(\xi)] \} \in D.$$

Then, there exist $F' \subseteq F$ *and* $D' \supseteq D$ *such that* $\mathrm{card}(F \setminus F') \leq \tau$ *and* (F', D') *is* τ-*consistent, and there is a function* $b \colon \lambda \to M$ *such that for every* $i < \tau$ *and every approximation* φ'_i *of* φ_i

$$\{ \xi < \lambda \mid \mathcal{M} \models \varphi'_i[b(\xi), a_{i,1}(\xi), \ldots, a_{i,n(i)}(\xi)] \} \in D'.$$

Proof. Similar to Lemma 5 of [She71] (or Lemma 6.1.14 of [CK90]). The Perturbation Lemma (Proposition 5.15) is used to ensure that the values of the function b come from the dense set M. Note that each positive bounded formula has only countably many approximations. $\square$

The Ultrapower Theorem

10.7 Theorem. *Let* L *be a signature and suppose* $\mathcal{M}$ *and* $\mathcal{N}$ *are normed space* L-*structures which satisfy* $\mathcal{M} \equiv_A \mathcal{N}$. *Then there exists an ultrafilter* D *such that* $(\mathcal{M})_D \cong (\mathcal{N})_D$.

Proof. Note that this result is the converse of Corollary 9.4.

Let κ be an infinite cardinal number such that the density character of each sort of $\mathcal{M}$ and $\mathcal{N}$ is at most κ and such that $\mathrm{card}(L) \leq 2^\kappa$.

In order to simplify notation, in the proof of Theorem 10.7 we will assume that $\mathcal{M}$ and $\mathcal{N}$ each have only one sort (in addition to the special sort $\mathbb{R}$) and for each structure we will refer to this sort as *the universe.*

Let M, N be dense subsets of the universes of $\mathcal{M}, \mathcal{N}$ respectively such that card(M), card(N) are both $\leq \kappa$.

Let $\lambda = 2^\kappa$ and let μ be the least cardinal α such that $\lambda^\mu > \lambda$. Note that $\kappa < \mu \leq \lambda$ and μ is regular. We use an inductive construction to find simultaneously an ultrafilter D on λ and an isomorphism between $(\mathcal{M})_D$ and $(\mathcal{N})_D$.

Consider a family $a = (a(\xi) \mid \xi < \lambda)$ of elements of M; we say that the family is *bounded* if the norms $(\|a(\xi)\| \mid \xi < \lambda)$ are uniformly bounded. We let $\ell_\infty(\lambda, M)$ denote the set of all such bounded families from M and define $\|a\|_\infty$ to be the supremum of all $\|a(\xi)\|$ for $\xi < \lambda$. We make a similar definition for N in place of M. Note that the cardinalities of $\ell_\infty(\lambda, M)$ and $\ell_\infty(\lambda, N)$ are at most 2^λ. We fix listings of the elements of $\ell_\infty(\lambda, M)$ and $\ell_\infty(\lambda, N)$ indexed over the ordinals $< 2^\lambda$.

We use Lemma 10.4 to fix a set F_0 of functions from λ into μ such that card(F_0) $= 2^\lambda$ and $(F_0, \{\lambda\})$ is μ-consistent (and hence it is λ-consistent). Let $D_0 = \{\lambda\}$. Let $(S_j \mid j < 2^\lambda)$ be a list of all the subsets of λ.

By induction on all ordinals $i < 2^\lambda$, we construct

· a decreasing family $(F_i \mid i < 2^\lambda)$ of sets of functions from λ into μ,
· an increasing family $(D_i \mid i < 2^\lambda)$ of proper filters on λ, and
· families $(a_i \mid i < 2^\lambda)$ and $(b_i \mid i < 2^\lambda)$ in $\ell_\infty(\lambda, M)$ and $\ell_\infty(\lambda, N)$, respectively,

such that the following conditions hold for all $i < 2^\lambda$.

(1)(a) card($F_0 \setminus F_i$) $\leq \lambda + $card($i$) (and therefore card($F_i$) $= 2^\lambda$);
   (b) (F_i, D_i) is $(\lambda + $card($i$))-consistent.
(2) If $i = j + 1$, then either $S_j \in D_i$ or $\lambda \setminus S_j \in D_i$.
(3) If $i = j + 1$ and j is an even ordinal, then there exists a filter D_j' on λ such that

$$D_j \subseteq D_j' \subseteq D_i,$$

and for every positive bounded L-formula $\varphi(x, x_1, \ldots, x_n)$ and every choice of $j_1, \ldots, j_n < j$:

(a) either the set

$$\{\xi < \lambda \mid \mathcal{M} \models \varphi[a_j(\xi), a_{j_1}(\xi), \ldots, a_{j_n}(\xi)]\}$$

or its complement is in D_j';

(b) for every $\psi > \varphi$,

$$\{\xi < \lambda \mid \mathcal{M} \models \varphi[a_j(\xi), a_{j_1}(\xi), \ldots, a_{j_n}(\xi)]\} \in D'_j$$

implies

$$\{\xi < \lambda \mid \mathcal{N} \models \psi[b_j(\xi), b_{j_1}(\xi), \ldots, b_{j_n}(\xi)]\} \in D_i.$$

(4) If $i = j + 1$ and j is an odd ordinal, then there exists a filter D'_j on λ such that

$$D_j \subseteq D'_j \subseteq D_i,$$

and for every positive bounded L-formula $\varphi(x, x_1, \ldots, x_n)$ and every choice of $j_1, \ldots, j_n < j$:

(a) either the set

$$\{\xi < \lambda \mid \mathcal{N} \models \varphi[b_j(\xi), b_{j_1}(\xi), \ldots, b_{j_n}(\xi)]\}$$

or its complement is in D'_j;

(b) for every $\psi > \varphi$,

$$\{\xi < \lambda \mid \mathcal{N} \models \varphi[b_j(\xi), b_{j_1}(\xi), \ldots, b_{j_n}(\xi)]\} \in D'_j$$

implies

$$\{\xi < \lambda \mid \mathcal{M} \models \psi[a_j(\xi), a_{j_1}(\xi), \ldots, a_{j_n}(\xi)]\} \in D_i.$$

(5) If i is a limit ordinal,

$$F_i = \bigcap_{j<i} F_j \qquad D_i = \bigcup_{j<i} D_j.$$

After this construction is completed, we will let $D = \bigcup_{i<2^\lambda} D_i$. Condition (2) ensures that D is an ultrafilter. The inductive construction will be carried out so that the families $(a_i \mid i < 2^\lambda)$ and $(b_i \mid i < 2^\lambda)$ exhaust $\ell_\infty(\lambda, M)$ and $\ell_\infty(\lambda, N)$ respectively. Therefore the family of equivalence classes $((a_i)_D \mid i < 2^\lambda)$ is norm dense in $(\mathcal{M})_D$ and $((b_i)_D \mid i < 2^\lambda)$ is dense in $(\mathcal{N})_D$.

The desired isomorphism T of $(\mathcal{M})_D$ onto $(\mathcal{N})_D$ is initially defined to take $(a_i)_D$ to $(b_i)_D$ for each $i < 2^\lambda$. To show that T has the desired properties, fix $\varphi(x_1, \ldots, x_n)$ and suppose

$$\{\xi < \lambda \mid \mathcal{M} \models \varphi[a_{j_1}(\xi), \ldots, a_{j_n}(\xi)]\} \in D,$$

where $j_1, \ldots, j_n < 2^\lambda$. Let j be the least even ordinal such that

$$j_1, \ldots, j_n < j.$$

By (3)(a) we have

$$\{\xi < \lambda \mid \mathcal{M} \models \varphi[a_{j_1}(\xi), \dots, a_{j_n}(\xi)]\} \in D'_j,$$

so by (3)(b) for every $\psi > \varphi$ we have

$$\{\xi < \lambda \mid \mathcal{N} \models \psi[b_{j_1}(\xi), \dots, b_{j_n}(\xi)]\} \in D_{j+1} \subseteq D.$$

It follows that T extends to an isomorphism from $(\mathcal{M})_D$ onto $(\mathcal{N})_D$.

We now describe the construction of $(F_i \mid i < 2^\lambda)$, $(D_i \mid i < 2^\lambda)$, $(a_i \mid i < 2^\lambda)$, and $(b_i \mid i < 2^\lambda)$. If i is a limit ordinal, we define F_i and D_i to make clause (5) true and use Remark 10.2(2) to verify clause (1)(b). All other conditions to be satisfied apply to the case where i is a successor ordinal, say $i = j + 1$. The order in which we define the next elements of our four families is: a_j, b_j and then D_i, F_i.

If j is even, we define a_j to be the first element of $\ell_\infty(\lambda, M)$ that does not appear in the list $(a_l \mid l < j)$ and then construct b_j and D_i, F_i in that order; if j is odd we define b_j to be the first element of $\ell_\infty(\lambda, N)$ that does not appear in the list $(b_l \mid l < j)$ and then construct a_j and D_i, F_i in that order. Because of the symmetry between these situations, we will discuss only the case when j is even.

For each positive bounded L-formula $\varphi(x, y_1, \dots, y_n)$ and each finite sequence of ordinals $j_1, \dots, j_n < j$, let

$$X(\varphi, j_1, \dots, j_n) = \{\xi < \lambda \mid \mathcal{M} \models \varphi[a_j(\xi), a_{j_1}(\xi), \dots, a_{j_n}(\xi)]\}.$$

There are at most $\lambda + \mathrm{card}(i)$ such sets X. Hence, since (F_j, D_j) is $(\lambda + \mathrm{card}(i))$-consistent, Lemma 10.5 provides $F'_j \subseteq F_j$ and $D'_j \supseteq D_j$ such that

· $\mathrm{card}(F_j \setminus F'_j) \leq \lambda + \mathrm{card}(i)$,
· (F'_j, D'_j) is $(\lambda + \mathrm{card}(i))$-consistent,
· for every choice of $\varphi(x, y_1, \dots, y_n)$ and $j_1, \dots, j_n < j$,

$$X(\varphi, j_1, \dots, j_n) \in D'_j \quad \text{or} \quad \lambda \setminus X(\varphi, j_1, \dots, j_n) \in D'_j,$$

· either $S_j \in D'_j$ or $\lambda \setminus S_j \in D'_j$.

Fix a formula $\varphi(x, y_1, \dots, y_n)$ and $j_1, \dots, j_n < j$, and suppose that $X(\varphi, j_1, \dots, j_n) \in D'_j$. Let u be a rational number such that $u > \|a_j\|_\infty$. Then the set $Y(u, \varphi, j_1, \dots, j_n)$ is in D'_j, where

$$Y(u, \varphi, j_1, \dots, j_n) =$$
$$\{\xi < \lambda \mid \mathcal{M} \models \exists_u x \, \varphi(x, y_1, \dots, y_n)[a_{j_1}(\xi), \dots, a_{j_n}(\xi)]\}.$$

We claim that for every rational number $w > u$ and every approximation ψ of φ, the set $Z(w, \psi, j_1, \ldots, j_n)$ is also in D'_j, where

$$Z(w, \psi, j_1, \ldots, j_n) =$$
$$\{ \xi < \lambda \mid \mathcal{N} \models \exists_w x \; \psi(x, y_1, \ldots, y_n)[b_{j_1}(\xi), \ldots, b_{j_n}(\xi)] \}.$$

To see this, suppose $Z(w, \psi, j_1, \ldots, j_n) \notin D'_j$. Then let $l = \max_k j_k$, and fix an approximation φ' of φ such that $\varphi < \varphi' < \psi$ and a rational number v such that $u < v < w$. We have $Z(w, \psi, j_1, \ldots, j_n) \notin D_l$ and hence $\lambda \setminus Z(w, \psi, j_1, \ldots, j_n) \in D_l$ by (4)(a). But then,

$$\{ \xi < \lambda \mid \mathcal{N} \models \forall_w x \; \mathrm{neg}(\psi)(x, y_1, \ldots, y_n)[b_{j_1}(\xi), \ldots, b_{j_n}(\xi)] \} \in D_l,$$

so by (4)(b),

$$\{ \xi < \lambda \mid \mathcal{M} \models \forall_v x \; \mathrm{neg}(\varphi')(x, y_1, \ldots, y_n)[a_{j_1}(\xi), \ldots, a_{j_n}(\xi)] \}$$
$$\in D_j \subseteq D'_j,$$

which contradicts the fact that D'_j is a proper filter.

There are two situations in which the description above needs to be clarified, at the point where we are showing that $Z(w, \psi, j_1, \ldots, j_n) \notin D'_j$ and we set $l = \max_k j_k$. When $j = 0$ we do not have $j_1, \ldots, j_n$ and the formulas involved have no free variables; here we use the hypothesis that $\mathcal{M} \equiv_A \mathcal{N}$ and recall that $D_0 = \{\lambda\}$. When $j > 0$ we may assume $n \geq 1$ without loss of generality and the argument proceeds as given.

We now apply Lemma 10.6 to obtain $F_i \subseteq F'_j$ and $D_i \supseteq D'_j$ such that $\mathrm{card}(F'_j \setminus F_i) \leq \lambda + \mathrm{card}(i)$ and (F_i, D_i) is $(\lambda + \mathrm{card}(i))$-consistent, and a function $b_j : \lambda \to N$ such that

$$\{ \xi < \lambda \mid \mathcal{N} \models \psi[b_j(\xi), b_{j_1}(\xi), \ldots, b_{j_n}(\xi)] \} \in D_i$$

for every positive bounded formula $\varphi(x, y_1, \ldots, y_n)$ such that

$$X(\varphi, j_1, \ldots, j_n) \in D'_j$$

and every approximation ψ of φ. In particular, we can assume $\|b_j\|_\infty \leq \|a_j\|_\infty$, so b_j is bounded.

Conditions (1)–(4) are immediate from the construction. $\qquad\square$

The Ultraproduct Theorem

10.8 Theorem. *Let κ be an infinite cardinal and let $\lambda = 2^\kappa$. Then there exists an ultrafilter D on λ satisfying the following properties: if L is a signature with $\mathrm{card}(L) \leq \lambda$ and $(\mathcal{M}_\xi \mid \xi < \lambda)$ and $(\mathcal{N}_\xi \mid \xi < \lambda)$ are* *families of normed space L-structures such that the density character of each sort of* $\mathcal{M}_\xi$ *and* $\mathcal{N}_\xi$ *is at most* κ *for all* $\xi < \lambda$, *then*

$$\left(\prod_{\xi<\lambda} \mathcal{M}_\xi\right)_D \equiv_A \left(\prod_{\xi<\lambda} \mathcal{N}_\xi\right)_D$$

implies

$$\left(\prod_{\xi<\lambda} \mathcal{M}_\xi\right)_D \cong \left(\prod_{\xi<\lambda} \mathcal{N}_\xi\right)_D.$$

Moreover, in each such situation the ultraproduct $\left(\prod_{\xi<\lambda} \mathcal{M}_\xi\right)_D$ *is* κ^+- *saturated.*

Proof. Let μ be the smallest cardinal number α such that $\lambda^\alpha > \lambda$, so $\kappa < \mu \leq \lambda$ and μ is regular.

To clarify the proof, we have divided it into five parts.

PART 1. SOME TECHNICAL SIMPLIFICATIONS. In order to simplify notation, we will explicitly treat only normed space structures which have only one sort in addition to the special sort $\mathbb{R}$. The calligraphic letters $\mathcal{M}$, $\mathcal{N}$, etc. will be used to denote normed space structures, and the corresponding Roman letters M, N, etc. will denote dense subsets of their respective universes with cardinality at most κ.

Without loss of generality, we fix a signature L of cardinality λ such that every signature of cardinality at most λ occurs as a subsignature of L. We also restrict our attention to structures $\mathcal{M}$ whose universe has a dense subset M that is a cardinal $\leq \kappa$ and whose universe is a cardinal $\leq \lambda$. Each normed space structure needing to be considered is isomorphic to one satisfying these restrictions. The number of subsignatures of L is 2^λ, and for each such subsignature L_0 and each $\chi \leq \kappa$, the number of L_0-structures whose universe is a cardinal $\leq \lambda$ that has χ as a dense subset is $(\lambda^\chi)^{\mathrm{card}(L_0)} \leq \lambda^\lambda = 2^\lambda$. Hence the number of normed space structures under consideration is 2^λ.

PART 2. A LIST OF STRUCTURES, FILTERS, AND ELEMENTS. Let

$$(G_\alpha \mid \alpha < 2^\lambda)$$

be a list of all the filters on λ which have a set of at most λ generators.

For each $\alpha < 2^\lambda$ let

$$\left((\mathcal{M}_\xi^{\alpha,\gamma}, \mathcal{N}_\xi^{\alpha,\gamma})_{\xi<\lambda} \mid \gamma < 2^\lambda\right)$$

be a list of all the families $(\mathcal{M}_\xi^{\alpha,\gamma}, \mathcal{N}_\xi^{\alpha,\gamma})_{\xi<\lambda}$ which satisfy the following conditions:

· There exists a subsignature L^α of L such that for every $\xi < \lambda$, $\mathfrak{M}_\xi^{\alpha,\gamma}$ and $\mathfrak{N}_\xi^{\alpha,\gamma}$ are L^α-structures.

· For every $\xi < \lambda$ the universes of $\mathfrak{M}_\xi^{\alpha,\gamma}$ and $\mathfrak{N}_\xi^{\alpha,\gamma}$ are cardinals $\leq \lambda$ and have dense subsets that are cardinals $\leq \kappa$.

· For every L^α-sentence θ and every $\theta' > \theta$, either the set

$$\{\xi < \lambda \mid \mathfrak{M}_\xi^{\alpha,\gamma} \models \theta'\}$$

or its complement is in G_α.

· For every L^α-sentence θ and every $\theta' > \theta$, either the set

$$\{\xi < \lambda \mid \mathfrak{N}_\xi^{\alpha,\gamma} \models \theta'\}$$

or its complement is in G_α.

· For every L^α-sentence θ and every $\theta' > \theta$,

$$\{\xi < \lambda \mid \mathfrak{M}_\xi^{\alpha,\gamma} \models \theta\} \in G_\alpha \quad \text{implies} \quad \{\xi < \lambda \mid \mathfrak{N}_\xi^{\alpha,\gamma} \models \theta'\} \in G_\alpha$$

and

$$\{\xi < \lambda \mid \mathfrak{N}_\xi^{\alpha,\gamma} \models \theta\} \in G_\alpha \quad \text{implies} \quad \{\xi < \lambda \mid \mathfrak{M}_\xi^{\alpha,\gamma} \models \theta'\} \in G_\alpha.$$

For each $\alpha, \gamma < 2^\lambda$ and $\xi < \lambda$ we let $M_\xi^{\alpha,\gamma}, N_\xi^{\alpha,\gamma}$ denote the cardinals $\leq \kappa$ that are dense subsets of the universes of $\mathfrak{M}_\xi^{\alpha,\gamma}, \mathfrak{N}_\xi^{\alpha,\gamma}$ respectively.

For $\alpha, \gamma < 2^\lambda$, let

$$\{a_\rho^{\alpha,\gamma} \mid \rho < 2^\lambda\}$$

be the set of bounded elements of $\prod_{\xi < \lambda} M_\xi^{\alpha,\gamma}$ and

$$\{b_\rho^{\alpha,\gamma} \mid \rho < 2^\lambda\}$$

the set of bounded elements of $\prod_{\xi < \lambda} N_\xi^{\alpha,\gamma}$.

Fix a family

$$C = (c_i \mid i < 2^\lambda)$$

of distinct constant symbols which do not occur in L, and for each $i < 2^\lambda$ let

$$C_i = \{c_j \mid j < i\}.$$

Using cardinality considerations, we can fix a function g defined on the set $\{i \mid i < 2^\lambda\}$ satisfying the properties listed below.

(I) For each $i < 2^\lambda$, one of the following conditions holds:

    (i) $g(i) = (0, S)$, where S is a subset of λ.

    (ii) $g(i) = (1, \alpha, \gamma, \rho, a_\rho^{\alpha,\gamma})$, where $\alpha, \gamma, \rho < 2^\lambda$;

   (iii) $g(i) = (2, \alpha, \gamma, \rho, b_\rho^{\alpha,\gamma})$, where $\alpha, \gamma, \rho < 2^\lambda$;

   (iv) $g(i) = (3, \alpha, \gamma, K, p, R)$, where K is a subset of C_i with card$(K) \leq \kappa$, $p = p(x)$ is a set of positive bounded $L(K)$-formulas which is closed under conjunction, and R is a non-negative rational number;

   (v) $g(i) = (4, \alpha, \gamma, K, p, R)$, where K is a subset of C_i with card$(K) \leq \kappa$, $p = p(x)$ is a set of positive bounded $L(K)$-formulas which is closed under conjunction, and R is a non-negative rational number.

(II) Each of the following elements can be expressed as $g(i)$ for a unique $i < 2^\lambda$:

   (i) Every pair of the form $(0, S)$, where S is a subset of λ;

   (ii) Every quintuple of the form $(1, \alpha, \gamma, \rho, a_\rho^{\alpha,\gamma})$, where $\alpha, \gamma, \rho < 2^\lambda$;

   (iii) Every quintuple of the form $(2, \alpha, \gamma, \rho, b_\rho^{\alpha,\gamma})$, where $\alpha, \gamma, \rho < 2^\lambda$;

   (iv) Every sextuple of the form $(3, \alpha, \gamma, K, p, R)$, where K is a subset of C_i with card$(K) \leq \kappa$, $p = p(x)$ is a set of positive bounded $L(K)$-formulas which is closed under conjunction, and R is a nonnegative rational number.

   (v) Every sextuple of the form $(4, \alpha, \gamma, K, p, R)$, where K is a subset of C_i with card$(K) \leq \kappa$, $p = p(x)$ is a set of positive bounded $L(K)$-formulas which is closed under conjunction, and R is a nonnegative rational number.

(III) For fixed α and γ, the cases given by (I)(ii) and (I)(iii) above alternate; more precisely,

   (i) If $g(i) = (1, \alpha, \gamma, \rho, a_\rho^{\alpha,\gamma})$ and

$$i^* = \min_{j>i} \left[\text{there exists } \sigma < 2^\lambda \text{ such that} \right.$$
$$g(j) = (1, \alpha, \gamma, \sigma, a_\sigma^{\alpha,\gamma}) \quad \text{or}$$
$$\left. g(j) = (2, \alpha, \gamma, \sigma, b_\sigma^{\alpha,\gamma}) \right],$$

then $g(i^*) = (2, \alpha, \gamma, \sigma, b_\sigma^{\alpha,\gamma})$ for some $\sigma < 2^\lambda$.

(ii) If $g(i) = (2, \alpha, \gamma, \rho, b_\rho^{\alpha,\gamma})$ and

$$i^* = \min_{j>i} \Big[\text{there exists } \sigma < 2^\lambda \text{ such that}$$

$$g(j) = (1, \alpha, \gamma, \sigma, a_\sigma^{\alpha,\gamma}) \quad \text{or}$$

$$g(j) = (2, \alpha, \gamma, \sigma, b_\sigma^{\alpha,\gamma}) \Big],$$

then $g(i^*) = (1, \alpha, \gamma, \sigma, a_\sigma^{\alpha,\gamma})$ for some $\sigma < 2^\lambda$.

PART 3. AN INDUCTIVE CONSTRUCTION. The construction is an elaboration of the ideas used in the proof of Theorem 10.7, with which we assume the reader is familiar. First, we use Lemma 10.4 to fix a set F_0 of 2^λ functions from λ into μ such that $(F_0, \{\lambda\})$ is μ-consistent (and hence λ-consistent). Then we let $D_0 = \{\lambda\}$, and for $\alpha, \gamma < 2^\lambda$ we let $A_0^{\alpha,\gamma}$ and $B_0^{\alpha,\gamma}$ be the empty functions.

By induction on all ordinals $i < 2^\lambda$, we will now construct:

· a decreasing family $(F_i \mid i < 2^\lambda)$ of sets of functions from λ into μ,
· an increasing family $(D_i \mid i < 2^\lambda)$ of proper filters on λ,
· for each choice of $\alpha, \gamma < 2^\lambda$ functions

$$A_i^{\alpha,\gamma} : C_i \to \ell_\infty(\lambda, (M_\xi^{\alpha,\gamma} \mid \xi < \lambda)),$$
$$B_i^{\alpha,\gamma} : C_i \to \ell_\infty(\lambda, (N_\xi^{\alpha,\gamma} \mid \xi < \lambda)),$$

where $A_i^{\alpha,\gamma}$ extends $A_j^{\alpha,\gamma}$ for $i > j$ and $B_i^{\alpha,\gamma}$ extends $B_j^{\alpha,\gamma}$ for $i > j$,

such that the following conditions hold:

(1)(a) $\operatorname{card}(F_0 \setminus F_i) \leq \lambda + \operatorname{card}(i)$,
   (b) (F_i, D_i) is $(\lambda + \operatorname{card}(i))$-consistent.
(2) If $i = j + 1$ and $g(j) = (0, S)$, then either $S \in D_i$ or $\lambda \setminus S \in D_i$.
(3) If $i = j + 1$, $g(j) = (1, \alpha, \gamma, \rho, a_\rho^{\alpha,\gamma})$, and the set $D_j \cup G_\alpha$ generates a proper filter on λ, then there exist a filter D_j' with

$$D_j \subseteq D_j' \subseteq D_i$$

and a bounded element

$$b \in \prod_{\xi < \lambda} N_\xi^{\alpha,\gamma}$$

such the functions $A_i^{\alpha,\gamma}, B_i^{\alpha,\gamma}$ are defined by

$$A_i^{\alpha,\gamma}(c) = \begin{cases} A_j^{\alpha,\gamma}(c), & \text{if } c \in C_j \\ a_\rho^{\alpha,\gamma}, & \text{if } c = c_j, \end{cases}$$

$$B_i^{\alpha,\gamma}(c) = \begin{cases} B_j^{\alpha,\gamma}(c), & \text{if } c \in C_j \\ b, & \text{if } c = c_j, \end{cases}$$

and the following conditions are satisfied: given a positive bounded formula $\varphi(x, x_1, \ldots, x_n)$ and ordinals $j_1, \ldots, j_n < j$, if

$$A_j^{\alpha,\gamma}(c_{j_k}) = a_k \qquad\qquad (k = 1, \ldots, n),$$
$$B_j^{\alpha,\gamma}(c_{j_k}) = b_k \qquad\qquad (k = 1, \ldots, n),$$

then:

(a) either the set

$$\left\{ \xi < \lambda \;\middle|\; \mathcal{M}_\xi^{\alpha,\gamma} \models \varphi[a_\rho^{\alpha,\gamma}(\xi), a_1(\xi), \ldots, a_n(\xi)] \right\}$$

or its complement is in D_j'; and

(b) for every approximation φ' of φ,

$$\left\{ \xi < \lambda \;\middle|\; \mathcal{M}_\xi^{\alpha,\gamma} \models \varphi[a_\rho^{\alpha,\gamma}(\xi), a_1(\xi), \ldots, a_n(\xi)] \right\} \in D_j'$$

implies

$$\left\{ \xi < \lambda \;\middle|\; \mathcal{N}_\xi^{\alpha,\gamma} \models \varphi'[b(\xi), b_1(\xi), \ldots, b_n(\xi)] \right\} \in D_i.$$

(4) If $i = j+1$, $g(j) = (2, \alpha, \gamma, \rho, b_\rho^{\alpha,\gamma})$, and the set $D_j \cup G_\alpha$ generates a proper filter on λ, then there exist a filter D_j' with

$$D_j \subseteq D_j' \subseteq D_i$$

and a bounded element

$$a \in \prod_{\xi < \lambda} M_\xi^{\alpha,\gamma}$$

such that the functions $A_i^{\alpha,\gamma}, B_i^{\alpha,\gamma}$ are defined by

$$A_i^{\alpha,\gamma}(c) = \begin{cases} A_j^{\alpha,\gamma}(c), & \text{if } c \in C_j \\ a, & \text{if } c = c_j, \end{cases}$$

$$B_i^{\alpha,\gamma}(c) = \begin{cases} B_j^{\alpha,\gamma}(c), & \text{if } c \in C_j \\ b_\rho^{\alpha,\gamma}, & \text{if } c = c_j, \end{cases}$$

and the following conditions are satisfied: given a positive bounded formula $\varphi(x, x_1, \ldots, x_n)$ and ordinals $j_1, \ldots, j_n < j$, if

$$A_j^{\alpha,\gamma}(c_{j_k}) = a_k \qquad\qquad (k = 1, \ldots, n),$$
$$B_j^{\alpha,\gamma}(c_{j_k}) = b_k \qquad\qquad (k = 1, \ldots, n),$$

then:

(a) either the set

$$\left\{ \xi < \lambda \ \middle| \ \mathcal{N}_\xi^{\alpha,\gamma} \models \varphi[b_\rho^{\alpha,\gamma}(\xi), b_1(\xi), \ldots, b_n(\xi)] \right\}$$

or its complement is in D_j'; and

(b) for every approximation φ' of φ,

$$\left\{ \xi < \lambda \ \middle| \ \mathcal{N}_\xi^{\alpha,\gamma} \models \varphi[b_\rho^{\alpha,\gamma}(\xi), b_1(\xi), \ldots, b_n(\xi)] \right\} \in D_j'$$

implies

$$\left\{ \xi < \lambda \ \middle| \ \mathcal{M}_\xi^{\alpha,\gamma} \models \varphi'[a(\xi), a_1(\xi), \ldots, a_n(\xi)] \right\} \in D_i.$$

(5) If $i = j + 1$, $g(j) = (3, \alpha, \gamma, K, p, R)$, $K \subseteq \operatorname{dom}(A_j^{\alpha,\gamma})$, and for every finite subset $p_0(x)$ of $(p(x))^+$ we have

$$\left\{ \xi < \lambda \ \middle| \ (\mathcal{M}_\xi^{\alpha,\gamma}, A_j^{\alpha,\gamma}(c)(\xi) \mid c \in K) \models \exists_R x \bigwedge p_0(x) \right\} \in D_j,$$

then there exists a bounded element a of $\prod_{\xi<\lambda} M_\xi^{\alpha,\gamma}$ such that

$$\left\{ \xi < \lambda \ \middle| \ (\mathcal{M}_\xi^{\alpha,\gamma}, A_j^{\alpha,\gamma}(c)(\xi) \mid c \in K) \models p[a] \right\} \in D_i.$$

(6) If $i = j + 1$, $g(j) = (4, \alpha, \gamma, K, p, R)$, $K \subseteq \operatorname{dom}(B_j^{\alpha,\gamma})$, and for every finite subset $p_0(x)$ of $(p(x))^+$ we have

$$\left\{ \xi < \lambda \ \middle| \ (\mathcal{N}_\xi^{\alpha,\gamma}, B_j^{\alpha,\gamma}(c)(\xi) \mid c \in K) \models \exists_R x \bigwedge p_0(x) \right\} \in D_j,$$

then there exists a bounded element b of $\prod_{\xi<\lambda} N_\xi^{\alpha,\gamma}$ such that

$$\left\{ \xi < \lambda \ \middle| \ (\mathcal{N}_\xi^{\alpha,\gamma}, B_j^{\alpha,\gamma}(c)(\xi) \mid c \in K) \models p[b] \right\} \in D_i.$$

(7) If i is a limit ordinal, then

- $F_i = \bigcap_{j<i} F_j$,
- $D_i = \bigcup_{j<i} D_j$,
- $A_i^{\alpha,\gamma} = \bigcup_{j<i} A_j^{\alpha,\gamma}$,
- $B_i^{\alpha,\gamma} = \bigcup_{j<i} B_j^{\alpha,\gamma}$.

PART 4. DESCRIPTION OF THE INDUCTIVE CONSTRUCTION. As in the proof of Theorem 10.7, the case that requires special attention is the successor case, $i = j + 1$.

If $g(j) = (0, S)$, we use Lemma 10.5 to find F_i and D_i as stated in (2), and define

$$A_i^{\alpha,\gamma}(c) = \begin{cases} A_j^{\alpha,\gamma}(c), & \text{if } c \in C_j \\ 0, & \text{if } c = c_j, \end{cases}$$

$$B_i^{\alpha,\gamma}(c) = \begin{cases} B_j^{\alpha,\gamma}(c), & \text{if } c \in C_j \\ 0, & \text{if } c = c_j. \end{cases}$$

If $g(j) = (1, \alpha, \gamma, \rho, a_\rho^{\alpha,\gamma})$ and the set $D_j \cup G_\alpha$ generates a proper filter on λ, we define $A_i^{\alpha,\gamma}$ and $B_i^{\alpha,\gamma}$ as given in (3) and proceed as in the proof of Theorem 10.7 to find F_i and D_i as stated in (3).

If $g(j) = (2, \alpha, \gamma, \rho, b_\rho^{\alpha,\gamma})$ and the set $D_j \cup G_\alpha$ generates a proper filter on λ, we define $A_i^{\alpha,\gamma}$ and $B_i^{\alpha,\gamma}$ as given in (4) and proceed as in the proof of Theorem 10.7 to find F_i and D_i as stated in (4).

If $i = j + 1$, $g(j) = (3, \alpha, \gamma, K, p, R)$, $K \subseteq \text{dom}(A_j^{\alpha,\gamma})$, and for every finite subset $p_0(x)$ of $(p(x))^+$ we have

$$\left\{ \xi < \lambda \;\middle|\; (\mathfrak{M}_\xi^{\alpha,\gamma}, A_j^{\alpha,\gamma}(c)(\xi) \mid c \in K) \models \exists_R x \bigwedge p_0(x) \right\} \in D_j,$$

we use Lemma 10.6 to find F_i and D_i as stated in (5), and let $A_i^{\alpha,\gamma}, B_i^{\alpha,\gamma}$ be as when $g(j) = (0, S)$.

If $i = j + 1$, $g(j) = (4, \alpha, \gamma, K, p, R)$, $K \subseteq \text{dom}(B_j^{\alpha,\gamma})$, and for every finite subset $p_0(x)$ of $(p(x))^+$ we have

$$\left\{ \xi < \lambda \;\middle|\; (\mathfrak{N}_\xi^{\alpha,\gamma}, B_j^{\alpha,\gamma}(c)(\xi) \mid c \in K) \models \exists_R x \bigwedge p_0(x) \right\} \in D_j,$$

we use Lemma 10.6 to find F_i and D_i as stated in (6), and let $A_i^{\alpha,\gamma}, B_i^{\alpha,\gamma}$ be as when $g(j) = (0, S)$.

If $i = j + 1$ and $g(j)$ is not of any of the preceding forms, we simply let $F_i = F_j$, $D_i = D_j$, and let $A_i^{\alpha,\gamma}, B_i^{\alpha,\gamma}$ be as when $g(j) = (0, S)$.

PART 5. HOW THE CONSTRUCTION YIELDS THE THEOREM. Let

$$D = \bigcup_{i < 2^\lambda} D_i.$$

By (2), D is an ultrafilter.

Suppose that L_0 is a subsignature of L and

$$(\mathfrak{M}_\xi \mid \xi < \lambda), \qquad (\mathfrak{N}_\xi \mid \xi < \lambda)$$

are families of normed space L_0-structures such that the density character of each sort of $\mathcal{M}_\xi$ and $\mathcal{N}_\xi$ is at most κ for all $\xi < \lambda$; suppose further that

$$\left(\prod_{\xi<\lambda} \mathcal{M}_\xi \right)_D \equiv_A \left(\prod_{\xi<\lambda} \mathcal{N}_\xi \right)_D.$$

Claim 1. There exist $\alpha, \gamma < 2^\lambda$ such that for all $\xi < \lambda$,

$$\mathcal{M}_\xi \cong \mathcal{M}_\xi^{\alpha,\gamma}, \qquad \mathcal{N}_\xi \cong \mathcal{N}_\xi^{\alpha,\gamma}.$$

Proof of Claim 1. For each L_0-sentence θ, let

$$X(\theta) = \{\, \xi < \lambda \mid \mathcal{M}_\xi \models \theta \,\},$$
$$Y(\theta) = \{\, \xi < \lambda \mid \mathcal{N}_\xi \models \theta \,\},$$

and let G be the filter generated by the sets

$$X(\theta') \text{ and } Y(\theta'), \qquad \text{where } \theta' > \theta \in \mathrm{Th}_A\left(\left(\prod_{\xi<\lambda} \mathcal{M}_\xi \right)_D \right).$$

Note that $G \subseteq D$. Moreover, G is generated by at most λ sets, so $G = G_\alpha$ for some $\alpha < 2^\lambda$. Further, for every L_0-sentence θ and every approximation θ' of θ we have

$$\{\xi < \lambda \mid \mathcal{M}_\xi \models \theta\} \in G \text{ implies } \{\xi < \lambda \mid \mathcal{N}_\xi \models \theta'\} \in G$$

and

$$\{\xi < \lambda \mid \mathcal{N}_\xi \models \theta\} \in G \text{ implies } \{\xi < \lambda \mid \mathcal{M}_\xi \models \theta'\} \in G,$$

so Claim 1 follows. $\square$

Fix α and γ as given by Claim 1, and let

$$A^{\alpha,\gamma} = \bigcup_{i<2^\lambda} A_i^{\alpha,\gamma},$$
$$A^{\alpha,\gamma} = \bigcup_{i<2^\lambda} B_i^{\alpha,\gamma},$$
$$C = \bigcup_{i<2^\lambda} C_i.$$

By (3) and (4) (and property (II) of the function g), every bounded element of $\prod_{\xi<\lambda} M_\xi^{\alpha,\gamma}$ is of the form $A^{\alpha,\gamma}(c)$ for some $c \in C$ and every bounded element of $\prod_{\xi<\lambda} N_\xi^{\alpha,\gamma}$ is of the form $B^{\alpha,\gamma}(c)$ for some $c \in C$.

Claim 2. Every element of $(\prod_{\xi<\lambda} \mathcal{M}_\xi^{\alpha,\gamma})_D$ is of the form $(a)_D$ for some bounded element a of $\prod_{\xi<\lambda} M_\xi^{\alpha,\gamma}$.

Proof of Claim 2. The set of equivalence classes of the form $(a)_D$, where a is a bounded element of $\prod_{\xi<\lambda} M_\xi^{\alpha,\gamma}$, is dense in $(\prod_{\xi<\lambda} M_\xi^{\alpha,\gamma})_D$, so we may use condition (5) to argue as in the proof of Proposition 9.21.   □

Claim 3. Every element of $(\prod_{\xi<\lambda} N_\xi^{\alpha,\gamma})_D$ is of the form $(b)_D$ for some bounded element b of $\prod_{\xi<\lambda} N_\xi^{\alpha,\gamma}$.

Proof of Claim 3. Similar to the proof of Claim 2, but using condition (6) in place of condition (5).   □

Claim 4. The structure $(\prod_{\xi<\lambda} M_\xi^{\alpha,\gamma})_D$ is κ^+-saturated; hence so is $(\prod_{\xi<\lambda} M_\xi)_D$.

Proof of Claim 4. By Claim 2 and condition (5).   □

Claim 5. If $\varphi(x_1,\ldots,x_n)$ is a positive bounded L_0-formula,

$$a_1,\ldots,a_n \in \ell_\infty(\lambda,(M_\xi^{\alpha,\gamma} \mid \xi < \lambda)),$$
$$b_1,\ldots,b_n \in \ell_\infty(\lambda,(N_\xi^{\alpha,\gamma} \mid \xi < \lambda)),$$

and $c_1,\ldots,c_n \in C$ are such that

$$A^{\alpha,\gamma}(c_k) = a_k \qquad\qquad (k = 1,\ldots,n),$$
$$B^{\alpha,\gamma}(c_k) = b_k \qquad\qquad (k = 1,\ldots,n),$$

then

$$\left\{ \xi < \lambda \;\middle|\; M_\xi^{\alpha,\gamma} \models \varphi[a_1(\xi),\ldots,a_n(\xi)] \right\} \in D$$

implies

$$\left\{ \xi < \lambda \;\middle|\; N_\xi^{\alpha,\gamma} \models \varphi'[b_1(\xi),\ldots,b_n(\xi)] \right\} \in D \quad \text{for every } \varphi' > \varphi.$$

Proof of Claim 5. Suppose

$$\left\{ \xi < \lambda \;\middle|\; M_\xi^{\alpha,\gamma} \models \varphi[a_1(\xi),\ldots,a_n(\xi)] \right\} \in D.$$

Let i be the least ordinal such that $g(i) = (1,\ldots)$ and $a_1,\ldots,a_n$ are in the range of $A_i^{\alpha,\gamma}$. Let j be such that $i = j+1$. The set $D_j \cup G_\alpha$ generates a proper filter on λ since $G_\alpha = G \subseteq D$. Hence, by condition (3)(a),

$$\left\{ \xi < \lambda \;\middle|\; M_\xi^{\alpha,\gamma} \models \varphi[a_1(\xi),\ldots,a_n(\xi)] \right\} \in D_j',$$

and by (3)(b),

$$\left\{ \xi < \lambda \;\middle|\; \mathcal{N}_\xi^{\alpha,\gamma} \models \varphi'[b_1(\xi),\dots,b_n(\xi)] \right\} \in D_i \subseteq D$$

for every $\varphi' > \varphi$. □

By Claims 2, 3, and 5,

$$(\prod_{\xi<\lambda} \mathcal{M}_\xi^{\alpha,\gamma})_D \cong (\prod_{\xi<\lambda} \mathcal{N}_\xi^{\alpha,\gamma})_D$$

and therefore

$$(\prod_{\xi<\lambda} \mathcal{M}_\xi)_D \cong (\prod_{\xi<\lambda} \mathcal{N}_\xi)_D.$$

by Claim 1. □

10.9 Remark. An important case of Theorems 10.7 and 10.8 and their consequences is when $\kappa = \aleph_0$. In this case $\mathrm{card}(L) \leq 2^{\aleph_0}$ and we are considering normed space L-structures $\mathcal{M}$ such that every sort of $\mathcal{M}$ is separable. In this situation, the ultrafilter D in Theorems 10.7 and 10.8 lives on an index set of cardinality equal to the continuum $c = 2^{\aleph_0}$. In the ultrapowers and ultraproducts considered in these results, each sort has density character at most 2^c.

10.10 Remark. For Banach spaces and Banach lattices (but not for normed space structures in general) Theorem 10.7 was first proved in [Hen75] (see especially Theorem 4) and was further treated in [Hen76] (see Theorem 1.13). See also [HM83a] where several characterizations of $\equiv_A$ for Banach spaces are discussed. Those proofs applied the Keisler-Shelah Theorem together with Lyndon's Theorem from ordinary model theory to give an indirect proof of the Banach space result, rather than constructing a direct proof as is done here. Stern [Ste78] (see also [Ste75]) applied the Keisler-Shelah Theorem in a similar way to prove that if X is a Banach space and U, V are ultrafilters, then there exists an ultrafilter D such that $((X)_U)_D$ and $((X)_V)_D$ are isometrically isomorphic. This is, of course, a special case of Theorem 10.7. Shelah's full result for ultraproducts, and its analogue Theorem 10.8 for normed spaced structures, are closely connected to the Isomorphism Properties [Hen74] [HM83b] used in nonstandard analysis. Another characterization of $\equiv_A$ can be given in terms of Ehrenfeucht-Fraissé Games. Discussions of this topic in the restricted setting of pure Banach spaces can be found in [HM83a], [HM83b], and [HH86]. For a general signature L, the results and proofs are essentially the same as the ones given in those papers.

11

Alternative Formulations of the Theory

It is possible to handle the role of the scalar field $\mathbb{R}$ in a number of different ways, without changing the overall expressive power of this logic. For example, we might have included constant symbols in every signature for all the real numbers, as opposed to only the rationals. (We chose to do the latter in order to avoid artificially increasing the cardinality of signatures, since this has an effect on such results as the Downward Löwenheim-Skolem Theorem (Proposition 9.13)).

Using the isomorphism theorem for ultrapowers (Theorem 10.7) we can show that adding all the reals as constants does not change approximate elementary equivalence of normed space structures, although it does increase the set of positive bounded formulas. (The Perturbation Lemma (Proposition 5.15) can also be used to prove this.)

11.1 Proposition. *Let L be a signature and let $C = \{c_r \mid r \in \mathbb{R}\}$ be a new set of constant symbols in 1-1 correspondence with $\mathbb{R}$; for each $r \in \mathbb{R}$ the constant symbol c_r has the sort $s_{\mathbb{R}}$. For each normed space L-structure $\mathcal{M}$ let $\mathcal{M}'$ be the $L(C)$-structure that expands $\mathcal{M}$ and interprets c_r as r for each $r \in \mathbb{R}$. Then, for any L-structures $\mathcal{M}$ and $\mathcal{N}$,*

$$\mathcal{M} \equiv_A \mathcal{N} \quad \text{if and only if} \quad \mathcal{M}' \equiv_A \mathcal{N}'.$$

and

$$\mathcal{M} \preceq_A \mathcal{N} \quad \text{if and only if} \quad \mathcal{M}' \preceq_A \mathcal{N}'.$$

Proof. Since $\mathcal{M}$ is a reduct of $\mathcal{M}'$ for each normed space L-structure $\mathcal{M}$, the right to left directions of these equivalences are immediate. We prove the left to right implications using the isomorphism theorem for ultrapowers. Let $\mathcal{M}$, $\mathcal{N}$ be L-structures such that $\mathcal{M} \equiv_A \mathcal{N}$. Take κ to be a cardinal number greater than the density character of each sort of $\mathcal{M}$ and of $\mathcal{N}$, and such that $\operatorname{card}(L) \leq \kappa$. By Theorem 10.7 there is an ultrafilter D on $\lambda = 2^\kappa$ such that $(\mathcal{M})_D \cong (\mathcal{N})_D$. It follows that $((\mathcal{M})_D)' \cong ((\mathcal{N})_D)'$. One easily sees that $(\mathcal{M}')_D \cong ((\mathcal{M})_D)'$ and the same for $\mathcal{N}$. Therefore $(\mathcal{M}')_D \cong (\mathcal{N}')_D$. By Corollary 9.4 it follows that $\mathcal{M}' \equiv_A \mathcal{N}'$.

The second implication follows from the first when it is applied to an expansion by constants of $\mathcal{M}$ in which every element of $\mathcal{M}$ is given a name. $\qquad\square$

Note that a similar argument shows that if we add every continuous function $f \colon \mathbb{R}^n \to \mathbb{R}$ to a normed space structure $\mathcal{M}$, then the positive bounded theory of the new structure is determined by $\mathrm{Th}_A(\mathcal{M})$; we would make this precise using the same language as in Proposition 11.1. This can sometimes be useful when characterizing properties of normed space structures explicitly by positive bounded sentences.

Similar considerations show that it does not change the expressive power of this logic to completely omit the real sort, as long as this is done carefully. To illustrate what is involved here, we consider only the example of Banach spaces with no additional structure. Let L_0 be the signature for such structures. (See Chapter 7.) We call an L_0-formula *restricted* if it does not contain any variables or constant symbols whose sort is $s_\mathbb{R}$, and if the only function symbols it contains are the norm and the 0 and $+$ of the normed space sort. Up to equivalence in every normed space L_0-structure, the restricted atomic formulas are $\|k_1 x_1 + \cdots + k_n x_n\| \leq r$ and $r \leq \|k_1 x_1 + \cdots + k_n x_n\|$, where $k_1, \ldots, k_n$ are integers, r is a rational number, and $x_1, \ldots, x_n$ are variables of the normed space sort. Restricted formulas in general are constructed from restricted atomic formulas using $\wedge, \vee$ and bounded quantifiers $\exists_r x, \forall_r x$ where r is a positive rational number and x is a variable of the normed space sort. Note that the weak negation and every approximation of a restricted L_0-formula are restricted.

The following result shows that the overall expressive power of restricted formulas is the same as that of the full class of positive bounded L_0-formulas.

11.2 Proposition. *Let X, Y be normed spaces, considered as normed space L_0-structures.*

(i) *Suppose that for every restricted L_0-sentence σ,*

$$X \models_A \sigma \quad \text{if and only if} \quad Y \models_A \sigma.$$

Then $X \equiv_A Y$.

(ii) *Assume $X \subseteq Y$. Suppose that for every $a_1, \ldots, a_n \in X$ and every restricted L_0-formula $\varphi(x_1, \ldots, x_n)$,*

$$X \models_A \varphi[a_1, \ldots, a_n] \quad \text{if and only if} \quad Y \models_A \varphi[a_1, \ldots, a_n].$$

Then $X \preceq_A Y$.

Proof. We prove (i); (ii) is proved similarly. Assume that X, Y satisfy the hypotheses of (i). We indicate how the proof of Theorem 10.7 can be modified to prove that there is an ultrafilter D such that $(X)_D \cong (Y)_D$. From this it follows that $X \equiv_A Y$, as desired, using Corollary 9.4.

Take κ to bound the cardinalities of X and Y, and then construct an ultrafilter D on $\lambda = 2^\kappa$ as in the proof of Theorem 10.7. The modification of the proof consists of replacing every reference to *positive bounded formula* in the inductive construction by *restricted positive bounded formula*. At the end of the construction one has a mapping $T \colon \ell_\infty(\lambda, X) \to \ell_\infty(\lambda, Y)$ such that for every restricted positive bounded L_0-formula $\varphi(x_1, \ldots, x_n)$, every approximation φ' of φ, and every $a_1, \ldots, a_n \in \ell_\infty(\lambda, X)$,

$$\{\xi < \lambda \mid X \models \varphi[a_1(\xi), \ldots, a_n(\xi)]\} \in D$$

implies

$$\{\xi < \lambda \mid Y \models \varphi'[T(a_1)(\xi), \ldots, T(a_n)(\xi)]\} \in D.$$

Using the expressive power of restricted atomic formulas, it follows immediately that T induces an isomorphism between $(X)_D$ and $(Y)_D$ as L_0-structures. □

The previous argument can be applied in a much more general setting, and this is only a limited example of how one can vary the handing of the scalar sort without changing the overall expressive power of the logic.

We note that restricted L_0-formulas are (up to equivalence in all normed space L_0-structures) among the positive bounded formulas for the normed space setting that were considered in earlier papers such as [Hen76] [HM83b] [HH86]. Moreover, the concept of approximate satisfaction considered in those papers is easily seen to be equivalent to the one used here. From Proposition 11.2 it follows that the equivalence relation $\equiv_A$ for the signature L_0 that is treated in this paper is the same as the one considered in those earlier papers (which only considered Banach spaces without additional structure).

12

Homogeneous Structures

12.1 Definition. Let $\mathcal{M}$ be a normed space L-structure and let κ be an infinite cardinal. We say that $\mathcal{M}$ is *strongly κ-homogeneous* if the following condition holds: whenever $L(C)$ is an extension of L by constants with $\mathrm{card}(C) < \kappa$ and $\mathbf{A}, \mathbf{B}$ are interpretations of C in $\mathcal{M}$ such that

$$(\mathcal{M}, \mathbf{A}) \equiv_A (\mathcal{M}, \mathbf{B})$$

one has

$$(\mathcal{M}, \mathbf{A}) \cong (\mathcal{M}, \mathbf{B}).$$

Note that an isomorphism from $(\mathcal{M}, \mathbf{A})$ onto $(\mathcal{M}, \mathbf{B})$ is an *automorphism* of $\mathcal{M}$ that takes $\mathbf{A}(c)$ to $\mathbf{B}(c)$ for each $c \in C$.

12.2 Proposition. *Let κ be an infinite cardinal number and let $\lambda = 2^\kappa$. Then the ultrafilter D on λ given by Theorem 10.8 satisfies the following property: Whenever we are given*

· *a signature L with $\mathrm{card}(L) \leq \lambda$, and*
· *a family of normed space L-structures $(\mathcal{M}_\xi \mid \xi < \lambda)$ such that the density character of each sort of $\mathcal{M}_\xi$ is at most κ for every $\xi < \lambda$,*

then the ultraproduct $(\prod_{\xi < \lambda} \mathcal{M}_\xi)_D$ is always κ^+-saturated and strongly λ^+-homogeneous.

Before proving Proposition 12.2, we note the following important corollary:

12.3 Corollary. *Let κ be an infinite cardinal number and let $\lambda = 2^\kappa$. Then there exists an ultrafilter D on λ satisfying the following property: whenever $\mathcal{M}$ is an L-structure such that $\mathrm{card}(L) \leq \lambda$ and the density character of each sort of $\mathcal{M}$ is at most κ, the ultrapower $(\mathcal{M})_D$ is κ^+-saturated and strongly λ^+-homogeneous.*

Proof. By Propositions 6.7 and 12.2.   □

Proof of Proposition 12.2. The κ^+-saturation of $(\prod_{\xi<\lambda}\mathcal{M}_\xi)_D$ is given
by the last part of Theorem 10.8. We now prove that $(\prod_{\xi<\lambda}\mathcal{M}_\xi)_D$ is
strongly λ^+-homogeneous. To this end, fix an extension by constants
$L(C)$ of L with $\mathrm{card}(C)\le\lambda$. Note that $\mathrm{card}(L(C))\le\lambda$.

Suppose that $\mathbf{A}$ is any interpretation of C in $(\prod_{\xi<\lambda}\mathcal{M}_\xi)_D$. The key
observation for this proof is that there exist interpretations $\mathbf{A}_\xi$ of C in
$\mathcal{M}_\xi$ for each $\xi<\lambda$ such that $\mathbf{A}(c)=((\mathbf{A}_\xi(c))_{\xi<\lambda})_D$ for every $c\in C$ and
hence that

$$\left(\left(\prod_{\xi<\lambda}\mathcal{M}_\xi\right)_D,\mathbf{A}\right)=\left(\prod_{\xi<\lambda}(\mathcal{M}_\xi,\mathbf{A}_\xi)\right)_D.$$

Suppose that $\mathbf{A},\mathbf{B}$ are interpretations of C in $(\prod_{\xi<\lambda}\mathcal{M}_\xi)_D$ such that

$$\left(\left(\prod_{\xi<\lambda}\mathcal{M}_\xi\right)_D,\mathbf{A}\right)\equiv_A\left(\left(\prod_{\xi<\lambda}\mathcal{M}_\xi\right)_D,\mathbf{B}\right).$$

Apply the previous remark to $\mathbf{A}$ and $\mathbf{B}$, obtaining $\mathbf{A}_\xi$ and $\mathbf{B}_\xi$ for each
$\xi<\lambda$ such that

$$\left(\left(\prod_{\xi<\lambda}\mathcal{M}_\xi\right)_D,\mathbf{A}\right)=\left(\prod_{\xi<\lambda}(\mathcal{M}_\xi,\mathbf{A}_\xi)\right)_D$$

and

$$\left(\left(\prod_{\xi<\lambda}\mathcal{M}_\xi\right)_D,\mathbf{B}\right)=\left(\prod_{\xi<\lambda}(\mathcal{M}_\xi,\mathbf{B}_\xi)\right)_D.$$

From this it follows that

$$\left(\prod_{\xi<\lambda}(\mathcal{M}_\xi,\mathbf{A}_\xi)\right)_D\equiv_A\left(\prod_{\xi<\lambda}(\mathcal{M}_\xi,\mathbf{B}_\xi)\right)_D.$$

Therefore,

$$\left(\left(\prod_{\xi<\lambda}\mathcal{M}_\xi\right)_D,\mathbf{A}\right)\cong\left(\left(\prod_{\xi<\lambda}\mathcal{M}_\xi\right)_D,\mathbf{B}\right)$$

by Theorem 10.8 and the equations above.   □

12.4 Remark. A technically easier proof that κ-saturated, strongly
κ-homogeneous approximate elementary extensions exist can be given
using the existence of highly saturated extensions and a union-of-chains
argument, as we now outline. We may assume κ is regular, increasing
κ if necessary. Given any normed space L-structure $\mathcal{M}$, we construct
an approximate elementary chain $(\mathcal{M}_\alpha\mid\alpha<\kappa)$ whose union has the
desired properties. Let $\mathcal{M}_0=\mathcal{M}$; for each $\alpha<\kappa$, let $\mathcal{M}_{\alpha+1}$ be an

approximate elementary extension of $\mathcal{M}_\alpha$ that is τ_α-saturated, where τ_α is a cardinal bigger than $\text{card}(L)$ and bigger than the cardinality of each sort of $\mathcal{M}_\alpha$; take unions at limit ordinals. Let $\mathcal{N}$ be the union of $(\mathcal{M}_\alpha \mid \alpha < \kappa)$. By Corollary 9.12, $\mathcal{M} \preceq_A \mathcal{N}$. An argument such as in the proof of Proposition 9.25 shows that $\mathcal{N}$ is κ-saturated. The fact that $\mathcal{N}$ is strongly κ-homogeneous follows from an inductive argument whose successor steps are based on the following easily proved fact:

Suppose $\mathcal{M} \preceq_A \mathcal{N}$ and $\mathcal{N}$ is τ-saturated, where τ is a cardinal bigger than $\text{card}(L)$ and bigger than the cardinality of each sort of $\mathcal{M}$. Let C be a set of new constants with $\text{card}(C) < \tau$. Suppose $\mathbf{A}, \mathbf{B}$ are interpretations of C in $\mathcal{M}$ such that $(\mathcal{M}, \mathbf{A}) \equiv_A (\mathcal{N}, \mathbf{B})$. Then there exists an approximate elementary embedding $T \colon \mathcal{M} \to \mathcal{N}$ such that for every $c \in C$, $T(\mathbf{A}(c)) = \mathbf{B}(c)$.

12.5 Remark. S. Buechler and O. Lessmann have developed an approach to geometric stability theory in the absence of full model theoretic compactness [BL] . We briefly indicate how this can be applied to normed space structures. Let L be any signature and let $\mathcal{M}_0$ be a normed space L-structure. Let λ be a cardinal number $\geq \aleph_1$. Applying Corollary 12.3 to $\mathcal{M}_0$ and a sufficiently large κ, let $\mathcal{M}$ be a Banach space ultrapower of $\mathcal{M}_0$ such that $\mathcal{M}$ is λ-saturated and strongly λ-homogeneous. We regard $\mathcal{M}$ as a logical structure $(M, \mathcal{R})$ in the sense of [BL, Definition 1.1] by taking $\mathcal{R}$ to be the collection of all finitary relations that are definable in $\mathcal{M}$ by positive bounded L-formulas (without parameters except for the ones named by constant symbols of L). (Recall that since $\mathcal{M}$ is at least $\aleph_1$-saturated, $\models_A$ is equivalent to $\models$ in $\mathcal{M}$ for positive bounded formulas; see Proposition 9.20.) It is easy to verify that $(M, \mathcal{R})$ satisfies the concept of strong λ-homogeneity used by Buechler and Lessmann (see [BL, Definition 1.3]) and therefore that their methods apply to this structure. Stability and simplicity of the original structure $\mathcal{M}_0$ can be studied by applying the definitions and methods from [BL] to $\mathcal{M}$. (It is not difficult to show that this is independent of $\mathcal{M}$, as long as λ is sufficiently large.)

In the mid-1990s, the second author developed a concept of stability for normed space structures in the context of the model theory developed here; see [Iov94] [Iov96] [Iov98] [Iov99a] [Iov99b]. Using what is presented in those papers (especially in [Iov99b, Section 4]) and in [BL], and adapting certain other arguments in the literature to the setting of positive bounded formulas and normed space structures, it can be shown that these two notions of stability for normed space structures are equivalent.

A number of interesting normed space structures are stable in this sense. For example, a type-counting argument based on simple features of Hilbert space geometry can be used to show that if $\mathcal{M}$ consists of a Hilbert space H together with a family $(T_i \mid i \in I)$ of bounded linear operators $T_i \colon H \to H$, then $\mathcal{M}$ is stable. Another example of a stable normed space structure is the Banach lattice $L_p(\mu)$, where $1 \leq p < \infty$ and μ is any measure.

An explanation of the equivalence of these two notions of stability for normed space structures and a treatment of these examples will be given elsewhere.

13

More Model Theory

Quantifier Complexity of Formulas

For every positive bounded L-formula $\varphi(x_1, \ldots, x_n)$ there exists a positive bounded L-formula $\psi(x_1, \ldots, x_n)$ such that ψ is of the form

$$Q_1 y_1 Q_2 y_2 \ldots Q_k y_k\, \theta(x_1, \ldots, x_n, y_1, \ldots, y_k)$$

where

· Q_i is a bounded quantifier of the form $\exists_{r_i}$ or $\forall_{r_i}$, for $i = 1, \ldots, k$;
· the positive bounded formula $\theta(x_1, \ldots, x_n, y_1, \ldots, y_k)$ is quantifier-free;
· $\psi(x_1, \ldots, x_n)$ is equivalent to $\varphi(x_1, \ldots, x_n)$ in the sense that for every normed space L-structure $\mathcal{M}$ and elements $a_1, \ldots, a_n$ of suitable sorts of $\mathcal{M}$,

$$\mathcal{M} \models \varphi[a_1, \ldots, a_n] \quad \text{if and only if} \quad \mathcal{M} \models \psi[a_1, \ldots, a_n].$$

The formula ψ is said to be in *prenex form*. We say that ψ is *existential* if all its quantifiers are existential. Similarly, we say that ψ is *universal* if all its quantifiers are universal.

Note that any approximation of a prenex form positive bounded formula will also be in prenex form, with exactly the same sequence of leading quantifier types ($\forall$ versus $\exists$).

Note also that the usual procedure for putting a given formula into prenex form, when applied to a positive bounded formula, commutes nicely with the process of forming approximations. If φ is a positive bounded formula and φ' is an approximation of φ, then the logic structure of φ' is identical to that of φ. Suppose ψ is in prenex form and is equivalent to φ, as above; let ψ' be obtained from φ' by applying the same sequence of prenex rules that was used to obtain ψ from φ. Then

ψ' is equivalent to φ' and ψ' is an approximation of ψ. This observation allows us to investigate the approximate satisfaction of the positive bounded formula φ by replacing φ with its equivalent prenex form ψ.

Axiomatizing Classes of Substructures

If φ is a universal sentence and $\mathcal{M}$ satisfies φ, then every substructure of $\mathcal{M}$ satisfies φ too. The following result shows that this is a characteristic way of describing uniform classes of L-structures that are closed under isomorphism and substructures:

13.1 Proposition. *Suppose that $\mathfrak{D}$ is a uniform class of normed space L-structures and let Γ be the set of universal L-sentences that are satisfied by every structure in $\mathfrak{D}$. Let $\mathfrak{C}$ be the class of L-structures that are isomorphic to a substructure of an ultraproduct of members of $\mathfrak{D}$. Then $\mathfrak{C}$ consists of the L-structures that satisfy Γ.*

Proof. Every structure in $\mathfrak{C}$ satisfies Γ, so we just need to prove the converse. Fix an L-structure $\mathcal{M}$ such that $\mathcal{M} \models \Gamma$. Let C be a set of new constants and $\mathbf{A}$ an interpretation of C in $\mathcal{M}$ such that every element of $\mathcal{M}$ is of the form $\mathbf{A}(c)$ for some $c \in C$. Let Δ be the set of all quantifier-free positive bounded sentences in $L(C)$ that are true in $(\mathcal{M}, \mathbf{A})$. Let Σ be the set of all finite subsets of Δ^+.

Consider $\xi = \{\sigma_1, \ldots, \sigma_n\} \in \Sigma$. There exist quantifier-free positive bounded L- formulas $\psi_j(x_1, \ldots, x_m)$ for $j = 1, \ldots, n$ and $c_1, \ldots, c_m \in C$ such that for each $j = 1, \ldots, n$, the formula σ_j is equal to $\psi_j(c_1, \ldots, c_m)$. Note that x_i and c_i must have the same sort for $i = 1, \ldots, m$. Moreover, for each $j = 1, \ldots, n$ there exists a positive bounded L-formula φ_j with $\varphi_j < \psi_j$ and $\varphi_j(c_1, \ldots, c_m) \in \Delta$. For each $i = 1, \ldots, m$ let r_i be a rational number with $\|\mathbf{A}(c_i)\| \le r_i$. Without loss of generality we may assume that each φ_j contains the formula $\|x_i\| \le r_i$ as a conjunct. Let $s_i > r_i$ be the rational number for which each approximation ψ_j contains $\|x_i\| \le s_i$ as a conjunct in the corresponding place.

We claim that there exist an L-structure $\mathcal{N}_\xi \in \mathfrak{D}$ and an interpretation $\mathbf{A}_\xi$ of C in $\mathcal{N}_\xi$ such that the $L(C)$-structure $(\mathcal{N}_\xi, \mathbf{A}_\xi)$ satisfies every element of ξ. Otherwise we would have

$$\forall_{s_1} x_1 \ldots \forall_{s_m} x_m \bigvee_{i=1}^{n} \mathrm{neg}(\psi_i) \in \Gamma,$$

and hence this universal sentence would be true in $\mathcal{M}$. However, this is impossible, since by our choice of formulas we have

$$\mathcal{M} \not\models \bigvee_{i=1}^{n} \text{neg}(\psi_i)[\mathbf{A}(c_1), \dots, \mathbf{A}(c_m)].$$

Let D be an ultrafilter on Σ that contains the set

$$\{\xi \in \Sigma \mid \psi \in \xi\}$$

whenever $\psi \in \Delta^+$. Then the D-ultraproduct of $(\mathcal{N}_\xi \mid \xi \in \Lambda)$ contains a substructure isomorphic to $\mathcal{M}$. Therefore $\mathcal{M}$ is in $\mathfrak{C}$. □

13.2 Definition. Suppose that $\mathfrak{C}$ is a uniform class of normed space L-structures and there exists a set Γ of universal positive bounded L-sentences such that $\mathfrak{C}$ is exactly the class of L-structures that satisfy Γ. In this situation we will say that $\mathfrak{C}$ *is axiomatizible by universal L-sentences.*

13.3 Corollary. *Let $\mathfrak{C}$ be a uniform class of L-structures. Then $\mathfrak{C}$ is axiomatizable by universal L-sentences if and only if it is closed under isomorphism, under ultraproducts, and under taking substructures.*

Proof. ($\Leftarrow$) Apply Proposition 13.1 with $\mathfrak{D} = \mathfrak{C}$. □

13.4 Examples. The literature contains a number of important examples of classes of normed space structures which can be shown to have axiomatizations by universal positive bounded sentences. In some cases the conditions in Corollary 13.3 are shown to be true of the class in question, and so one has a "soft" proof that the axiomatization exists. In other cases explicit conditions for membership in the class are given that can be expressed using universal (positive bounded) sentences (sometimes easily, sometimes with effort).

- Fix p in the interval $1 \le p < \infty$. [BDCK66] [DCK72] show that the class of Banach lattices isomorphic to $L_p(\mu)$ for some measure μ is axiomatizable by universal sentences in the signature of Banach lattices. (Also see the discussion of abstract L_p-spaces in [MN91] and [Lac74].)
- [Hen76] [Ste76] show that the class of Banach lattices (with a designated element) isomorphic to $C(K)$ (with constant function 1) for some compact Hausdorff space K is axiomatizable by universal sentences in the signature of Banach lattices with a designated element. (Also see the discussion of abstract M-spaces in [MN91] and [Lac74].)

- Fix p in the interval $1 \leq p < \infty$. [BDCK65] [BDCK66] [Kri67] show that the class of Banach spaces embedable into $L_p(\mu)$ for some measure μ is axiomatizable by universal sentences in the signature of pure Banach spaces.
- [DCK70] considers a class $\mathfrak{D}$ of Banach spaces closed under ultraproducts, and shows that the class of Banach spaces λ-embedable in a member of $\mathfrak{D}$ can be axiomatized by conditions expressible by universal sentences in the signature of pure Banach spaces. See also [DC72a] (especially Theorem 1) and [DCK72].
- [DC72b] gives a characterization due to Krivine of embedability in classes of "ψ-spaces" that are closed under ultraproduct and "ψ-sums", using conditions that can be expressed by universal sentences in the expansion of the signature of pure Banach spaces obtained by adding a function symbol for the function ψ.

Axiomatizability of Classes of Structures

13.5 Definition. Suppose that $\mathfrak{C}, \mathfrak{D}$ are classes of normed space L-structures such that $\mathfrak{C} \subseteq \mathfrak{D}$. We say that $\mathfrak{C}$ is *axiomatizable in $\mathfrak{D}$ by positive bounded sentences* if there exists a set Γ of positive bounded L-sentences such that the structures in $\mathfrak{C}$ are exactly those normed space L-structures in $\mathfrak{D}$ which approximately satisfy Γ. When this condition holds for Γ, we say that Γ is a set of *positive bounded axioms* for $\mathfrak{C}$ in $\mathfrak{D}$.

In this section we characterize axiomatizability by positive bounded sentences in terms of ultraproducts. The ideas are patterned after a well known characterization of axiomatizability in first order logic due to Keisler [Kei61]. (See Corollary 6.1.16 in [CK90].)

13.6 Proposition. *Suppose that $\mathfrak{C}, \mathfrak{D}$ are classes of normed space L-structures with $\mathfrak{C} \subseteq \mathfrak{D}$ such that $\mathfrak{D}$ uniform and is closed under isomorphisms and ultraproducts. The following conditions are equivalent:*

(i) *$\mathfrak{C}$ is axiomatizable in $\mathfrak{D}$ by positive bounded sentences;*
(ii) *$\mathfrak{C}$ is closed under isomorphisms and ultraproducts, and $\mathfrak{D} \setminus \mathfrak{C}$ is closed under ultrapowers.*

Proof. (i)$\Rightarrow$(ii) follows from Propositions 5.14 and 9.3.

To prove (ii)$\Rightarrow$(i), we let Γ be the set of positive bounded L-sentences that are satisfied by every structure in $\mathfrak{C}$. We claim that Γ is a set of positive bounded axioms for $\mathfrak{C}$ in $\mathfrak{D}$. To prove this, suppose that $\mathcal{M}$ is a structure in $\mathfrak{D}$ such that $\mathcal{M} \models_A \Gamma$. Let

$$\Sigma = \big\{\, \{\psi_1, \ldots, \psi_n\} \mid \psi_1, \ldots, \psi_n \in (\mathrm{Th}_A(\mathcal{M}))^+,\ n \in \mathbb{N} \,\big\}.$$

If $\xi = \{\psi_1, \ldots, \psi_n\} \in \Sigma$, then there exists a structure in $\mathfrak{C}$ which satisfies every element of ξ; otherwise we would have $\bigvee_{i=1}^n \mathrm{neg}(\psi_i) \in \Gamma$, so $\mathcal{M} \models_A \bigvee_{i=1}^n \mathrm{neg}(\psi_i)$; but this is impossible because if $\varphi_1, \ldots, \varphi_n \in \mathrm{Th}_A(\mathcal{M})$ are such that $\varphi_i < \psi_i$ for $i = 1, \ldots, n$, we have $\mathcal{M} \models_A \bigwedge_{i=1}^n \varphi_i$, which is a contradiction.

For each $\xi = \{\psi_1, \ldots, \psi_n\} \in \Sigma$ let $\mathcal{N}_\xi$ be a structure in $\mathfrak{C}$ which satisfies every sentence in ξ. Let D be an ultrafilter on Σ that contains the set

$$\{\xi \in \Sigma \mid \psi \in \xi\}$$

for each $\psi \in \mathrm{Th}_A(\mathcal{M}))^+$. Let $\mathcal{K} = (\prod_{\xi \in \Lambda} \mathcal{N}_\xi)_D$. Then $\mathcal{K} \models_A \mathrm{Th}_A(\mathcal{M})$, so we have $\mathcal{K} \equiv_A \mathcal{M}$. Since $\mathfrak{C}$ is closed under ultraproducts, $\mathcal{K}$ is in $\mathfrak{C}$. By Theorem 10.7, there exists an ultrafilter D such that $(\mathcal{K})_D$ and $(\mathcal{M})_D$ are isomorphic. Now $(\mathcal{K})_D$ is in $\mathfrak{C}$, so $(\mathcal{M})_D$ is in $\mathfrak{C}$, since $\mathfrak{C}$ is closed under isomorphisms. Therefore, since $\mathfrak{D} \setminus \mathfrak{C}$ is closed under ultrapowers, $\mathcal{M}$ is also in $\mathfrak{C}$. $\square$

13.7 Remark. The proof of Proposition 13.6 contains the following useful elementary result: let $\mathfrak{C}$ be a uniform class of L-structures and let Γ be the set of all positive bounded sentences σ such that $\mathcal{M} \models_A \sigma$ holds for all $\mathcal{M} \in \mathfrak{C}$. Then, every approximate model of Γ is approximately elementarily equivalent to some ultraproduct of structures from $\mathfrak{C}$.

13.8 Examples. The literature contains a number of important examples of classes of normed space structures with axiomatizations by positive bounded sentences.

· [HM74a] [Ste76] [Ste78] contain results from which one can show that the class of Banach spaces isometric to $L_p(\mu)$ for some measure μ can be axiomatized by positive bounded sentences in the signature of pure Banach spaces (p is fixed, in the interval $1 \le p < \infty$). The key idea is that these spaces are the same as the spaces that are $\mathcal{L}_{p,\lambda}$-spaces for all $\lambda > 1$. A full analysis of the complete positive bounded theories of Banach spaces in these classes is discussed in [Hen76]; the added information needed to axiomatize a complete theory is the "number" of atoms of the measure μ (this "number" is either an integer ≥ 0 or ∞). (For a discussion of $\mathcal{L}_{p,\lambda}$-spaces, needed in this paragraph and the next, see Chapter II.5 in [LT73].)

· The papers cited in the previous item also cover the case $p = \infty$. The axiomatizable class here consists of all Banach spaces whose dual space is isometric to an $L_1(\mu)$-space; these are the same as the spaces that are $\mathcal{L}_{\infty,\lambda}$-spaces for all $\lambda > 1$. The classification of these Banach spaces under $\equiv_A$ is more or less completely open.

· The results in [Hei81] show that the class of Banach spaces isometric to $C(K)$ for some compact Hausdorff space K can be axiomatized by positive bounded sentences in the signature of pure Banach spaces. The key result in this paper is that the class of $C(K)$ Banach spaces is closed under ultraroots. An application of Proposition 13.6 yields that the required set of positive bounded sentences exists; no explicit set of axioms is known. Certain properties of K are known to be axiomatizable in this sense, such as connectedness and total disconnectedness (see [Hen76]), but a complete classification of $C(K)$ Banach spaces under $\equiv_A$ is far from being known. Some partial results are given in [HHM83] [HHM86] [HHM87].

13.9 Remark. There are a number of uniform classes of normed space structures that have been shown to be closed under ultraproducts, but for which the corresponding closure under "ultraroots" is either false or not known to be true. These provide open problems as to the (optimal) applicability of Proposition 13.6. Among the many papers in which the structure of ultraproducts of spaces of interest in functional analysis is studied, we mention the following: [DCK72] [Ray97] for Orlicz and Köthe function spaces; [LR84a] [LR84b] [HLR85] [Ray86] [HLR91]for vector valued L_p-spaces, in particular generalized $L_p(L_q)$-spaces; [Ray91] [Ray98b] for interpolation spaces; and [Gro84] [Ray98a] [Ray02] for noncommutative L_p-spaces.

Quantifier Elimination

13.10 Definition. Let Γ be a positive bounded theory in a signature L and let $\varphi(x_1, \ldots, x_n)$ be a positive bounded L-formula. We say that φ is *approximable in Γ by quantifier-free formulas* if the following condition holds: for every approximation φ' of φ and every positive real number r there exist a quantifier-free positive bounded L-formula $\theta(x_1, \ldots, x_n)$ and an approximation θ' of θ such that whenever $\mathcal{M}$ is an approximate model of Γ and $a_1, \ldots, a_n$ are elements of suitable sorts of $\mathcal{M}$ with $\|a_j\| \leq$

r for $j = 1, \ldots, n$,

$$\mathcal{M} \models \varphi[a_1, \ldots, a_n] \quad \text{implies} \quad \mathcal{M} \models \theta[a_1, \ldots, a_n],$$
$$\mathcal{M} \models \theta'[a_1, \ldots, a_n] \quad \text{implies} \quad \mathcal{M} \models \varphi'[a_1, \ldots, a_n].$$

13.11 Remark. In Definition 13.10, one may replace $\models$ by $\models_{\mathcal{A}}$ without changing the concept being defined.

13.12 Proposition. *Let Γ be a positive bounded theory in a signature L such that the class of approximate models of Γ is uniform, and let $\varphi(x_1, \ldots, x_n)$ be a positive bounded L-formula, where x_j is a variable of sort s_j, for $j = 1, \ldots, n$. The following conditions are equivalent.*

(i) *φ is approximable in Γ by quantifier-free formulas;*
(ii) *Whenever we are given*

   · *two approximate models $\mathcal{M}$ and $\mathcal{N}$ of Γ,*
   · *substructures $\mathcal{M}_0 \subseteq \mathcal{M}$ and $\mathcal{N}_0 \subseteq \mathcal{N}$,*
   · *an isomorphism $(T^{(s)} \mid s \in S)$ of $\mathcal{M}_0$ onto $\mathcal{N}_0$, and*
   · *elements $a_1, \ldots, a_n$ of suitable sorts of $\mathcal{M}_0$,*

   we have

   $$\mathcal{M} \models_{\mathcal{A}} \varphi[a_1, \ldots, a_n] \quad \text{implies} \quad \mathcal{N} \models_{\mathcal{A}} \varphi[T^{(s_1)}(a_1), \ldots, T^{(s_1)}(a_n)].$$

Moreover, for the implication (ii)$\Rightarrow$(i) it suffices to assume (ii) only when $\mathcal{M}_0$ and $\mathcal{N}_0$ are finitely generated substructures.

Proof. We omit the straightforward proof of (i)$\Rightarrow$(ii). To prove (ii)$\Rightarrow$(i), fix an approximation φ' of φ and $r > 0$ in order to find a pair θ, θ' satisfying the implications in Definition 13.10.

Let $\Sigma(x_1, \ldots, x_n)$ be the set of all quantifier-free L-formulas of the form $\mathrm{neg}(\sigma)(x_1, \ldots, x_n)$, where σ satisfies the following condition: there exists an approximation τ of σ such that whenever $\mathcal{N}$ is an approximate model of Γ and $b_1, \ldots, b_n$ are elements of suitable sorts of $\mathcal{N}$ with $\|b_j\| \leq r$ for $j = 1, \ldots, n$,

$$\mathcal{N} \models \tau[b_1, \ldots, b_n] \quad \text{implies} \quad \mathcal{N} \models \varphi'[b_1, \ldots, b_n].$$

We claim that the set

$$\{ \bigwedge_{j=1}^{n} \|x_j\| \leq r \} \cup \{ \varphi(x_1, \ldots, x_n) \} \cup \Sigma(x_1, \ldots, x_n)$$

is not satisfiable in any model of Γ. To show this, suppose that $\mathcal{M}$ is an approximate model of Γ and $a_1, \ldots, a_n$ are elements of suitable sorts of $\mathcal{M}$ with $\|a_j\| \leq r$ for $j = 1, \ldots, n$ such that $\mathcal{M} \models \varphi[a_1, \ldots, a_n]$ and $\mathcal{M} \models \Sigma[a_1, \ldots, a_n]$. Let $p(x_1, \ldots, x_n)$ be the set of all quantifier-free L-formulas $\sigma(x_1, \ldots, x_n)$ such that $\mathcal{M} \models \sigma[a_1, \ldots, a_n]$, and suppose that $\mathcal{N}$ is an approximate model of Γ and $b_1, \ldots, b_n$ are elements of suitable sorts of $\mathcal{N}$ with $\|b_j\| \leq r$ for $j = 1, \ldots, n$ such that $\mathcal{N} \models p[b_1, \ldots, b_n]$. Then the map $a_j \mapsto b_j$ determines an isomorphism between the substructure of $\mathcal{M}$ generated by $\{a_1, \ldots, a_n\}$ and the substructure of $\mathcal{N}$ generated by $\{b_1, \ldots, b_n\}$. Therefore, by (ii), we have $\mathcal{N} \models \varphi'[b_1, \ldots, b_n]$.

Thus, by the Compactness Theorem (Proposition 9.5), there exist $\sigma \in p$ and an approximation τ of σ such that if $\mathcal{N}$ is an approximate model of Γ and $b_1, \ldots, b_n$ are elements of suitable sorts of $\mathcal{N}$ with $\|b_j\| \leq r$ for $j = 1, \ldots, n$, we have

$$\mathcal{N} \models \tau[b_1, \ldots, b_n] \quad \text{implies} \quad \mathcal{N} \models \varphi'[b_1, \ldots, b_n].$$

Take an approximation σ' of σ such that $\sigma < \sigma' < \tau$. By the definition of Σ, we have $\mathrm{neg}(\sigma') \in \Sigma$ and therefore $\mathcal{M} \models \mathrm{neg}(\sigma')[a_1, \ldots, a_n]$; but this contradicts the fact that $\sigma \in p$ (and the definition of p), so the claim is proved.

By the preceding claim and the Compactness Theorem, we find formulas $\chi_1, \ldots, \chi_k \in \Sigma$ such that whenever $\mathcal{M}$ is an approximate model of Γ and $a_1, \ldots, a_n$ are elements of suitable sorts of $\mathcal{M}$ with $\|a_j\| \leq r$ for $j = 1, \ldots, n$, we have

$$\mathcal{M} \models \varphi[a_1, \ldots, a_n] \quad \text{implies} \quad \mathcal{M} \models \mathrm{neg}(\chi_1) \vee \cdots \vee \mathrm{neg}(\chi_k)[a_1, \ldots, a_n].$$

Each χ_i is of the form $\mathrm{neg}(\sigma_i)$, where there exists $\tau_i > \sigma_i$ such that whenever $\mathcal{M}$ is an approximate model of Γ and $a_1, \ldots, a_n$ are elements of suitable sorts of $\mathcal{M}$ with $\|a_j\| \leq r$ for $j = 1, \ldots, n$,

$$\mathcal{M} \models \tau_i[a_1, \ldots, a_n] \quad \text{implies} \quad \mathcal{M} \models \varphi'[a_1, \ldots, a_n].$$

Thus, Condition (i) follows by taking as θ the formula $\sigma_1 \vee \cdots \vee \sigma_k$ and as θ' the formula $\tau_1 \vee \cdots \vee \tau_k$. $\square$

13.13 Definition. A positive bounded theory Γ in a signature L admits *quantifier elimination* if every L-formula is approximable in Γ by quantifier-free formulas.

13.14 Remarks.

(i) Let Γ be a positive bounded theory in a signature L, and let $L(C)$ be an extension of L by constants. If Γ admits quantifier elimination in L then Γ admits quantifier elimination in $L(C)$.

(ii) Let $\Gamma \subseteq \Gamma'$ be positive bounded theories in a signature L. If Γ admits quantifier elimination in L, then Γ' admits quantifier elimination in L.

13.15 Lemma. *Suppose that Γ is a positive bounded theory in a signature L and ψ is a positive bounded L formula. If every formula φ with $\varphi < \psi$ is approximable in Γ by quantifier-free formulas, then $\mathrm{neg}(\psi)$ is approximable in Γ by quantifier-free formulas.*

Proof. If $\psi(x_1, \ldots, x_n)$ be a positive bounded L-formula, a typical approximation of $\mathrm{neg}(\psi)$ is a positive bounded formula of the form $\mathrm{neg}(\varphi)$, where $\varphi < \psi$. Assume that such φ is approximable in Γ by quantifier-free formulas and take $r > 0$ and an approximation φ' of φ with $\varphi < \varphi' < \psi$. Fix a quantifier-free positive bounded formula $\theta(x_1, \ldots, x_n)$ and an approximation θ' of θ such that whenever $\mathcal{M}$ is an approximate model of Γ and $a_1, \ldots, a_n$ are elements of suitable sorts of $\mathcal{M}$ with $\|a_j\| \leq r$ for $j = 1, \ldots, n$,

$$\mathcal{M} \models \varphi[a_1, \ldots, a_n] \quad \text{implies} \quad \mathcal{M} \models \theta[a_1, \ldots, a_n],$$
$$\mathcal{M} \models \theta'[a_1, \ldots, a_n] \quad \text{implies} \quad \mathcal{M} \models \varphi'[a_1, \ldots, a_n].$$

Fix also an approximation θ^* of θ such that $\theta < \theta^* < \theta'$. Then, if $\mathcal{M}$ and $a_1, \ldots, a_n$ are as above, we have

$$\mathcal{M} \models \mathrm{neg}(\psi)[a_1, \ldots, a_n] \quad \text{implies} \quad \mathcal{M} \models \mathrm{neg}(\theta')[a_1, \ldots, a_n]$$
$$\mathcal{M} \models \mathrm{neg}(\theta^*)[a_1, \ldots, a_n] \quad \text{implies} \quad \mathcal{M} \models \mathrm{neg}(\varphi)[a_1, \ldots, a_n]$$

by the basic properties of neg (see Remark 5.13). $\qquad\square$

13.16 Lemma. *Suppose that Γ is a positive bounded theory in a signature L and that every existential positive bounded L-formula with one existential quantifier is approximable in Γ by quantifier-free formulas. Then Γ admits quantifier elimination.*

Proof. Note that by Lemma 13.15, the hypothesis implies that every positive bounded formula in prenex normal form with one universal bounded quantifier is approximable in Γ by quantifier-free formulas. The proof that every positive bounded L-formula φ is approximable in Γ by quantifier-free formulas is by induction on the complexity of φ. $\qquad\square$

13.17 Proposition. *Let Γ be a positive bounded theory in a signature L such that the class of approximate models of Γ is uniform. Then the following conditions are equivalent:*

(i) *Γ admits quantifier elimination;*

(ii) *If $\mathcal{M}$ and $\mathcal{N}$ are approximate models of Γ, then every embedding of a substructure of $\mathcal{M}$ into $\mathcal{N}$ can be extended to an embedding of $\mathcal{M}$ into an approximate elementary extension of $\mathcal{N}$.*

Moreover, if $\operatorname{card}(L) \leq \kappa$, then in condition (ii) it suffices to consider approximate models $\mathcal{M}$ such that every sort of $\mathcal{M}$ has density character $\leq \kappa$.

Proof. Let us first prove that (ii) implies (i). By Lemma 13.16 it suffices to show that each existential L-formula with one existential quantifier is approximable in Γ by quantifier-free formulas. Consider such a formula $\exists_r y \varphi(x_1, \ldots, x_n, y)$ where φ is a quantifier-free positive bounded L-formula and x_j is a variable of sort s_j, for $j = 1, \ldots, n$. We show that this formula satisfies condition (ii) of Proposition 13.12. Therefore, suppose we are given approximate models $\mathcal{M}$ and $\mathcal{N}$ of Γ, substructures $\mathcal{M}_0 \subseteq \mathcal{M}$ and $\mathcal{N}_0 \subseteq \mathcal{N}$, an isomorphism $(T^{(s)} \mid s \in S)$ of $\mathcal{M}_0$ into $\mathcal{N}_0$, and elements $a_1, \ldots a_n$ of suitable sorts of $\mathcal{M}_0$, such that

$$\mathcal{M} \models_A \exists_r y \varphi(x_1, \ldots, x_n, y)[a_1, \ldots, a_n].$$

We need to prove that

$$\mathcal{N} \models_A \exists_r y \varphi(x_1, \ldots, x_n, y)[T^{(s_1)}(a_1), \ldots, T^{(s_1)}(a_n)].$$

Fix $s > r$ and any approximation φ' of φ, and let b be an element of a suitable sort of $\mathcal{M}$ such that $\|b\| \leq s$ and

$$\mathcal{M} \models \varphi'(x_1, \ldots, x_n, y)[a_1, \ldots, a_n, b].$$

Applying the assumed condition (ii), we may extend T to an embedding of $\mathcal{M}$ into an approximate elementary extension $\mathcal{N}'$ of $\mathcal{N}$. Let c be the image of b under this embedding. This implies that $\|c\| \leq s$ and, since φ is quantifier-free, we also have

$$\mathcal{N}' \models_A \varphi'(x_1, \ldots, x_n, y)[T^{(s_1)}(a_1), \ldots, T^{(s_1)}(a_n), c],$$

and hence

$$\mathcal{N} \models_A \exists_s y \varphi'(x_1, \ldots, x_n, y)[T^{(s_1)}(a_1), \ldots, T^{(s_1)}(a_n)].$$

Since s and φ' were arbitrary, we may conclude that

$$\mathcal{N} \models_A \exists_r y \varphi(x_1,\ldots,x_n,y)[T^{(s_1)}(a_1),\ldots,T^{(s_1)}(a_n)].$$

To prove (i)$\Rightarrow$(ii), assume that Γ admits quantifier elimination, and suppose that $\mathcal{M}$ and $\mathcal{N}$ are approximate models of Γ and $(T^{(s)} \mid s \in S)$ is an isomorphism of a substructure $\mathcal{M}_0$ of $\mathcal{M}$ into $\mathcal{N}$. Then, by the quantifier elimination assumption, whenever $\varphi(x_1,\ldots,x_n)$ is an existential L-formula with x_j of sort s_j for $j = 1,\ldots,n$ and $a_1,\ldots,a_n$ are elements of suitable sorts of $\mathcal{M}_0$, we have

$$\mathcal{M} \models \varphi[a_1,\ldots,a_n] \quad \text{implies} \quad \mathcal{N} \models_A \varphi[T^{(s_1)}(a_1),\ldots,T^{(s_1)}(a_n)].$$

Therefore (ii) follows by an application of the Compactness Theorem (Proposition 9.22). □

13.18 Example. Fix p in the interval $1 \leq p < \infty$ and let $\mathfrak{C}_p$ be the class of all Banach lattices that are isomorphic to $L_p(\mu)$ for some *atomless* measure μ. There is a set Γ_p of positive bounded sentences in the signature for Banach lattices such that $\mathfrak{C}_p$ is the set of all Banach lattices that are approximate models of Γ_p. The Downward Löwenheim-Skolem Theorem (Proposition 9.13) shows that Γ_p is a complete positive bounded theory, since any two separable members of $\mathfrak{C}_p$ are isomorphic.

We indicate briefly a proof that Γ_p *admits quantifier elimination*. This is done by verifying condition (ii) of Proposition 13.17; for this we use the structure theory of abstract L_p-spaces as found in the proof of Theorem 3 in [BDCK66] and in Section 15 in [Lac74].

Let $\mathcal{M},\mathcal{N}$ be members of $\mathfrak{C}_p$ and let T be an isomorphism from a substructure $\mathcal{M}_0$ of $\mathcal{M}$ into $\mathcal{N}$. We must extend T to an embedding of $\mathcal{M}$ into some approximate elementary extension of $\mathcal{N}$. Since the signature of Banach lattices is countable, we may assume that $\mathcal{M}$ is separable and it is evidently no loss of generality to assume that $\mathcal{M}$ is complete.

Let $\mathcal{N}'$ be an $\aleph_1$-saturated approximate elementary extension of $\mathcal{N}$. (It suffices to consider the case where $\mathcal{N}'$ is an ultrapower of $L_p(\lambda)$, where λ is Lebesgue measure on the unit interval. This may simplify some of the arguments below for some readers.) We will prove that every embedding T of a substructure of $\mathcal{M}$ into $\mathcal{N}'$ extends to an embedding of $\mathcal{M}$ into $\mathcal{N}'$. Using a Zorn's Lemma argument, it suffices to prove that if T is an embedding of a proper substructure $\mathcal{M}_0$ of $\mathcal{M}$ into $\mathcal{N}'$, then there exists *some* $\mathcal{M}_1$ with $\mathcal{M}_0 \subset \mathcal{M}_1 \subseteq \mathcal{M}$ and an extension of T to an embedding of $\mathcal{M}_1$ into $\mathcal{N}'$.

So, let $\mathcal{M}_0 \subset \mathcal{M}$ and let T be an embedding of $\mathcal{M}_0$ into $\mathcal{N}'$. First we may assume that $\mathcal{M}_0$ is not $\{0\}$; otherwise, take e to be any positive element of $\mathcal{M}$ having norm 1, and extend T so it takes e to any similar element of $\mathcal{N}'$. The one-dimensional space spanned by e is a substructure (sublattice) of $\mathcal{M}$ and T is an embedding. Next we may assume that $\mathcal{M}_0$ is closed; otherwise take $\mathcal{M}_1$ to be the closure and extend T by uniform continuity.

With these reductions, $\mathcal{M}_0$ is a non-zero closed vector sublattice of $\mathcal{M}$, and therefore it is an $L_p(\nu)$-space in its own right (but the measure ν need not be atomless). By separability, there is a norm-one, positive element e of $\mathcal{M}_0$ which has full support; that is, the band generated by e in $\mathcal{M}_0$ is all of $\mathcal{M}_0$.

Next consider the case where the band generated by e in $\mathcal{M}$ is *not* all of $\mathcal{M}$. Then there must exist a norm-one, positive element f of $\mathcal{M}$ that is orthogonal to e. Since $\mathcal{N}'$ is $\aleph_1$-saturated, we can find a norm-one, positive element f' of $\mathcal{N}'$ which is orthogonal to $T(e)$. In this situation we let $\mathcal{M}_1$ be the substructure of $\mathcal{M}$ that is generated by $\mathcal{M}_0$ and f; our choice of f ensures that $\mathcal{M}_1$ is the Banach lattice direct sum of $\mathcal{M}_0$ and the one dimensional sublattice generated by f. Thus we extend T to an embedding of $\mathcal{M}_1$ into $\mathcal{N}'$ by mapping f to f'.

Finally, we must deal with the case where the band generated by e in $\mathcal{M}$ *is* all of $\mathcal{M}$. In this case we use the structure theory for abstract L_p-spaces to pass to an isomorphic situation in which $\mathcal{M}$ is L_p (where the measure is Lebesgue measure on the unit interval $([0,1], \Sigma, \lambda)$) and $\mathcal{M}_0$ is the sublattice of all functions in L_p that are measurable with respect to Σ_0, where Σ_0 is a sub-σ-algebra of Σ. Let A be a set of positive measure in Σ that is not in Σ_0; such a set exists because $\mathcal{M}_0$ is a proper closed sublattice of $\mathcal{M}$. We want to extend T to an embedding of the sublattice of $\mathcal{M}$ that is generated by $\mathcal{M}_0$ and χ_A (the characteristic function of A).

Again we use the structure theory of abstract L_p-spaces, this time applied to $\mathcal{N}'$. Namely, we may represent $\mathcal{N}'$ as $L_p(\mu)$ for some (necessarily atomless) measure μ, and we may do this in such a way that for every set B in Σ_0, $T(\chi_B)$ is the characteristic function $\chi_{\sigma(B)}$ of some μ-measurable set $\sigma(B)$. It is a routine saturation argument to show that there exists a μ-measurable set A' such that for any $B \in \Sigma_0$, we have $\lambda(B \cap A) = \mu(\sigma(B) \cap A')$ and $\lambda(B \setminus A) = \mu(\sigma(B) \setminus A')$. The desired extension of T can then be taken to map χ_A to $\chi_{A'}$.

We close this example by pointing out that it can be paralleled for Banach spaces without lattice structure using the equimeasurability results of Rudin [Rud76] and Plotkin [Plo71]. The result is that the class of $L_p(\mu)$-spaces, where μ is atomless, admits quantifier elimination *in the language of pure Banach spaces*, as long as $1 \leq p < \infty$ and $p \neq 4, 6, 8, \ldots$. As above, the proof goes via Proposition 13.17, and we need to verify condition (ii); however, in this situation it is linear subspaces and linear embeddings that need to be considered. The results of Rudin and Plotkin are used to extend an embedding from an arbitrary linear subspace to the closed sublattice that it generates, after which the argument given above takes over.

Approximability by Formulas in a Given Class

13.19 Definition. Suppose that L is a signature and Θ is a class of positive bounded L-formulas that is closed under approximations. If Γ is a positive bounded theory in L and $\varphi(x_1, \ldots, x_n)$ is a positive bounded L-formula, we will say that φ is *approximable in Γ by formulas from Θ* if the following condition holds. For every approximation φ' of φ and every positive real number r there exists $\theta(x_1, \ldots, x_n) \in \Theta$ and an approximation θ' of θ such that whenever $\mathcal{M}$ is an approximate model of Γ and $a_1, \ldots, a_n$ are elements of suitable sorts of $\mathcal{M}$ with $\|a_j\| \leq r$ for $j = 1, \ldots, n$,

$$\mathcal{M} \models \varphi[a_1, \ldots, a_n] \quad \text{implies} \quad \mathcal{M} \models \theta[a_1, \ldots, a_n],$$
$$\mathcal{M} \models \theta'[a_1, \ldots, a_n] \quad \text{implies} \quad \mathcal{M} \models \varphi'[a_1, \ldots, a_n].$$

Notice that the concept of approximability by quantifier-free formulas introduced in Definition 13.10 is a particular instance of the concept defined above, namely, when Θ is the class of all quantifier-free L-formulas.

13.20 Proposition. *Suppose that L is a signature and Θ is a class of positive bounded L-formulas that is closed under conjunction, disjunction, and approximations. Let Γ be a positive bounded theory in L such that the class of approximate models of Γ is uniform, and let $\varphi(x_1, \ldots, x_n)$ be a positive bounded L-formula. Then the following conditions are equivalent:*

(i) *Either φ is approximable in Γ by formulas from Θ, or φ is satisfiable in every approximate model of Γ, or φ is satisfiable in no approximate model of Γ;*

(ii) *If $\mathcal{M}$ and $\mathcal{N}$ are approximate models of Γ and $a_1, \ldots, a_n$ and $b_1, \ldots, b_n$ are elements of suitable sorts of $\mathcal{M}$ and $\mathcal{N}$, respectively, such that for every $\theta(x_1, \ldots, x_n) \in \Theta$,*

$$\mathcal{M} \models_A \theta[a_1, \ldots, a_n] \quad \text{implies} \quad \mathcal{N} \models_A \theta[b_1, \ldots, b_n],$$

then we have

$$\mathcal{M} \models_{\mathcal{A}} \varphi[a_1, \ldots, a_n] \quad implies \quad \mathcal{N} \models_{\mathcal{A}} \varphi[b_1, \ldots, b_n].$$

Proof. Analogous to the proof of Proposition 13.12. □

13.21 Example. As an example of how Proposition 13.20 can be applied, we consider one of the alternative formulations discussed in Chapter 11; the others can be handled similarly.

Let L_0 be the signature for Banach spaces with no additional structure and let Θ be the set of *restricted* positive bounded L_0-formulas. Recall that these are built up from atomic formulas of the form

$$\|k_1 x_1 + \cdots + k_n x_n\| \leq r,$$
$$\|k_1 x_1 + \cdots + k_n x_n\| \geq r,$$

where $x_1, \ldots, x_n$ are variables of the normed space sort, $k_1, \ldots, k_n$ are positive integers, and r is a rational number, using $\wedge, \vee$ and bounded quantifiers $\exists_r x, \forall_r x$ where r is a positive rational number and x is a variable of the normed space sort.

It is clear that Θ is closed under conjunction, disjunction, and approximations. Moreover, Proposition 11.2 shows that Θ satisfies condition (ii) of Proposition 13.20 when applied to the class of all Banach spaces and to any positive bounded L_0-formula φ. Therefore, every positive bounded L_0-formula is approximable in the class of all Banach spaces by restricted L_0 formulas.

Banach Spaces with Isomorphic Ultrapowers

Here we work in the pure Banach space signature L_0 and we identify every Banach space L_0-structure with its underlying Banach space. (See Chapter 7.)

13.22 Definition. Let φ be a positive bounded L_0-formula. For every rational number $\lambda > 1$ we define an approximation φ_λ of φ. The definition is by induction on the complexity of φ:

If φ is:	φ_λ is:
$\|t\| \leq r$	$\|t\| \leq \lambda r$
$\|t\| \geq r$	$\|t\| \geq \frac{r}{\lambda}$
$(\psi \wedge \theta)$	$(\psi_\lambda \wedge \theta_\lambda)$
$(\psi \vee \theta)$	$(\psi_\lambda \vee \theta_\lambda)$
$\exists_r x\, \psi$	$\exists_{\lambda r} x\, \psi_\lambda$
$\forall_r x\, \psi$	$\forall_{\frac{r}{\lambda}} x\, \psi_\lambda$

If X and Y are Banach spaces and λ is a real number with $\lambda \geq 1$, we say that X and Y are λ-*isomorphic* if there is a linear isomorphism $T\colon X \to Y$ such that $\|T\|, \|T^{-1}\| \leq \lambda$. In this case we say that the map T is a λ-isomorphism.

A concept of approximate elementary equivalence suitably adapted to the "λ-isomorphic" context was developed in [HH86]. Here we summarize briefly what was done in the first part of that paper, and prove a version of the Ultraproduct Theorem (Theorem 10.8) for λ-isomorphisms between ultraproducts of Banach spaces. The corresponding result for ultrapowers of Banach spaces is Theorem 4 in [HH86].

13.23 Remarks. Let φ be a positive bounded L_0-formula. Then:

(i) For each $\lambda > 1$, φ_λ is an approximation of φ;
(ii) for every approximation φ' of φ there exists a rational number $\lambda > 1$ such that $\varphi < \varphi_\lambda < \varphi'$;
(iii) if $\lambda, \mu > 1$, then $(\varphi_\lambda)_\mu$ is $\varphi_{\lambda\mu}$;
(iv) if $\lambda < \mu$, then $(\mathrm{neg}(\varphi_\mu))_\lambda$ is $\mathrm{neg}(\varphi_{\mu/\lambda})$.

By Proposition 6.4 and Remark 13.23(i)(ii), for every pair of Banach spaces X and Y, $X \equiv_A Y$ if and only if for every positive bounded L_0-formula φ we have $X \models \varphi$ implies $Y \models \varphi_\lambda$, for every $\lambda > 1$. It is therefore natural to generalize the concept $\equiv_A$ of approximate elementary equivalence for L_0-structures as follows. (See [HH86, Definition 2 and Lemma 3].)

13.24 Definition. If X, Y are Banach spaces and λ is a real number with $\lambda \geq 1$, we write $X \equiv_A^\lambda Y$ if for every positive bounded L_0-sentence φ,

$$X \models \varphi \quad \text{implies} \quad Y \models \varphi_\mu, \quad \text{for every } \mu > \lambda.$$

13.25 Proposition. *If X, Y, and Z are Banach spaces and $\lambda, \mu \geq 1$, then*

(i) $X \equiv_{\mathcal{A}}^{\lambda} X$;

(ii) *If $X \equiv_{\mathcal{A}}^{\lambda} Y$, then $Y \equiv_{\mathcal{A}}^{\lambda} X$;*

(iii) *If $X \equiv_{\mathcal{A}}^{\lambda} Y$ and $Y \equiv_{\mathcal{A}}^{\mu} Z$, then $X \equiv_{\mathcal{A}}^{\lambda\mu} Z$.*

Proof. Part (i) is immediate and Part (ii) follows from Remark 13.23-(ii). To prove (iii), assume that $X \equiv_{\mathcal{A}}^{\lambda} Y$ and φ is a positive bounded L_0-sentence such that $Y \models \varphi$ and $X \models \mathrm{neg}(\varphi_\nu)$ for some $\nu > \lambda$. Then, by Remark 13.23-(iii), for every μ with $1 < \mu < \nu$ we have $Y \models \mathrm{neg}(\varphi_\mu)$; but this contradicts the assumption that $Y \models \varphi$. □

The main result of this section is the following version of the Ultraproduct Theorem (Theorem 10.8). (Also, see [HH86, Remark, page 313].)

13.26 Theorem. *Let κ be an infinite cardinal. The ultrafilter D on 2^κ given by Theorem 10.8 satisfies the following property: if λ is a real number with $\lambda \geq 1$ and $(X_\xi \mid \xi < \Lambda)$ and $(Y_\xi \mid \xi < \Lambda)$ are families of Banach spaces such that $\mathrm{density}(X_\xi)$ and $\mathrm{density}(Y_\xi)$ are uniformly bounded by κ and*

$$\left(\prod_{\xi < \Lambda} X_\xi \right)_D \equiv_{\mathcal{A}}^{\lambda} \left(\prod_{\xi < \Lambda} Y_\xi \right)_D,$$

then $(\prod_{\xi < \Lambda} X)_D$ and $(\prod_{\xi < \Lambda} Y)_D$ are λ-isomorphic.

One way to prove this theorem would be to make the appropriate changes in the proof of Theorem 10.8; that is, one would suitably change the approximations of positive bounded formulas in the inductive construction by μ-approximations, for rational numbers $\mu > \lambda$. This approach would have the advantage that one could apply it to signatures more general than L_0. Below we present an alternative proof which invokes Theorem 10.8 as well as the following result proved in [HH86, Theorem 4]. (Note that the argument at [HH86, lines 12–17, page 308] uses a special case of what is given here in Theorem 10.7.)

13.27 Theorem. *Let X and Y be separable Banach spaces and let λ be a real number with $\lambda \geq 1$. The following conditions are equivalent:*

(i) $X \equiv_{\mathcal{A}}^{\lambda} Y$;

(ii) *There exist Banach spaces $\widehat{X}$ and $\widehat{Y}$ such that $X \preceq_A \widehat{X}$ and $Y \preceq_A \widehat{Y}$ and $\widehat{X}, \widehat{Y}$ are λ-isomorphic.*

Proof of Theorem 13.26. Let D be the ultrafilter on 2^κ given by Theorem 10.8, and let

$$X = (\prod_{\xi < \Lambda} X_\xi)_D$$

$$Y = (\prod_{\xi < \Lambda} Y_\xi)_D.$$

The Downward Löwenheim-Skolem Theorem (Proposition 9.13) gives us separable spaces X' and Y' such that $X \equiv_A X'$ and $Y \equiv_A Y'$. Then Theorem 13.27 gives λ-isomorphic Banach spaces $\widehat{X'}$ and $\widehat{Y'}$ such that $X' \preceq_A \widehat{X'}$ and $Y' \preceq_A \widehat{Y'}$. By the Downward Löwenheim-Skolem Theorem (applied to a structure that contains $\widehat{X'}$ and $\widehat{Y'}$ as sorts and the given λ-isomorphism between them as a distinguished function), we obtain separable Banach spaces X'' and Y'' such that $X'' \equiv_A X$ and $Y'' \equiv_A Y$, and a λ-isomorphism $T : X'' \to Y''$. Now $(T)_D : (X'')_D \to (Y'')_D$ is also a λ-isomorphism. By the choice of D, X is isometric to $(X'')_D$ and Y is isometric to $(Y'')_D$. Composing $(T)_D$ with these isometries, we obtain the desired λ-isomorphism between X and Y. $\square$

14

Types

14.1 Definition. Suppose that $\mathcal{M}$ is a normed space L-structure, $L(C)$ is an extension of L by constants, and $\mathbf{A}$ is an interpretation of C in $\mathcal{M}$ (see Definition 3.7). Let $e_1, \ldots, e_n$ be elements of $\mathcal{M}$. Fix distinct variables $x_1, \ldots, x_n$ of L such that x_j and e_j have the same sort for each $j = 1, \ldots, n$. We define the *type of* $(e_1, \ldots, e_n)$ *over* $\mathbf{A}$, denoted

$$\mathrm{tp}_{\mathcal{M}}(e_1, \ldots, e_n/\mathbf{A}),$$

to be the set of all positive bounded $L(C)$-formulas $\varphi(x_1, \ldots, x_n)$ such that

$$(\mathcal{M}, \mathbf{A}) \models_A \varphi[e_1, \ldots, e_n].$$

When it can be done without confusion, we omit the subscript $\mathcal{M}$ from this notation and denote the type by

$$\mathrm{tp}(e_1, \ldots, e_n/\mathbf{A}).$$

If C is empty, we write $\emptyset$ in place of $\mathbf{A}$ in this notation, or omit it entirely.

14.2 Remarks.

(i) Let $\mathcal{M}$, $\mathbf{A}$ be as above, and let $e_1, \ldots, e_n$ and $e_1', \ldots, e_n'$ be elements of $\mathcal{M}$ such that e_j and e_j' are in the same sort of $\mathcal{M}$ for $j = 1, \ldots, n$. Then

$$\mathrm{tp}_{\mathcal{M}}(e_1, \ldots, e_n/\mathbf{A}) = \mathrm{tp}_{\mathcal{M}}(e_1', \ldots, e_n'/\mathbf{A})$$

if and only if

$$(\mathcal{M}, \mathbf{A}, e_1, \ldots, e_n) \equiv_A (\mathcal{M}, \mathbf{A}, e_1', \ldots, e_n').$$

96

(ii) If $\mathrm{tp}(e_1,\ldots,e_n/\mathbf{A}) = \mathrm{tp}(e_1',\ldots,e_n'/\mathbf{A})$, then $\|e_j\| = \|e_j'\|$, for $j = 1,\ldots,n$.

(iii) If $(\mathcal{M},\mathbf{A})$ is an $L(C)$-structure and $\mathcal{M} \preceq_A \mathcal{N}$, then

$$\mathrm{tp}_{\mathcal{M}}(e_1,\ldots,e_n/\mathbf{A}) = \mathrm{tp}_{\mathcal{N}}(e_1,\ldots,e_n/\mathbf{A}).$$

14.3 Proposition. *Suppose that $\mathcal{M}$ is a normed space L-structure, $L(C)$ is an extension of L by constants, and $\mathbf{A}$ is an interpretation of C in $\mathcal{M}$. If $r_1,\ldots,r_n$ are nonnegative rational numbers and $t(x_1,\ldots,x_n)$ is a set of $L(C)$-formulas, then the following conditions are equivalent:*

(i) *There exists a structure $\mathcal{N}$ such that $\mathcal{M} \preceq_A \mathcal{N}$ and elements $e_1,\ldots,e_n$ of suitable sorts of $\mathcal{N}$ with $\|e_j\| \le r_j$ for $j = 1,\ldots,n$ such that*

$$t(x_1,\ldots,x_n) = \mathrm{tp}_{\mathcal{N}}(e_1,\ldots,e_n/\mathbf{A}).$$

(ii)(a) *The formula $\bigwedge_{j=1}^{n} \|x_j\| \le r_j$ is in $t(x_1,\ldots,x_n)$;*

  (b) *Every finite subset of t^+ is satisfiable in $(\mathcal{M},\mathbf{A})$;*

  (c) *Whenever $\varphi(x_1,\ldots,x_n)$ is a positive bounded $L(C)$-formula and φ' is an approximation of φ, we have $\varphi \in t$ or $\mathrm{neg}(\varphi') \in t$.*

Proof. The implication (i)$\Rightarrow$(ii) is immediate from the definitions. To prove (ii)$\Rightarrow$(i) let $B = \{\,b_1,\ldots,b_n\,\}$ be a set of new constant symbols, where we take the sort of b_j to be the same as the sort of x_j for $j = 1,\ldots,n$. Let $\mathfrak{C}$ be the class of $L(C \cup B)$-structures of the form

$$(\mathcal{M},\mathbf{A},\mathbf{B}),$$

where $\mathbf{B}$ is an interpretation of $\{\,b_1,\ldots,b_n\,\}$ in $(\mathcal{M},\mathbf{A})$, and $\|\mathbf{B}(b_j)\| \le r_j$ for $j = 1,\ldots,n$. Then $\mathfrak{C}$ is a uniform class of $L(C \cup B)$-structures. By (ii), every finite subset of $(t(b_1,\ldots,b_n))^+$ is satisfied by a structure in $\mathfrak{C}$, so (i) follows from the Compactness Theorem (Proposition 9.5) and Corollary 9.7. □

14.4 Definition. Let $\mathcal{M}$, $L(C)$, and $\mathbf{A}$ be as above. A set $t(x_1,\ldots,x_n)$ of positive bounded $L(C)$-formulas satisfying the (equivalent) conditions of Proposition 14.3 is called a *type over* $\mathbf{A}$. If $\mathcal{N}$ and $(e_1,\ldots,e_n)$ are as in (i) of the proposition, we say that $(e_1,\ldots,e_n)$ *realizes t in* $\mathcal{N}$ or that $(e_1,\ldots,e_n)$ is a *realization of t in* $\mathcal{N}$.

14.5 Remarks. Let L, C, $\mathcal{M}$, and $\mathbf{A}$ be as in the statement of Proposition 14.3.

(i) If $t(x_1, \ldots, x_n)$ is a type over **A**, the structure $\mathcal{N}$ and the elements $e_1, \ldots, e_n$ given by Proposition 14.3 can be taken so that

$$\mathcal{N} \models t[e_1, \ldots, e_n],$$

not just $\mathcal{N} \models_A t[e_1, \ldots, e_n])$; this may be achieved by taking $\mathcal{N}$ to be $\aleph_1$-saturated (see Propositions 9.20 and 9.18).

(ii) Suppose that $(t_i \mid i \in I)$ is a family of types over **A** such that t_i is realized in $\mathcal{N}_i$, for every $i \in I$ (so $\mathcal{M} \preceq_A \mathcal{N}_i$). Using the Compactness Theorem (Proposition 9.5) one can find a structure $\mathcal{N}$ such that $\mathcal{M} \preceq_A \mathcal{N}$ and for every $i \in I$, there is an approximate elementary embedding of $\mathcal{N}_i$ into $\mathcal{N}$ whose restriction to $\mathcal{M}$ is the identity; hence every t_i is realized in $\mathcal{N}$. Thus, any given family of types over **A** can be realized in a single approximate elementary extension of $\mathcal{M}$. This observation will be used heavily. (See also Proposition 14.8 below.)

(iii) Suppose that $\mathcal{M}$ is an L-structure and $(t_i \mid i \in I)$ is a family of types over **A** as in the previous paragraph. Let κ be an infinite cardinal number such that $\mathrm{card}(L) \leq \kappa$ and the density character of each sort of $\mathcal{M}$ is at most κ. Then, using the Downward Löwenheim-Skolem Theorem (Proposition 9.13), the approximate elementary extension $\mathcal{N}$ of $\mathcal{M}$ in the previous paragraph can be found such that the density character of each sort of $\mathcal{N}$ is at most $\mathrm{card}(I) + \kappa$.

14.6 Definition. If $\Gamma(x_1, \ldots, x_n)$ is a set of L-formulas and Γ_0 is a set of L-formulas such that $\Gamma_0 \subseteq \Gamma$, let us say that Γ is *determined by* Γ_0 (or that Γ_0 *determines* Γ) if the following condition holds: whenever $\mathcal{M}$ is an L-structure and $a_1, \ldots, a_n$ are elements of suitable sorts of $\mathcal{M}$,

$$\mathcal{M} \models_A \Gamma_0[a_1, \ldots, a_n] \quad \text{implies} \quad \mathcal{M} \models_A \Gamma[a_1, \ldots, a_n].$$

14.7 Proposition. *Suppose that $\mathcal{M}$ is a normed space L-structure and we have the following situation:*

· *$L(C)$ is an extension of L by constants and **A** is an interpretation of C in $\mathcal{M}$,*

· *$L(C \cup D)$ is an extension of $L(C)$ by constants and **B** is an interpretation of D in $(\mathcal{M}, \mathbf{A})$, and*

· *every element of $\mathcal{M}$ in the range of **B** is in the closed L-substructure of $\mathcal{M}$ generated by the range of **A**.*

Then, whenever $\mathcal{N}$ is an approximate elementary extension of $\mathcal{M}$ and $e_1, \ldots, e_n$ are elements of $\mathcal{N}$, the type $\mathrm{tp}(e_1, \ldots, e_n / \mathbf{A} \cup \mathbf{B})$ is determined by $\mathrm{tp}(e_1, \ldots, e_n / \mathbf{A})$.

Proof. By the Perturbation Lemma (Proposition 5.15). □

The language of types allows us to rephrase the concept of saturation.

14.8 Proposition. *Suppose* $\mathcal{M}$ *is a normed space L-structure and κ is an infinite cardinal. Then the following conditions are equivalent:*

(i) $\mathcal{M}$ *is κ-saturated;*

(ii) *If $L(C)$ is an extension of L by constants with* $\mathrm{card}(C) < \kappa$ *and* $\mathbf{A}$ *is an interpretation of C in* $\mathcal{M}$*, then every type over* $\mathbf{A}$ *is realized in* $\mathcal{M}$*.*

Spaces of Types

In this section, the signature L will be fixed, and $\mathcal{M}$ will denote a fixed normed space L-structure. Throughout the section, $L(C)$ will denote an extension of L by constants and $\mathbf{M}$ will be an interpretation of C in $\mathcal{M}$ such that every element of $\mathcal{M}$ is in the range of $\mathbf{M}$. The letters $\mathbf{A}$, $\mathbf{B}$, etc. will denote restrictions of $\mathbf{M}$ to subsets of C. We write $\mathbf{A} \subseteq \mathbf{B}$ to indicate that $\mathbf{A}$ is a restriction of $\mathbf{B}$.

Given $\mathbf{A} \subseteq \mathbf{M}$ as above, the set of types over $\mathbf{A}$ is denoted $S(\mathbf{A})$. If $x_1, \ldots, x_n$ are distinct variables, the set of types $t(x_1, \ldots, x_n)$ over $\mathbf{A}$ will be denoted $S_{(x_1, \ldots, x_n)}(\mathbf{A})$. The word "type" will mean "type over $\mathbf{A}$, where $\mathbf{A} \subseteq \mathbf{M}$". Note that $S(\mathbf{A})$ and $S_{(x_1, \ldots, x_n)}(\mathbf{A})$ depend only on $\mathrm{Th}_{\mathcal{A}}(\mathcal{M}, \mathbf{A})$.

If $t(x_1, \ldots, x_n)$ is a type, we denote by $\|t\|$ the real number $\max_j \|e_j\|$, where $(e_1, \ldots e_n)$ is a realization of t. (Note that $\|t\|$ depends only on t and not on the particular realization chosen.)

The Logic Topology on Types

If $\varphi(x_1, \ldots, x_n)$ is an $L(C)$-formula, we let

$$[\varphi] = \{\, t(x_1, \ldots, x_n) \mid t \text{ is a type and } \varphi \in t \,\}.$$

14.9 Definition. The *logic topology* on types is defined as follows. If t is a type, the basic neighborhoods of t are the sets of the form $[\varphi]$ with $\varphi \in t^+$. If $x_1, \ldots, x_n$ are variables, then the *logic topology on* $S_{(x_1, \ldots, x_n)}(\mathbf{A})$ is the restriction to $S_{(x_1, \ldots, x_m)}(\mathbf{A})$ of the logic topology.

Note that the logic topology is Hausdorff. Each basic neighborhood $[\varphi]$ is closed in the logic topology; indeed, if t is a type not in $[\varphi]$, then for some $\varphi' > \varphi$ we have that $t \in [\mathrm{neg}(\varphi')]$ while $[\varphi]$ and $[\mathrm{neg}(\varphi')]$ are always disjoint.

14.10 Proposition. *For any distinct variables $x_1, \ldots, x_n$ and any real number $r > 0$, the set of types $t \in S_{(x_1,\ldots,x_n)}(\mathbf{A})$ such that $\|t\| \leq r$ is compact with respect to the logic topology.*

Proof. Proposition 14.10 is in fact a restatement of the Compactness Theorem. Below we present an alternative proof using ultrapowers.

Suppose that $(t_\xi)_{\xi \in \Lambda}$ is a family of types in $S_{(x_1,\ldots,x_n)}(\mathbf{A})$ such that $\|t_\xi\| \leq r$ for every $\xi \in \Lambda$ and that D is an ultrafilter on Λ. Let $\bar{e}_\xi$ be a realization of t_ξ in an approximate elementary extension $\mathcal{N}_\xi$ of $(\mathcal{M}, \mathbf{A})$. Then $(\mathcal{N}_\xi \mid \xi \in \Lambda)$ is a uniform family of structures and $(\bar{e}_\xi)_{\xi \in \Lambda} \in \ell_\infty(\Lambda, \mathcal{N}_\xi \mid \xi \in \Lambda))$. It follows that

$$\mathrm{tp}\left(((\bar{e}_\xi)_{\xi \in \Lambda})_D \,/\, \mathbf{A} \right) = \lim_{\xi, D} (t_\xi)_{\xi \in \Lambda}$$

using Proposition 9.2. □

14.11 Corollary. *The logic topology on $S_{(x_1,\ldots,x_n)}(\mathbf{A})$ is locally compact and σ-compact.*

Proof. For every type $t(x_1, \ldots, x_n) \in S_{(x_1,\ldots,x_n)}(\mathbf{A})$, there exists a rational number r such that the formula $\bigwedge_{j=1}^n \|x_j\| \leq r$ is in t^+. By the preceding proposition $[\bigwedge_{j=1}^n \|x_j\| \leq r]$ is a compact neighborhood of t. □

14.12 Remark. The logic topology on $S_{(x_1,\ldots,x_n)}(\mathbf{A})$ is not compact $(n > 0)$.

THE METRIC d ON TYPES

Let $\mathcal{K}$ be an approximate elementary extension of $\mathcal{M}$ such that each type in $S(\mathbf{A})$ is realized in $\mathcal{K}$. (See Remark 14.5(ii) for the existence of $\mathcal{K}$.) Define d on $S_{(x_1,\ldots,x_n)}(\mathbf{A})$ as follows.

$$d(t, t') = \inf \left\{ \max_j \|e_j - e_j'\| \;\middle|\; \mathcal{K} \models_A t[\bar{e}], \; \mathcal{K} \models_A t'[\bar{e}'] \right\}.$$

It follows easily that d is a pseudometric. The fact that d is a metric follows from the Perturbation Lemma. Note that d does not depend on $\mathcal{K}$, since $\mathcal{K}$ realizes every type of a $2n$-tuple $(e_1, \ldots, e_n, e_1', \ldots, e_n')$ over $\mathbf{A}$.

Note that if $t, t' \in S_{(x_1,\ldots,x_n)}(\mathbf{A})$, then by the Compactness Theorem there exist realizations $(e_1, \ldots, e_n)$ of t and $(e_1', \ldots, e_n')$ of t' in the same approximate elementary extension of $\mathcal{M}$, such that $\max_j \|e_j - e_j'\| = d(t, t')$.

14.13 Proposition. *The d-topology on $S_{(x_1,\ldots,x_n)}(\mathbf{A})$ is finer than the logic topology on $S_{(x_1,\ldots,x_n)}(\mathbf{A})$.*

Proof. By the Perturbation Lemma (Proposition 5.15). □

14.14 Proposition. *The metric space $(S_{(x_1,\ldots,x_n)}(\mathbf{A}), d)$ is complete.*

Proof. Similar to the proof of Proposition 9.21. □

14.15 Proposition. *The following conditions are equivalent for a pair of types $t, t' \in S_{(x_1,\ldots,x_n)}(\mathbf{A})$ and a rational $\delta > 0$:*

(i) $d(t, t') \leq \delta$;

(ii) *Every finite subset of the set*

$$t(x_1, \ldots, x_n)^+ \cup t'(x_1', \ldots, x_n')^+ \cup \{ \bigwedge_{j=1}^{n} \|x_j - x_j'\| \leq \epsilon \mid \epsilon > \delta \}$$

is satisfiable in $(\mathcal{M}, \mathbf{A})$.

(iii) *If κ is an infinite cardinal with $\kappa > \mathrm{card}(L(C))$ and $\mathcal{N}$ is a κ-saturated approximate elementary extension of $\mathcal{M}$, then in $\mathcal{N}$ there exist realizations $(e_1, \ldots, e_n)$ of t and $(e_1', \ldots, e_n')$ of t', such that*

$$\max_j \|e_j - e_j'\| \leq \delta;$$

(iv) *For every $L(C)$-formula $\varphi(x_1, \ldots, x_n)$, every approximation φ' of φ, and every rational $\epsilon > \delta$,*

$$\forall_\epsilon z_1 \ldots \forall_\epsilon z_n \, \varphi(x_1 + z_1, \ldots, x_n + z_n) \in t$$

implies

$$\forall_{\epsilon-\delta} z_1 \ldots \forall_{\epsilon-\delta} z_n \, \varphi'(x_1 + z_1, \ldots, x_n + z_n) \in t'.$$

Proof. Each implication between two of the conditions follows quickly from the definitions and (occasionally) the Compactness Theorem. □

14.16 Remark. In general there are many important topologies on spaces of types in the functional analysis setting; the more useful ones are given by uniform structures defined using positive bounded formulas. A general formulation of these topologies and preliminary discussion of their properties can be found in [Iov99a] and [Iov96].

Quantifier-Free Types

14.17 Definition. Suppose that $\mathcal{M}$ is a normed space L-structure, $L(C)$ is an extension of L by constants, and $\mathbf{A}$ is an interpretation of C in $\mathcal{M}$. Let $e_1, \ldots, e_n$ be elements of $\mathcal{M}$, and fix distinct variables $x_1, \ldots, x_n$ of L such that x_j and e_j have the same sort for each $j = 1, \ldots, n$. We define the *type of* $(e_1, \ldots, e_n)$ *over* $\mathbf{A}$, denoted

$$\mathrm{tpqf}_{\mathcal{M}}(e_1, \ldots, e_n / \mathbf{A}),$$

to be the set of all quantifier-free positive bounded $L(C)$-formulas $\varphi(x_1, \ldots, x_n)$ such that

$$(\mathcal{M}, \mathbf{A}) \models_A \varphi[e_1, \ldots, e_n].$$

When it can be done without confusion, we omit the subscript $\mathcal{M}$ from this notation and denote the type by

$$\mathrm{tpqf}(e_1, \ldots, e_n / \mathbf{A}).$$

If C is empty, we write $\emptyset$ in place of $\mathbf{A}$ in this notation, or omit it entirely.

14.18 Remarks.

(i) Let $\mathcal{M}, \mathbf{A}$ be as above, and let $e_1, \ldots, e_n$ and $e_1', \ldots, e_n'$ be elements of $\mathcal{M}$ such that e_j and e_j' are in the same sort of $\mathcal{M}$ for all $j = 1, \ldots, n$. Let $\mathcal{M}_0, \mathcal{M}_0'$ be the substructures of $\mathcal{M}$ generated by the range of $\mathbf{A}$ and $\{e_1, \ldots, e_n\}$, $\{e_1', \ldots, e_n'\}$, respectively. Then

$$\mathrm{tpqf}_{\mathcal{M}}(e_1, \ldots, e_n / \mathbf{A}) = \mathrm{tpqf}_{\mathcal{M}}(e_1', \ldots, e_n' / \mathbf{A})$$

if and only if there is an isomorphism from $\mathcal{M}_0$ onto $\mathcal{M}_1$ that is the identity on the range of $\mathbf{A}$ and takes e_j to e_j' for each $j = 1, \ldots, n$.

(ii) If $\mathrm{tpqf}(e_1, \ldots, e_n / \mathbf{A}) = \mathrm{tpqf}(e_1', \ldots, e_n' / \mathbf{A})$, then $\|e_j\| = \|e_j'\|$, for $j = 1, \ldots, n$.

(iii) If $(\mathcal{M}, \mathbf{A})$ is an $L(C)$-structure and $\mathcal{M} \subseteq \mathcal{N}$, then

$$\mathrm{tpqf}_{\mathcal{M}}(e_1, \ldots, e_n / \mathbf{A}) = \mathrm{tpqf}_{\mathcal{N}}(e_1, \ldots, e_n / \mathbf{A}).$$

14.19 Proposition. *Suppose that $\mathcal{M}$ is a normed space L-structure, $L(C)$ is an extension of L by constants, and $\mathbf{A}$ is an interpretation of C in $\mathcal{M}$. Let $r_1, \ldots, r_n$ be nonnegative rational numbers and $t(x_1, \ldots, x_n)$ a set of quantifier-free positive bounded $L(C)$-formulas. The following conditions are equivalent:*

(i) *There exists a structure* $\mathcal{N}$ *such that* $\mathcal{M} \preceq_A \mathcal{N}$ *and elements* $e_1, \ldots, e_n$ *of suitable sorts of* $\mathcal{N}$ *with* $\|e_j\| \le r_j$ *for* $j = 1, \ldots, n$ *such that*

$$t(x_1, \ldots, x_n) = \mathrm{tpqf}_{\mathcal{N}}(e_1, \ldots, e_n/\mathbf{A}).$$

(ii)(a) *The formula* $\bigwedge_{j=1}^{n} \|x_j\| \le r_j$ *is in* $t(x_1, \ldots, x_n)$;
  (b) *Every finite subset of* t^+ *is satisfiable in* $(\mathcal{M}, \mathbf{A})$;
  (c) *Whenever* $\varphi(x_1, \ldots, x_n)$ *is a quantifier-free positive bounded* $L(C)$-*formula and* φ' *is approximation of* φ, *we have* $\varphi \in t$ *or* $\mathrm{neg}(\varphi') \in t$.

14.20 Definition. Let $\mathcal{M}$, $L(C)$, and $\mathbf{A}$ be as in the preceding proposition. A set $t(x_1, \ldots, x_n)$ of quantifier free $L(C)$-formulas satisfying the (equivalent) conditions in Proposition 14.19 is called a *quantifier-free type* over $\mathbf{A}$. If $\mathcal{N}$ and $(e_1, \ldots, e_n)$ are as in (i) of the proposition, we say that $(e_1, \ldots, e_n)$ *realizes* t *in* $\mathcal{N}$ or that $(e_1, \ldots, e_n)$ is a *realization of* t *in* $\mathcal{N}$.

The following result characterizes quantifier elimination for a complete positive bounded theory in terms of types.

14.21 Proposition. *Suppose that L is a signature and Γ is a complete positive bounded theory in L. The following conditions are equivalent:*

(i) Γ *admits quantifier elimination;*
(ii) *Whenever $\mathcal{M}$ is an approximate model of Γ and $e_1, \ldots, e_n$ are elements of $\mathcal{M}$, the type*

$$\mathrm{tp}_{\mathcal{M}}(e_1, \ldots, e_n/\emptyset)$$

is determined by

$$\mathrm{tpqf}_{\mathcal{M}}(e_1, \ldots, e_n/\emptyset).$$

Proof. (i)$\Rightarrow$(ii) follows from the definitions. To prove (ii)$\Rightarrow$(i) one verifies condition (ii) of Proposition 13.12 simultaneously for all positive bounded formulas $\varphi(x_1, \ldots, x_n)$. $\qquad\square$

Quantifier-Free Types Over Banach Spaces

In this section we work in the pure Banach space signature L_0 and we identify every Banach space L_0-structure with its underlying Banach space. (See Chapter 7.)

The model theoretic concept of *type* was brought into Banach space theory by Krivine and Maurey [KM81]. For analysts, a type over a Banach space X is a function from X into $\mathbb{R}$ that characterizes a possible way to add an element to X while staying inside an ultrapower of X. Namely, given e in an ultrapower of X, one considers the function τ_e defined for $a \in X$ by $\tau_e(a) = \|a + e\|$. Note that τ_e and the norm on X determine the norm on the linear span of X and e. Together with Remark 14.18(i), this makes it clear that as e ranges over an ultrapower of X, there is a 1-1 correspondence between the quantifier-free type $\mathrm{tpqf}(e/X)$ and the function τ_e. (The definition of both objects involves a specific ultrapower of X in the background, from which the elements e are taken and in which norms are calculated.)

From the perspective of model theory, it is natural to generalize this in two ways: first, to consider n-tuples $(e_1, \ldots, e_n)$ in place of single elements e; second, to allow the base space X to be replaced by any subset A of X (including the empty set, in fact). Both of these generalizations yield benefits.

Throughout this section X is a fixed Banach space and A, B denote subsets of X. Let A be such a set, and consider a set $C = \{c_a \mid a \in A\}$ of new constant symbols that is in 1-1 correspondence with A. Define an interpretation $\mathbf{A}$ of C in X by setting $\mathbf{A}(c_a) = a$ for all $a \in A$. Take $e_1, \ldots, e_n$ in any approximate elementary extension of X and consider the quantifier-free type over $\mathbf{A}$ realized by $(e_1, \ldots, e_n)$ over $\mathbf{A}$. These are the objects that will be discussed in this section.

To simplify notation we will ignore the set of constants C and the interpretation $\mathbf{A}$; as is usual in model theory we will write $\mathrm{tpqf}(e_1, \ldots, e_n/A)$ instead of $\mathrm{tpqf}(e_1, \ldots, e_n/\mathbf{A})$ and we will speak of quantifier-free types over A rather than over $\mathbf{A}$. Also, we will write $L(A)$ for $L(C)$ and $(X, a)_{a \in A}$ for the $L(C)$-structure $(X, \mathbf{A})$. In this section, "quantifier-free type" will mean "quantifier-free type over A, realized in an approximate elementary extension of X, where A is a subset of X".

Let A be the closed linear span of B in X. Using Remark 14.18 we see that for any $e_1, \ldots, e_n$ in an approximate elementary extension of X, the quantifier-free type $\mathrm{tpqf}(e_1, \ldots, e_n/A)$ is determined by $\mathrm{tpqf}(e_1, \ldots, e_n/B)$. Hence, we lose no generality by focusing attention on types over closed linear subspaces of X.

14.22 Definition. Let A be a closed subspace of X. A *quantifier-free n-type over A* is a quantifier-free type of the form $\mathrm{tp}(e_1, \ldots, e_n/A)$, where $e_1, \ldots, e_n$ are in some approximate elementary extension of X.

Note that we have chosen to suppress the role of X in this terminology. Also, by Theorem 10.7, any approximate elementary extension of X has an ultrapower that is linearly isometric to an ultrapower of X.

Therefore, every quantifier-free n-type over a subspace of X is realized in some ultrapower of X.

14.23 Proposition. *Let A be a subspace of X. Let $r_1, \ldots, r_n$ be nonnegative rational numbers and $t(x_1, \ldots, x_n)$ a set of quantifier-free positive bounded $L(A)$-formulas. The following conditions are equivalent:*

(i) *There exists a Banach space $\mathfrak{X}$ such that $X \preceq_A \mathfrak{X}$ and elements $e_1, \ldots, e_n$ of $\mathfrak{X}$ with $\|e_j\| \leq r_j$ for $j = 1, \ldots, n$ such that*

$$t(x_1, \ldots, x_n) = \operatorname{tpqf}(e_1, \ldots, e_n/A).$$

(ii)(a) *The formula $\bigwedge_{j=1}^n \|x_j\| \leq r_j$ is in $t(x_1, \ldots, x_n)$;*
   (b) *Every finite subset of t^+ is satisfiable in $(X, a)_{a \in A}$;*
   (c) *Whenever $\varphi(x_1, \ldots, x_n)$ is a quantifier-free positive bounded $L(A)$-formula and φ' is approximation of φ, we have $\varphi \in t$ or $\operatorname{neg}(\varphi') \in t$.*

Proof. This is a special case of Proposition 14.19. □

If A is a subspace of X and $e_1, \ldots, e_n$ are in an approximate elementary extension of X, the quantifier-free type $\operatorname{tpqf}(e_1, \ldots, e_n/A)$ is determined by the truth values of the formulas

$$s_1 \leq \|x + r_1 e_1 + \cdots + r_n e_n\| \leq s_2,$$

as x ranges over A, where $r_1, \ldots, r_n, s_1, s_2$ are rational numbers. Therefore, $\operatorname{tpqf}(e_1, \ldots, e_n/A)$ provides the same information as the function

$$\tau_{e_1, \ldots, e_n, A} \colon A \times \mathbb{R}^n \to \mathbb{R}$$

defined for $a \in A$ and $r_1, \ldots, r_n \in \mathbb{R}$ by

$$\tau_{e_1, \ldots, e_n, A}(a, r_1, \ldots, r_n) = \|a + r_1 e_1 + \cdots + r_n e_n\|.$$

Since A is a subspace, homogeneity of the norm permits us to restrict $\tau_{e_1, \ldots, e_n, A}$ to $A \times \Delta$, where

$$\Delta = \{(r_1, \ldots, r_n) \in \mathbb{R} \mid \sum |r_i| = 1 \text{ and } \sum r_i > 0\},$$

without losing any information.

14.24 Proposition. *Let A be a subspace of X and $\tau \colon A \times \mathbb{R}^n \to \mathbb{R}$ any function. The following conditions are equivalent:*

(i) *There exists a Banach space $\mathfrak{X}$ such that $X \preceq_A \mathfrak{X}$ and elements $e_1, \ldots, e_n$ of $\mathfrak{X}$ such that for all $a \in A$ and $r_1, \ldots, r_n \in \mathbb{R}$*

$$\tau(a, r_1, \ldots, r_n) = \|a + r_1 e_1 + \cdots + r_n e_n\|.$$

(ii) *There exists a family $(u_1(i), \ldots, u_n(i))_{i \in I}$ in X^n and an ultrafilter D
on I such that for all $a \in A$ and $r_1, \ldots, r_n \in \mathbb{R}$*

$$\tau(a, r_1, \ldots, r_n) = \lim_{i, D} \| a + r_1 u_1(i) + \cdots + r_n u_n(i) \|.$$

Proof. (ii)$\Rightarrow$(i) follows from Proposition 14.25-(i) and Proposition 14.10.
To prove (i)$\Rightarrow$(ii) fix $e_1, \ldots, e_n$ as in (i) and note that, since $e_1, \ldots, e_n$
live in an approximate elementary extension of X and $A \subseteq X$, for every
formula $\varphi(x_1, \ldots, x_n) \in \mathrm{tpqf}(e_1, \ldots, e_n/A)$ of the form

$$s_1 \leq \| a + r_1 x_1 + \ldots r_n x_n \| \leq s_2$$

(where $a \in A$) and every approximation φ' of φ there exist elements
$u_1, \ldots, u_n \in X$ such that $X \models \varphi'[u_1, \ldots, u_n]$. $\qquad\square$

It is natural to define quantifier-free analogs of the topologies on types
defined earlier in this chapter. The *quantifier-free logic topology* is the
topology on quantifier-free types in which the basic neighborhoods of
a quantifier-free type t are the sets of the form $[\varphi]$ with $\varphi \in t^+$. (See
Definition 14.9). Let $\mathcal{X}$ be an approximate elementary extension of X
in which every quantifier-free type over every subspace of X is realized.
The *quantifier-free d-topology* is the topology on quantifier-free types
given by the metric d_{qf}, where for two quantifier-free types t, t',

$$d_{\mathrm{qf}}(t, t') = \inf \left\{ \max_j \| e_j - e'_j \| \ \mid \ \mathcal{X} \models_A t[\bar{e}], \ \mathcal{X} \models_A t'[\bar{e}'] \right\}.$$

This definition of the quantifier-free d-topology is independent of $\mathcal{X}$.

14.25 Proposition. *Let A be a subspace of X and regard quantifier-free
1-types over A as real-valued functions via the correspondence*

$$\mathrm{tpqf}(e/A) \rightsquigarrow \tau_{e,A}.$$

Then,

(i) *The quantifier-free logic topology corresponds to the topology of pointwise convergence on A;*

(ii) *The quantifier-free d-topology is finer than the topology of uniform
convergence on A.*

Proof. (i) is immediate from the definitions; (ii) is a consequence of the
triangle inequality. $\qquad\square$

References

[ACH97] L. O. Arkeryd, N. J. Cutland, and C. W. Henson, editors. *Nonstandard Analysis: Theory and Applications*, Dordrecht, 1997. Kluwer Academic Publishers Group.

[BDCK65] J. Bretagnolle, D. Dacunha-Castelle, and J.-L. Krivine. Fonctions de type positif sur les espaces L^p. *C. R. Acad. Sci. Paris*, 261:2153–2156, 1965.

[BDCK66] J. Bretagnolle, D. Dacunha-Castelle, and J.-L. Krivine. Lois stables et espaces L^p. *Ann. Inst. H. Poincaré Sect. B (N.S.)*, 2:231–259, 1965/1966.

[BL] S. Buechler and O. Lessmann. Simple homogeneous models. Submitted.

[CK90] C. C. Chang and H. J. Keisler. *Model theory*, volume 73 of *Studies in Logic and the Foundations of Mathematics*. North-Holland Publishing Co., Amsterdam, third edition, 1990.

[DC72a] D. Dacunha-Castelle. Applications des ultraproduits à la théorie des plongements des espaces de Banach. In *Actes du Colloque d'Analyse Fonctionnelle (Univ. Bordeaux, Bordeaux, 1971)*, pages 117–125. Bull. Soc. Math. France, Mém. No. 31–32. Soc. Math. France, Paris, 1972.

[DC72b] D. Dacunha-Castelle. Sur un théorème de J. L. Krivine concernant la caractérisation des classes d'espaces isomorphes à des espaces d'Orlicz généralisés et des classes voisines. *Israel J. Math.*, 13:261–276 (1973), 1972. Proceedings of the International Symposium on Partial Differential Equations and the Geometry of Normed Linear Spaces (Jerusalem, 1972).

[DCK70] D. Dacunha-Castelle and J.-L. Krivine. Ultraproduits d'espaces d'Orlicz et applications géométriques. *C. R. Acad. Sci. Paris Sér. A-B*, 271:A987–A989, 1970.

[DCK72] D. Dacunha-Castelle and J.-L. Krivine. Applications des ultraproduits à l'étude des espaces et des algèbres de Banach. *Studia Math.*, 41:315–334, 1972.

[Gro84] Ulrich Groh. Uniform ergodic theorems for identity preserving Schwarz maps on W^*-algebras. *J. Operator Theory*, 11(2):395–404, 1984.

[Hei80] S. Heinrich. Ultraproducts in Banach space theory. *J. Reine Angew. Math.*, 313:72–104, 1980.

[Hei81] S. Heinrich. Ultraproducts of L_1-predual spaces. *Fund. Math.*, 113(3):221–234, 1981.

[Hen74] C. W. Henson. The isomorphism property in nonstandard analysis and its use in the theory of Banach spaces. *J. Symbolic Logic*, 39:717–731, 1974.

[Hen75] C. W. Henson. When do two Banach spaces have isometrically isomorphic nonstandard hulls? *Israel J. Math.*, 22(1):57–67, 1975.

[Hen76] C. W. Henson. Nonstandard hulls of Banach spaces. *Israel J. Math.*, 25(1-2):108–144, 1976.

[Hen97] C. W. Henson. Foundations of nonstandard analysis: a gentle introduction to nonstandard extensions. In *Nonstandard analysis (Edinburgh, 1996)*, pages 1–49. Kluwer Acad. Publ., Dordrecht, 1997.

[HH86] S. Heinrich and C. W. Henson. Banach space model theory. II. Isomorphic equivalence. *Math. Nachr.*, 125:301–317, 1986.

[HHM83] S. Heinrich, C. W. Henson, and L. C. Moore, Jr. Elementary equivalence of L_1-preduals. In *Banach space theory and its applications (Bucharest, 1981)*, pages 79–90. Springer, Berlin, 1983.

108   *C. W. Henson and J. Iovino*

[HHM86] S. Heinrich, C. W. Henson, and L. C. Moore, Jr. Elementary
equivalence of $C_\sigma(K)$ spaces for totally disconnected, compact
Hausdorff K. *J. Symbolic Logic*, 51(1):135–146, 1986.

[HHM87] S. Heinrich, C. W. Henson, and L. C. Moore, Jr. A note on
elementary equivalence of $C(K)$ spaces. *J. Symbolic Logic*,
52(2):368–373, 1987.

[HL85] A. E. Hurd and P. A. Loeb. *An Introduction to Nonstandard Real
Analysis*. Academic Press, San Diego, 1985.

[HLR85] R. Haydon, M. Levy, and Y. Raynaud. On the local structure of
$L_p(X)$. In *Banach spaces (Columbia, Mo., 1984)*, pages 74–79. Springer,
Berlin, 1985.

[HLR91] R. Haydon, M. Levy, and Y. Raynaud. *Randomly normed spaces*.
Hermann, Paris, 1991.

[HM74a] C. W. Henson and L. C. Moore, Jr. Nonstandard hulls of the
classical Banach spaces. *Duke Math. J.*, 41:277–284, 1974.

[HM74b] C. W. Henson and L. C. Moore, Jr. Subspaces of the nonstandard
hull of a normed space. *Trans. Amer. Math. Soc.*, 197:131–143, 1974.

[HM83a] C. W. Henson and L. C. Moore, Jr. The Banach spaces $l_p(n)$ for
large p and n. *Manuscripta Math.*, 44(1-3):1–33, 1983.

[HM83b] C. W. Henson and L. C. Moore, Jr. Nonstandard analysis and the
theory of Banach spaces. In *Nonstandard analysis—recent developments
(Victoria, B.C., 1980)*, pages 27–112. Springer, Berlin, 1983.

[Iov94] J. Iovino. *Stable Theories in Functional Analysis*. PhD thesis,
University of Illinois at Urbana-Champaign, 1994.

[Iov96] J. Iovino. The Morley rank of a Banach space. *J. Symbolic Logic*,
61(3):928–941, 1996.

[Iov98] J. Iovino. Types on stable Banach spaces. *Fund. Math.*, 157(1):85–95,
1998.

[Iov99a] J. Iovino. Stable Banach spaces and Banach space structures. I.
Fundamentals. In *Models, algebras, and proofs (Bogotá, 1995)*, pages
77–95. Dekker, New York, 1999.

[Iov99b] J. Iovino. Stable Banach spaces and Banach space structures. II.
Forking and compact topologies. In *Models, algebras, and proofs
(Bogotá, 1995)*, pages 97–117. Dekker, New York, 1999.

[Iov01] J. Iovino. On the maximality of logics with approximations. *J.
Symbolic Logic*, 66(4):1909–1918, 2001.

[Kei61] H. J. Keisler. Ultraproducts and elementary classes. *Nederl. Akad.
Wetensch. Proc. Ser. A 64 = Indag. Math.*, 23:477–495, 1961.

[KM81] J.-L. Krivine and B. Maurey. Espaces de Banach stables. *Israel J.
Math.*, 39(4):273–295, 1981.

[Kri67] J.-L. Krivine. *Sous-espaces et Cones Convexes dans les Espaces L^p*.
Université de Paris, 1967. Thèse.

[Kri72] J.-L. Krivine. Théorie des modèles et espaces L^p. *C. R. Acad. Sci.
Paris Sér. A-B*, 275:A1207–A1210, 1972.

[Kri74] J.-L. Krivine. Langages à valeurs réelles et applications. *Fund. Math.*,
81:213–253, 1974. Collection of articles dedicated to Andrzej Mostowski
on the occasion of his sixtieth birthday, III.

[Lac74] H. E. Lacey. *The Isometric Theory of Classical Banach Spaces*.
Springer-Verlag, Berlin, 1974.

[Lin88] T. Lindstrom. An Invitation to Nonstandard Analysis. In
Nonstandard Analysis and its Applications, pages 1–105. Cambridge

university Press, Cambridge, 1988.

[Łoś55] J. Łoś. Quelques remarques, théorèmes et problèmes sur les classes définissables d'algèbres. In *Mathematical interpretation of formal systems*, pages 98–113. North-Holland Publishing Co., Amsterdam, 1955.

[LR84a] M. Levy and Y. Raynaud. Ultrapuissances de $L^p(L^q)$. In *Seminar on functional analysis, 1983/1984*, pages 69–79. Univ. Paris VII, Paris, 1984.

[LR84b] M. Levy and Y. Raynaud. Ultrapuissances des espaces $L^p(L^q)$. *C. R. Acad. Sci. Paris Sér. I Math.*, 299(3):81–84, 1984.

[LT73] J. Lindenstrauss and L. Tzafriri. *Classical Banach spaces.* Springer-Verlag, Berlin, 1973. Lecture Notes in Mathematics, Vol. 338.

[Lux69a] W. A. J. Luxemburg, editor. *Applications of model theory to algebra, analysis, and probability.* Holt, Rinehart and Winston, New York, 1969.

[Lux69b] W. A. J. Luxemburg. A general theory of monads. In *Applications of Model Theory to Algebra, Analysis, and Probability (Internat. Sympos., Pasadena, Calif., 1967)*, pages 18–86. Holt, Rinehart and Winston, New York, 1969.

[MN91] P. Meyer-Nieberg. *Banach Lattices.* Springer-Verlag, Berlin, 1991.

[Plo71] A. I. Plotkin. Continuation of L^p-isometries. *Zap. Naučn. Sem. Leningrad. Otdel. Mat. Inst. Steklov. (LOMI)*, 22:103–129, 1971.

[Ray86] Y. Raynaud. Sur les sous-espaces de $L^p(L^q)$. In *Séminaire d'Analyse Fonctionelle 1984/1985*, pages 49–71. Univ. Paris VII, Paris, 1986.

[Ray91] Y. Raynaud. Some remarks on ultrapowers and superproperties of the sum and interpolation spaces of Banach spaces. *Compositio Math.*, 79(3):295–319, 1991.

[Ray97] Y. Raynaud. Ultrapowers of Köthe function spaces. *Collect. Math.*, 48(4-6):733–742, 1997. Fourth International Conference on Function Spaces (Zielona Góra, 1995).

[Ray98a] Y. Raynaud. On ultrapowers of non commutative L_1 spaces. *Colloquium del Departamento de Analisis Matematico, Univ. Complutense Madrid,*, 43, 1998.

[Ray98b] Y. Raynaud. Ultrapowers of Calderón-Lozanovskii interpolation spaces. *Indag. Math. (N.S.)*, 9(1):65–105, 1998.

[Ray02] Y. Raynaud. On ultrapowers of non commutative L_p spaces. To appear, 2002.

[Rob66] A. Robinson. *Non-standard analysis.* North-Holland Publishing Co., Amsterdam, 1966.

[Rud76] W. Rudin. L^p-isometries and equimeasurability. *Indiana Univ. Math. J.*, 25(3):215–228, 1976.

[She71] S. Shelah. Every two elementarily equivalent models have isomorphic ultrapowers. *Israel J. Math.*, 10:224–233, 1971.

[Sim82] B. Sims. *"Ultra"-techniques in Banach space theory.* Queen's University, Kingston, Ont., 1982.

[Ste74] J. Stern. Sur certaines classes d'espaces de Banach caractérisées par des formules. *C. R. Acad. Sci. Paris Sér. A*, 278:525–528, 1974.

[Ste75] J. Stern. Propriétés locales et ultrapuissances d'espaces de Banach. In *Séminaire Maurey-Schwartz 1974-1975: Espaces L^p, applications radonifiantes et géométrie des espaces de Banach, Exp. Nos. VI et VII*, pages 30 pp. (erratum, p. 2). Centre Math., École Polytech., Paris, 1975.

[Ste76] J. Stern. Some applications of model theory in Banach space theory.

Ann. Math. Logic, 9(1-2):49–121, 1976.

[Ste77] J. Stern. Exemples d'application de la théorie des modèles à la théorie des espaces de Banach. In *Colloque International de Logique (Clermont-Ferrand, 1975)*, pages 19–37. Colloques Internat. CNRS, 249. CNRS, Paris, 1977.

[Ste78] J. Stern. Ultrapowers and local properties of Banach spaces. *Trans. Amer. Math. Soc.*, 240:231–252, 1978.

Index of Notation

$(F)_D$, 18
$(\mathcal{M})_D$, 19
$(X)_D$, 18
$((x_\xi)_{\xi \in \Lambda})_D$, 18
$[\varphi]$, 99

$\bar{a} \sim_\lambda \bar{b}$, 34

$\mathcal{B}_r(X)$, 18

$\mathrm{card}(L)$, 14

$d(t, t')$, 100

$f \colon s_1 \times \cdots \times s_m \to s_0$, 13
$f^{\mathcal{M}}$, 13
$\varphi(x_1, \ldots, x_n)$, 23
$\varphi < \varphi'$, 24
φ_λ, 92

$\Gamma(x_1, \ldots, x_n)$, 23
Γ^+, 25

$L(C)$, 14, 16
$L \subseteq L'$, 14
L_0, 34
$\ell_\infty(\Lambda, (X_\xi \mid \xi \in \Lambda))$, 17
$\lim_{\xi, D}$, 17

$(\mathcal{M}, \mathbf{A})$, 16
$\mathcal{M} \models \varphi[a_1, \ldots, a_n]$, 23
$\mathcal{M} \models_{\mathcal{A}} \varphi[a_1, \ldots, a_n]$, 25
$\mathcal{M} \models \Gamma[a_1, \ldots, a_n]$, 24
$\mathcal{M} \models_{\mathcal{A}} \Gamma[a_1, \ldots, a_n]$, 25
$\mathcal{M} \equiv_{\mathcal{A}} \mathcal{N}$, 31
$\mathcal{M} \preceq_{\mathcal{A}} \mathcal{N}$, 31
$\mathcal{M} \cong \mathcal{N}$, 15
$\mathcal{M} \subseteq \mathcal{N}$, 14

neg, 26

$(\prod_{\xi \in \Lambda} X_\xi)_D$, 18
$(\prod_{\xi \in \Lambda} \mathcal{M}_\xi)_D$, 41

$\exists_r x\, \varphi$, 22
$\forall_r x\, \varphi$, 22

$S_{(x_1, \ldots, x_n)}(\mathbf{A})$, 99
$s_{\mathbb{R}}$, 10, 13

$t(x_1, \ldots, x_n)$, 23
$t^{\mathcal{M}}$, 23
$\|t\|$, 99
$\mathrm{Th}_{\mathcal{A}}(\mathcal{M})$, 31
$\mathrm{tp}(e_1, \ldots, e_n / \mathbf{A})$, 96
$\mathrm{tp}_{\mathcal{M}}(e_1, \ldots, e_n / \mathbf{A})$, 96
$\mathrm{tpqf}(e_1, \ldots, e_n / \mathbf{A})$, 102
$\mathrm{tpqf}_{\mathcal{M}}(e_1, \ldots, e_n / \mathbf{A})$, 102

$\bigcup_{\xi \in \Lambda} \mathcal{M}_\xi$, 46

111

Index

approximable
>   by formulas in a given class, 91
>   by quantifier-free formulas, 84

approximate
>   elementary chain, 47
>   elementary embedding, 32
>   elementary extension, 32
>   elementary substructure, 31
>   model, 31
>   satisfiability, 49
>   truth, *see* approximate satisfaction

approximate satisfaction
>   of a formula, 25
>   of a set of formulas, 25

approximately elementarily equivalent,
>   structures, 31

approximation, of a positive bounded
>   formula, 24

automorphism, of a structure, 15

axiomatizable
>   by positive bounded sentences, 82
>   by universal sentences, 81

axioms, *see* positive bounded, axioms

Banach space structure, 10

cardinality, of a signature, 14

chain, of structures, 46
>   approximate elementary, 47

closure, of a structure, 15, 33

Compactness Theorem, 45
>   second version, 52

completion, of a structure, 15, 33

connective, positive, 22

constant symbol, 13

constant, of a structure, 10

countably incomplete, ultrafilter, 50

D-limit, *see* limit

d-metric, 100

density character, of a topological
>   space, 47

determined, 98

diagonal embedding
>   for normed spaces, 18
>   for structures, 19, 44

domain, of a function symbol, 13

element, of a structure, 11

embedding, 15
>   approximate elementary, 32

enlargement, 53

existential formula, 79

expansion, of a structure, 16
>   by constants, 16

extension
>   approximate elementary, 32

extension, of a signature, 14
>   by constants, 14

extension, of a structure, 14

finitely
>   generated, structure, 14

formula, *see* positive bounded formula
>   atomic, 22

function symbol, 13

homogeneous, *see* strongly
> κ-homogeneous

interpretation
>   of a function symbol, 13
>   of a set of constants, 16

isometric isomorphism, *see* isomorphism

isometric automorphism, *see*
>   automorphism

isometric embedding, *see* embedding

isomorphism, 15

112

κ-saturated structure, 49

L-structure, 13
L-term, 21
Löwenheim-Skolem Theorem
    Downward, 47
    Upward, 48
λ-approximation, 92
λ-equivalent, tuples of elements, 34
λ-isomorphic, 93
limit, with respect to an ultrafilter, 17
logic topology, 99
Łoś Theorem, 44

model, approximate, 31
modulus of uniform continuity, 18
    uniform, 38

nonstandard hull, 19, 41
normed space structure, 10
    based on, 10
    complete, 10
    finitely generated, 14
    κ-saturated, 49
    strongly κ-homogeneous, 75

operations
    of a structure, 10
    required, 10
    vector space, 10

Perturbation Lemma
    for a structure, 27
    for uniform classes, 43
positive bounded
    axioms, 82
    formula, 22
positive bounded formula
    existential, 79
    quantifier-free, 34
    universal, 79
positive bounded theory, 31
    complete, 45
    of a structure, 31
prenex form, 79

quantifier elimination
    for formulas, 84
    for theories, 86
quantifier, bounded, 22
quantifier-free
    formula, 34
    type, 102
        n-type, 104
        realization of, 103

range

of a function symbol, 13
of a term, 21
reduct, of a structure, 16

satisfaction
    of a formula, 23
    of a set of formulas, 24
satisfiable, 49
saturated, *see* κ-saturated structure
sentence, 22
signature, 13
    cardinality of, 14
    countable, 14
sort, of a variable, 21
sorts, of a structure, 10
    special, 10
strongly κ-homogeneous, structure, 75
structure, *see* normed space structure
subformula, of a formula, 22
subsignature, 14
substructure, 14
suitable sorts, elements of, 26

τ-consistent, 56
Tarski-Vaught Test, 33
term, 21
    evaluation of, 23
topology
    d-topology, *see* d-metric
    logic, *see* logic topology
type, 97
    n-type, 104
    norm of a, 99
    of a tuple, 96
    quantifier-free, *see* quantifier-free
        type
    realization of, 97

ultrapower
    of a normed space, 18
    of a normed space structure, 19
ultraproduct
    of a family of normed spaces, 18
    of a family of structures, 41
uniform
    bound for a family of functions, 38
    class of structures, 39
    family of functions, 38
    family of structures, 39
    modulus of uniform continuity, 38
universal formula, 79

variable, 21
    free, 22

weak negation, of a formula, 26

Actions of Polish Groups and Classification Problems

Alexander S. Kechris

Department of Mathematics
California Institute of Technology
Pasadena, California 91125 USA
kechris@caltech.edu
http://www.math.caltech.edu/people/kechris.html

1

Introduction

We will discuss in this paper some aspects of a general program whose goal is the development of the theory of definable actions of Polish groups, the structure and classification of their orbit spaces, and the closely related study of definable equivalence relations. This work is motivated by basic foundational questions, like understanding the nature of complete classification of mathematical objects up to some notion of equivalence by invariants, and creating a mathematical framework for measuring the complexity of such classification problems. This theory, which has been growing rapidly over the last few years, is developed within the context of descriptive set theory, which provides the basic underlying concepts and methods. On the other hand, in view of the broad scope of this theory, there are natural interactions of it with other areas of mathematics, such as the theory of topological groups, topological dynamics, ergodic theory and its relationships with the theory of operator algebras, model theory, and recursion theory.

Classically, in various branches of dynamics one studies actions of the groups of integers $\mathbb{Z}$, reals $\mathbb{R}$, Lie groups, or even more generally (second countable) locally compact groups. One of the goals of the theory is to expand this scope by considering the more comprehensive class of *Polish groups* (separable completely metrizable topological groups), which seems to be the widest class of well-behaved (for our purposes) groups and which includes practically every type of topological group we are interested in. One of the main problems concerning a given definable action of a Polish group G on a Polish space X is the complete classification of members of X up to orbit equivalence by invariants. (*Orbit equivalence* being the equivalence relation induced by the orbits of the action.) This is a special case of the more general problem of completely classifying elements of a given Polish space X up to some definable equivalence relation E on that space. This means finding a set of invariants I and a map $c : X \to I$ such that $xEy \Leftrightarrow c(x) = c(y)$, where for this to have any meaning, both I, c must be "explicit" or "definable" too. A typical example of this kind of problem is the classification of countable models of a theory up to isomorphism, the classification of the irreducible unitary representations of a locally compact group up to unitary equivalence, the classification of measure preserving transformations up to conjugacy, etc.

In measuring the complexity of the classification problem and the nature of the possible complete invariants for a given equivalence relation E, the following notion is important. Let E, E' be two equivalence relations on Polish spaces X, X'. We say that E is *Borel reducible* to E', in symbols

$$E \leq_B E',$$

if there is a Borel map $f : X \to X'$ such that $xEy \Leftrightarrow f(x)E'f(y)$. Letting then $\tilde{f}([x]_E) = [f(x)]_{E'}$, it is clear that $\tilde{f} : X/E \to X'/E'$ is an embedding of X/E into X'/E'. Intuitively, $E \leq_B E'$ can be interpreted as meaning any one of the following:

(i) E has a simpler classification problem than E': any complete invariants for E' work as well for E (after composing with f).

(ii) One can completely classify E-equivalence classes by invariants which are E'-equivalence classes.

(iii) The quotient space X/E "Borel embeds" into X'/E', so X/E has "Borel cardinality" less than or equal to that of X'/E'.

Also let

$$E \sim_B E' \Leftrightarrow E \leq_B E' \ \& \ E' \leq_B E.$$

This means that E, E' have equivalent classification problems or X/E, X'/E' have the same "Borel cardinality". Finally, let

$$E <_B E' \Leftrightarrow E \leq_B E' \ \& \ E' \leq_B E.$$

To illustrate these notions let us mention a couple of classical examples:

(i) The *Vitali* equivalence relation on $\mathbb{R}$ is defined by

$$x E_V y \Leftrightarrow x - y \in \mathbb{Q}$$

(so $\mathbb{R}/E_V = \mathbb{R}/\mathbb{Q}$). Denoting for any set X ambiguously also by X the equality relation on X, it is easily seen that $\mathbb{R} \leq_B E_V$. But it is also not hard to prove that $E_V \not\leq_B \mathbb{R}$. (Notice that this is a consequence of the following well-known fact: if $A \subseteq \mathbb{R}$ is Borel and invariant under $\mathbb{Q}$-translation, then A is either *meager*, i.e., of the first category, or *comeager*, i.e., its complement is of the first category. This is a special case of a general Topological 0-1 Law, see Kechris [95, 8.46].)

So $\mathbb{R} <_B E_V$. Thus $\mathbb{R}/\mathbb{Q}$ has bigger "Borel cardinality" than $\mathbb{R}$, although classically $\mathbb{R}/\mathbb{Q}$ has the same cardinality as $\mathbb{R}$.

(ii) If E denotes unitary equivalence of normal operators on a separable (complex) Hilbert space and $E' = \sim$ denotes measure equivalence of probability Borel measures on an uncountable Polish space X ($\mu \sim \nu \Leftrightarrow \mu << \nu \ \& \ \nu << \mu$), then the *Spectral Theorem* implies that $E \sim_B E'$.

This paper essentially consists of two parts. The first, which contains Sections 2-7, is a survey of certain aspects of the program discussed in this introduction and very few technical details are given here. The second, which contains Sections 8-13, gives a somewhat detailed technical exposition of Hjorth's recent theory of turbulence, which is first introduced in Section 7.

Acknowledgment. Research and preparation of this paper have been partially supported by NSF Grants DMS 96-19880 and DMS 99-87437.

2

The General Glimm-Effros Dichotomy

The Vitali equivalence relation plays a special role in the hierarchy of classification problems in view of a theorem known as the General Glimm-Effros Dichotomy that we will now explain.

Definition 2.1. An equivalence relation E on a Polish space X is called *concretely classifiable* or *smooth* if there is a Borel map $f : X \to Y, Y$ some Polish space, such that

$$xEy \Leftrightarrow f(x) = f(y).$$

So elements of X can be completely classified up to E-equivalence by invariants which are members of a Polish space, thus fairly "concrete."

Equivalently, E is concretely classifiable iff $E \leq_B \mathbb{R}$ iff $E \leq_B Y$ for some Polish space Y. In particular, if $E \leq_B \mathbb{N}$, i.e., E is Borel with only countably many equivalence classes, then E is concretely classifiable. If E has a *Borel selector*, i.e., a Borel function which chooses exactly one element out of each equivalence class, then E is concretely classifiable. The converse fails in general, e.g., every closed equivalence relation E is concretely classifiable but may not have a Borel selector (see Kechris [95, 18.D]). However, it is true in most natural examples.

Here are some examples of concretely classifiable E:

(i) E is Borel with every equivalence class finite. (This is because in this case we have a Borel selector.)

(ii) E is the equivalence relation of similarity on the $n \times n$ complex matrices. (This follows from the Jordan Canonical Form, which gives a Borel selector.)

(iii) Let G be a Polish group and $H \subseteq G$ a closed subgroup, and consider the equivalence relation on G:

$$x E_H y \Leftrightarrow x^{-1} y \in H.$$

(Again we have a Borel selector, see Kechris [95, 12.17]).

(iv) Let G be a type I Polish locally compact group and let E be the unitary equivalence relation on the irreducible unitary representations of G (see Mackey [78]). (This class of groups contains the compact, abelian, semi-simple Lie groups; etc.)

The following are examples of non-concretely classifiable (Borel) equivalence relations:

(v) The Vitali equivalence relation E_V. (This has been essentially proved in Section 1, Ex. (i).)

(vi) Consider the shift on $2^{\mathbb{Z}}$ and the corresponding equivalence relation E_S induced by the orbits of the shift. Or, consider an irrational rotation R on $\mathbb{T}$ and its associated orbit equivalence relation E_R. (Both are non-concretely classifiable, by an argument similar to that used for the Vitali equivalence relation.)

(vii) The unitary equivalence relation on the irreducible unitary representations of non-type I groups, e.g., F_2, the free group with 2 generators (see again Mackey [78]).

As in Examples (v), (vi) above, one way to show non-concrete classifiability is by using the following general fact:

Fact 2.2. *If E is an equivalence relation on a Polish space X such that every equivalence class is meager and every Borel E-invariant set is either meager or comeager, then E is not concretely classifiable.*

There is an analogous fact involving measure theoretic as opposed to topological notions. We first need the following definition.

Definition 2.3. Let X be a Polish space and denote by $P(X)$ the set of Borel probability measures on X. If E is an equivalence relation on X and $\mu \in P(X)$ we say that μ is E-*ergodic* if every E-invariant Borel set has μ-measure 0 or 1. We say that μ is E-*nonatomic* if every equivalence class has measure 0.

We now have:

Fact 2.4. *Let E be an equivalence relation. If E admits an E-ergodic, nonatomic measure, E is not concretely classifiable.*

The basic phenomenon now is that there is a "smallest" non-concretely classifiable Borel equivalence relation, namely the Vitali equivalence relation. For convenience, we will replace it, without any harm, by a combinatorial reformulation.

Consider the equivalence relation E_0 on $2^{\mathbb{N}}$ defined by

$$x E_0 y \Leftrightarrow \exists n \forall m \geq n (x_m = y_m).$$

Then it can be seen that $E_0 \sim_B E_V$ (Mycielski, see Mauldin-Ulam [87]), so these are equivalent for our purposes.

We now have

Theorem 2.5 (The General Glimm-Effros Dichotomy; Harrington-Kechris-Louveau [90]). *Let E be a Borel equivalence relation on a Polish space X. Then exactly one of the following holds:*

(I) *E is concretely classifiable.*

(II) *$E_0 \sqsubseteq_c E$, i.e., there is a continuous embedding $f : 2^{\mathbb{N}} \to X$ such that $x E_0 y \Leftrightarrow f(x) E f(y)$ (so that in particular $E_0 \leq_B E$).*

Moreover, (II) *is equivalent to*

(II)' *There exists an E-ergodic, non-atomic measure.*

Thus one also has a converse to 2.4 and this provides a useful existence theorem.

In Harrington, Kechris and Louveau [90] the reader can find more background on the history of this type of result and its origins in the theory of operator algebras.

3

Actions of Polish Groups

From now on we will be primarily interested in equivalence relations induced by actions of Polish groups.

Definition 3.1. Let G be a Polish group. A *Polish G-space* is a Polish space X together with a continuous action $(g, x) \mapsto g \cdot x$ of G into X. A *Borel G-space* is a Polish space with a Borel action.

It turns out that these two notions are essentially equivalent for our purposes, in view of the following:

Theorem 3.2 (Becker-Kechris [96]). *Any Borel G-space is Borel isomorphic to a Polish G-space.*

For any G-space X we denote by E_G^X the associated *orbit equivalence relation*

$$xE_G^Xy \Leftrightarrow \exists g(g \cdot x = y).$$

In general E_G^X is *analytic* but not Borel. Here are some examples:

(i) The isomorphism relation on countable structures of the language with one binary relation symbol and with standard universe $\mathbb{N}$ can be viewed, as explained in Section 6 below, as induced by a continuous action of a Polish group. It is not Borel (see, e.g., Kechris [95, 27.D]).

(ii) Let $I = [0, 1], \lambda =$ Lebesgue measure, and consider the Polish group $Aut(I, \lambda)$ of all measure preserving automorphisms on I (see, e.g., Kechris [95, 17.46]). Consider the conjugation action of this group into itself. The associated equivalence relation is of course the classical notion of isomorphism or conjugacy of measure preserving automorphisms in ergodic theory. Recently Hjorth [97] has shown that this equivalence relation is not Borel.

(iii) On the other hand, consider the unitary group $U(H)$ of a separable infinite-dimensional complex Hilbert space H, which is a Polish group, as explained, e.g., in Kechris [95, 9.B]. If we look at the conjugation action on this group and the corresponding orbit equivalence relation, i.e., the classical notion of unitary equivalence of unitary operators, the Spectral Theorem implies that it is Borel. (This is explained in Example (ii) of Section 1.)

The dichotomy theorem 2.5 does not hold for analytic equivalence relations. However, we can obtain an appropriate generalization by allowing a more liberal notion of invariants than that required by concrete classifiability.

Below by $2^{<\omega_1}$ we denote the set of transfinite sequences $(a_\xi)_{\xi<\theta}$ with $a_\xi \in \{0,1\}$ and θ a countable ordinal.

Definition 3.3. An equivalence relation E on a Polish space X is *Ulm-classifiable* if there is a "definable" map $f : X \to 2^{<\omega_1}$ such that $xEy \Leftrightarrow f(x) = f(y)$.

The concept of "definable" here can of course be made precise – it means "C-measurable in the codes" in the technical logical jargon. The reader can consult Hjorth-Kechris [95] for more details.

If E is Borel, then E is Ulm-classifiable iff E is concretely classifiable. An interesting example of an analytic equivalence relation induced by a Polish group action where these concepts differ is the isomorphism relation on countable abelian p-groups, which is Ulm-classifiable. See Hjorth-Kechris [95] again for a discussion of this example. It is of course the classical Ulm Theorem on the classification of such groups and the nature of the associated invariants that motivates our terminology.

We now have

Theorem 3.4 (Hjorth-Kechris [95], Becker). *For any Borel G-space X exactly one of the following holds:*

(I) E_G^X *is Ulm-classifiable.*

(II) $E_0 \sqsubseteq_c E_G^X$.

Actually in Hjorth-Kechris [95] this result is appropriately extended to arbitrary analytic equivalence relations, under the hypothesis of the existence of large cardinals.

An example where alternative (I) holds is the isomorphism relation of countable torsion abelian groups and an example where (II) holds is the isomorphism relation of countable torsion-free abelian groups (on which there is more in Section 6 below).

We will now consider the problem of classification of various classes of equivalence relations of the form E_G^X. The simplest case is when G is compact. It is not hard to see then that all E_G^X are concretely classifiable, so there is not much more to say here. The next simplest case is when G is countable (i.e., a discrete Polish group).

4

Actions of Countable Groups

Let G be a countable group and X a Borel G-space. Then it is clear that E_G^X is a Borel equivalence relation and every one of its equivalence class is countable.

Definition 4.1. A Borel equivalence relation E is *countable* if every equivalence class is countable.

We now have

Theorem 4.2 (Feldman-Moore [77]). *The following are equivalent for each Borel equivalence relation E on a Polish space X:*

(i) *E is countable.*

(ii) *$E = E_G^X$ for some countable group G and a Borel G-space X.*

Countable Borel equivalence relations have long been studied in ergodic theory and its relationship to the theory of operator algebras. One important observation is that many concepts of ergodic theory such as invariance, quasi-invariance (null set preservation), and ergodicity of measures depend only on the orbit equivalence relation and not the action inducing it. Also a countable Borel equivalence relation with an associated quasi-invariant probability measure gives rise to a canonical von Neumann algebra and this has important implications to classification problems of von Neumann algebras. See, for example, the survey Schmidt [90].

The simplest examples of countable Borel equivalence relations are those induced by Borel actions of the group of integers $\mathbb{Z}$, i.e., by the orbits of a single Borel automorphism. These are also called *hyperfinite* in view of the following result.

Theorem 4.3 (Slaman-Steel [88], Weiss [84]). *For each countable Borel equivalence relation E on a Polish space X the following are equivalent:*

(i) $E = E_{\mathbb{Z}}^X$, *for some Borel $\mathbb{Z}$-space X.*

(ii) $E = \bigcup_n E_n$, *where each E_n is a finite Borel equivalence relation (i.e., has finite equivalence classes) and $E_0 \subseteq E_1 \subseteq E_2 \subseteq \ldots$.*

Remark. The condition that $\{E_n\}$ forms an increasing sequence is important. If it is dropped, one obtains *all* countable Borel equivalence relations and these are not necessarily hyperfinite (see, e.g., Dougherty-Jackson-Kechris [94])).

Here are some examples (for which more details can be found in Dougherty-Jackson-Kechris [94]):

$$E_0, \quad E_V, \quad E_R, \quad E_S$$

are all hyperfinite, and is so is the *tail equivalence* relation E_t on $2^{\mathbb{N}}$ defined by:

$$x E_t y \Leftrightarrow \exists k \exists \ell \forall m (x_{m+k} = y_{m+\ell}).$$

We now have two main results concerning the classification of hyperfinite Borel equivalence relations.

Theorem 4.4 (Dougherty-Jackson-Kechris [94]). *If E, F are Borel hyperfinite but not concretely classifiable, then $E \sim_B F$.*

Thus we have the following exact picture concerning the ordering $<_B$ of hyperfinite E:

$$1 <_B 2 <_B 3 <_B \ldots <_B \mathbb{N} <_B \mathbb{R} <_{\mathbb{R}} E_0.$$

It is of interest also to classify here the hyperfinite E up to a much stricter notion of equivalence, namely *Borel isomorphism*. One motivation comes from an analogous problem in ergodic theory. The objects here are triples (X, E, μ), with X a Polish space, E a hyperfinite Borel equivalence relation and μ an E-quasi-invariant, ergodic probability measure on X. (Quasi-invariance simply means that any Borel automorphism inducing E leaves the μ-null sets invariant.) Isomorphisms of two such triples (X, E, μ), (X', E', μ') means a Borel isomorphism f of E, E', modulo Borel invariant null sets, which sends μ to a measure $f\mu \sim \mu'$. One distinguishes such triples into types

I_n $(n = 1, 2, \ldots)$, I_∞, II_n, II_∞, III_λ $(0 \leq \lambda \leq 1)$. The Dye-Krieger Classification Theory shows that, up to isomorphism, there is exactly one system of each type, except III_0. For the case III_0 a complete invariant of isomorphism is the so-called *Poincaré flow* associated with the system. Again the reader can consult Schmidt [90] for more on this.

Returning to the descriptive context we have the following answer which is quite different from the one in the measure theoretic context. Below we call an equivalence relation *aperiodic* if all its equivalence classes are infinite. For each hyperfinite Borel equivalence relation E we denote by $\mathcal{E}(E)$ the set of E-ergodic, invariant probability measures. (Again E-invariance means invariance under any Borel automorphism inducing E.)

It is clear that one can easily classify hyperfinite E which are concretely classifiable by simple, cardinality-type invariants. Also E can be canonically written as a direct sum $E = E_1 \oplus E_2$, where E_1 is a finite Borel equivalence relation and E_2 is aperiodic. Since finite Borel equivalence relations are concretely classifiable, we can restrict ourselves to aperiodic, non-concretely classifiable E. We now have, denoting by $\cong_B$ the relation of *Borel isomorphism*.

Theorem 4.5 (Dougherty-Jackson-Kechris [94]). *Let E, F be aperiodic, non-concretely classifiable hyperfinite Borel equivalence relations. Then*

$$E \cong_B F \Leftrightarrow \operatorname{card}(\mathcal{E}(E)) = \operatorname{card}(\mathcal{E}(F)).$$

Since $\mathcal{E}(E)$ is a Borel set in the Polish space of probability Borel measures, $\operatorname{card}(\mathcal{E}(E))$ can only take the values

$$0, 1, 2, \ldots, \aleph_0, 2^{\aleph_0},$$

so there are only countably many Borel isomorphism classes. They are realized by the following equivalence relations (in the same order)

$$E_t \,(\cong_B E_V), E_0 \,(\cong_B E_V|[0,1]), 2 \times E_0, 3 \times E_0, \ldots,$$
$$\mathbb{N} \times E_0, \mathbb{R} \times E_0 \,(\cong_B E_S^*)$$

(where E_S^* is the aperiodic part of the shift-equivalence relation E_S on $2^{\mathbb{Z}}$).

The proof of 4.5 uses 4.4 and the important result of Nadkarni [91].

It turns out that there are many other countable groups G for which *all* equivalence relations E_G^X (induced by Borel G-spaces X) are hyperfinite, so they fall under the context of the previous theory. It is not hard to see though that any such group G must be necessarily *amenable*, i.e., carry a left-invariant finitely additive invariant probability measure defined on all its subsets); see, for example, Kechris [91]. The following problem has been raised by Weiss [84].

Problem 4.6. *Let G be a countable amenable group, and X any Borel G-space. Is E_G^X hyperfinite?*

A positive almost everywhere answer is known in ergodic theory.

Theorem 4.7 (Ornstein-Weiss [80], Connes-Feldman-Weiss [81]). *For any countable amenable group G, any Borel G-space X and any probability Borel measure μ on X, E_G^X is hyperfinite on an invariant Borel set of μ-measure 1.*

Remark. A much stronger result is known in the case of category. By a result of Sullivan-Weiss-Wright [86], as strengthened subsequently by Woodin and Hjorth-Kechris [96], we have that for *any* countable Borel equivalence relation E on a Polish space X there is an invariant Borel comeager set $C \subseteq X$ with $E_G^X|C$ hyperfinite. Such a strong result is false for measure instead of category (see, e.g., again Kechris [91], §2).

The strongest known result concerning Weiss' problem to date is the following:

Theorem 4.8 (Weiss for $\mathbb{Z}^n$ (unpublished), Jackson-Kechris-Louveau [00] in general). *If G is a finitely generated group of polynomial growth, then every E_G^X is hyperfinite.*

There are, however, countable groups inducing non-hyperfinite equivalence relations, i.e., there are non-hyperfinite countable Borel equivalence relations. For example, any free action ($g \cdot x \neq x$, if $g \neq 1$) of the free group F_2 of two generators which has an invariant probability Borel measure induces a non-hyperfinite equivalence relation (see, e.g., Kechris [91], §2).

Concerning now general countable Borel equivalence relations we have the following fact:

Proposition 4.9 (Dougherty-Jackson-Kechris [94]). *There exists a universal countable Borel equivalence relation E_∞, i.e., $E \leq_B E_\infty$ for any countable Borel equivalence relation E.*

This is clearly uniquely determined up to $\sim_B$. It is not hyperfinite by our previous remarks and the fact that if F is hyperfinite and $E \leq_B F$, E is also hyperfinite. Thus $E_0 <_B E_\infty$.

One of the standard realizations of E_∞ is the following:

Consider the shift-action of F_2 on 2^{F_2} $(g \cdot p(h) = p(g^{-1}h))$ and denote by $E(F_2, 2)$ the corresponding orbit equivalence relation. Then $E(F_2, 2) \sim_B E_\infty$ (see Dougherty-Jackson-Kechris [94, 1.8]).

It follows from 2.5 that all non-concretely classifiable countable Borel E fall in the interval

$$E_0 \leq_B E \leq_B E_\infty.$$

It is known that there are strictly intermediate relations between E_0 and E_∞. One of the most interesting ones is $E^*(F_2, 2)$, the restriction of $E(F_2, 2)$ to the free part of 2^{F_2}, i.e., the set of $p \in 2^{F_2}$ for which $g \cdot p \neq p$ for any $g \neq 1$. This is denoted by $E_{\infty T}$, up to $\sim_B$ equivalence, because it has the following universality property: Call a countable Borel equivalence E on X relation *treeable* if there is a Borel acyclic graph on X whose connected components are the E-equivalence classes. Thus restricted on each equivalence class, this becomes a *tree*, i.e., a connected and acyclic graph. It is not hard to see that hyperfinite $\Rightarrow$ treeable and that $E^*(F_2, 2)$ is treeable (use the Cayley graph of F_2). It turns out now that $E^*(F_2, 2)$ is the universal treeable Borel equivalence relation, i.e., $E \leq_B E^*(F_2, 2)$ for any treeable Borel E. We now have

$$E_0 <_B E_{\infty T} <_B E_\infty.$$

These facts are proved in Jackson-Kechris-Louveau [00] and are based on results of Adams [88] in ergodic theory.

In general the structure of $\leq_B$ in the interval $[E_0, E_\infty]$ remains mysterious. For example, it is still open whether there are countable Borel equivalence relations which are incomparable with respect to $\leq_B$, although it should be safe to conjecture that they do exist. (This has now been proved; see Adams-Kechris [00].)

5

Actions of Locally Compact Groups

The next most complex class of Polish groups are the locally compact ones, whose actions have long been studied in ergodic theory, e.g., in the case of Lie groups. It can be seen again that for G locally compact every E_G^X is Borel (see Kechris [95, 35.49]).

We now have the following main result:

Theorem 5.1 (Kechris [92]). *Let G be a locally compact group and X a Borel G-space. Then E_G^X has a complete discrete section, i.e., there is a Borel set $S \subseteq X$ meeting every orbit and there is nbhd U of the identity $1 \in G$ such that for $x \in S$, $U \cdot x \cap S = \{x\}$.*

In particular, the intersection of S with every orbit is countable and $E_G^X|S$ is a countable Borel equivalence relation on S, with $E_G^X \sim_B E_G^X|S$. Thus we have:

Corollary 5.2. *If G is Polish locally compact and X is a Borel G-space, then $E_G^X \sim_B E$, for a countable Borel equivalence relation E.*

Thus we see that orbit equivalence relations of Borel actions of locally compact groups fall, for our purposes, within the context of Section 4.

Remark. Theorem 5.1 is a descriptive strengthening of results of Ambrose [41] and Feldman-Hahn-Moore [79] in the measure theoretic context. The case $G = \mathbb{R}$ of 5.1 was earlier proved by Wagh [88].

We will now discuss an interesting application of Theorem 5.1 and the ideas explained in Section 4 to determine the precise complexity of the classification of Riemann surfaces, i.e., one-dimensional complex manifolds, and domains (open connected sets) in $\mathbb{C}$ (which are special examples of Riemann surfaces) up to isomorphism, i.e., *conformal equivalence*. This result is due to Hjorth-Kechris [00].

One can parametrize in a standard way Riemann surfaces so that the parameter space, call it R, is a Polish space. Every element $r \in R$ represents a Riemann surface S_r and for each Riemann surface S there is $r \in R$ with $S_r = S$. Let $\cong_R$ be the equivalence relation of isomorphism on the parameter space R, i.e., $r \cong_R r'$ iff $S_r, S_{r'}$, are isomorphic. Similarly one can define a parameter space D for domains in $\mathbb{C}$, where each $d \in D$ corresponds canonically to a domain D_d and any domain is of the form D_d for some $d \in D$ and we let $d \cong_D d'$ iff D_d, D_d are isomorphic.

Using the uniformization theory for Riemann surfaces, it can be shown that $\cong_R, \cong_D$ are $\sim_B$ to equivalence relations induced by Borel actions of locally compact groups. In particular, they are $\sim_B$ to countable Borel equivalence relations, so $\cong_R, \cong_D \leq_B E_\infty$. In fact, the following result computes the exact complexity of $\cong_R, \cong_D$.

Theorem 5.3 (Hjorth-Kechris [00]).

$$(\cong_R) \sim_B (\cong_D) \sim_B E_\infty.$$

Looking at this problem was motivated by the work of Becker-Henson-Rubel [80] on conformal invariants for domains, and recent correspondence with Ward Henson, who prompted a rethinking of the issues discussed in that paper in the context of the theory explained here.

Theorem 5.3 implies for example that $\cong_D$ is not Ulm-classifiable. It also solves problem Q10 raised in Becker-Henson-Rubel [80, p. 176], by showing that it is indeed possible to assign a complete system of conformal invariants which take the form of countable subsets of $\mathbb{C}$.

6

Actions of the Infinite Symmetric Group

We denote by S_∞ the infinite symmetric group, i.e., the group of permutations of $\mathbb{N}$ with the pointwise convergence topology in which it is a Polish group. Actions of this group and its closed subgroups are of particular interest to logicians in view of the following facts (which can all be found, for example, in Becker-Kechris [96]).

Consider a countable language $L = \{f, g, \ldots, R, S, \ldots\}$ consisting of function symbols $f, g, \ldots$ and relation symbols $R, S, \ldots$, and countable structures

$$\mathcal{A} = < A, f^{\mathcal{A}}, g^{\mathcal{A}}, \ldots, R^{\mathcal{A}}, S^{\mathcal{A}}, \ldots >$$

for L. Since we will be mainly interested in infinite structures (i.e., A infinite) we will assume that $A = \mathbb{N}$. Then the automorphism group, $Aut(\mathcal{A})$, of $\mathcal{A}$ is a closed subgroup of S_∞ and every closed subgroup of S_∞ is of the form $Aut(\mathcal{A})$.

Next we can form the usual space of all countable L-structures (with universe $\mathbb{N}$), which we denote by X_L. For example, if $L = \{f, R\}$, with f k-ary and R m-ary, then $X_L = \mathbb{N}^{(\mathbb{N}^k)} \times 2^{(\mathbb{N}^m)}$. This is clearly a Polish space. There is a canonical continuous action of S_∞ on X_L, called the *logic action*. An element $g \in S_\infty$ acts on $\mathcal{A}$ by simply replacing $\mathcal{A}$ by its isomorphic copy $g \cdot \mathcal{A}$, obtained by applying g. We thus have a Polish S_∞-space whose associated orbit equivalence relation is clearly the isomorphism relation $\cong$ between structures.

If σ is a sentence in the infinitary language $L_{\omega_1 \omega}$, obtained by extending first-order logic by allowing countable conjunctions and disjunctions, then by $Mod(\sigma)$ we denote the set of all $\mathcal{A} \in X_L$ which satisfy σ, i.e.,

$$Mod(\sigma) = \{\mathcal{A} \in X_L : \mathcal{A} \models \sigma\}.$$

133

Then $Mod(\sigma)$ is an isomorphism invariant Borel subset of X_L and by a theorem of Lopez-Escobar every such set is of the form $Mod(\sigma)$. We denote the isomorphism relation restricted to $Mod(\sigma)$ by $\cong_\sigma$, i.e.,

$$\mathcal{A} \cong_\sigma \mathcal{B} \Leftrightarrow \mathcal{A}, \mathcal{B} \in Mod(\sigma) \ \& \ \mathcal{A} \cong \mathcal{B}.$$

We now have

Theorem 6.1 (Becker-Kechris [96]). *Let L be any language containing relation symbols of unbounded arity. Then the logic action on X_L is universal for all the Borel S_∞-actions, i.e., if X is a Borel S_∞-space there is a Borel injection $\pi : X \to X_L$ preserving the action: $\pi(g \cdot x) = g \cdot \pi(x)$.*

In particular, every Borel S_∞-space is Borel isomorphic to the logic action on some $Mod(\sigma)$, $\sigma \in L_{\omega_1\omega}$, and any given $E^X_{S_\infty}$ is Borel isomorphic to $\cong_\sigma$, for some such σ.

Since the regular representation shows that every countable group is a closed subgroup of S_∞ and since, in general, for any closed subgroup G of a Polish group H and any Borel G-space X, there is a Borel H-space Y with $E^X_G \sim_B E^Y_H$ (Mackey; see Becker-Kechris [96,2.3.5]), it follows that every countable Borel equivalence relation is $\sim_B$ ($\cong_\sigma$) for some $\sigma \in L_{\omega_1\omega}$. Thus the orbit equivalence relations induced by Borel S_∞-spaces are more general than those discussed in the previous two sections.

In general $\cong_\sigma$ is analytic but not Borel. For example if $\gamma =$ axioms for graphs, then $\cong_\gamma$ is not Borel. Moreover $\cong_\gamma$ has the following universality property:

$$(\cong_\sigma) \leq_B (\cong_\gamma),$$

i.e., $\cong_\gamma$ is universal among all orbit equivalence relations induced by Borel S_∞-spaces.

In the recent papers Hjorth-Kechris [96] and Hjorth-Kechris-Louveau [98], the case when $\cong_\sigma$ is Borel was studied in some detail. This case occurs often in practice. It was shown in these papers that the descriptive complexity of $\cong_\sigma$ essentially determines the types of complete invariants for $\cong_\sigma$. The appropriate notion of descriptive complexity here is that of the potential class of $\cong_\sigma$.

Definition 6.2. Let Γ be a Borel class such as $\boldsymbol{\Sigma}^0_1$ (open), $\boldsymbol{\Pi}^0_1$ (closed), $\boldsymbol{\Sigma}^0_2$ (F_σ), $\boldsymbol{\Pi}^0_2$ (G_δ), $\boldsymbol{\Sigma}^0_3$ ($G_{\delta\sigma}$), $\boldsymbol{\Pi}^0_3$ ($F_{\sigma\delta}$), etc. We say that an equivalence relation E on a Polish space X is of *potential class* Γ if $E \leq_B F$, where F is in the class Γ.

The following results are part of the general analysis carried out in the above two papers (except for the first two which are earlier and actually hold for arbitrary Borel equivalence relations; this is easy for (i) and was shown in Harrington-Kechris-Louveau [90] for (ii)).

Theorem 6.3. *Let $\sigma \in L_{\omega_1\omega}$. Then we have*

(i) *(folklore)* $\cong_\sigma$ *is potentially* $\mathbf{\Sigma}_1^0$ *iff it has countably many equivalence classes.*

(ii) *(Burgess [79])* $\cong_\sigma$ *is potentially* $\mathbf{\Pi}_2^0$ *iff* $\cong_\sigma$ *is potentially* $\mathbf{\Pi}_1^0$ *iff cong$_\sigma$ is concretely classifiable.*

(iii) *(Hjorth-Kechris [96], Hjorth-Kechris-Louveau [98])* $\cong_\sigma$ *is potentially* $\mathbf{\Sigma}_3^0$ *iff* $\cong_\sigma$ *is potentially* $\mathbf{\Sigma}_2^0$ *iff* $(\cong_\sigma) \leq_B E_\infty$.

(iv) *(Hjorth-Kechris-Louveau [98])* $\cong_\sigma$ *is potentially* $\mathbf{\Pi}_3^0$ *iff* $(\cong_\sigma) \leq_B E_{\mathrm{ctble}}$ *where E_{ctble} is the equivalence relation on $\mathbb{R}^{\mathbb{N}}$ given by*

$$(x_n) E_{\mathrm{ctble}}(y_n) \Leftrightarrow \{x_n : n \in \mathbb{N}\} = \{y_n : n \in \mathbb{N}\},$$

so that $\mathbb{R}^{\mathbb{N}}/E_{\mathrm{ctble}}$ is canonically identified with $\{A \subseteq \mathbb{R} : A$ is nonempty countable$\}$. (Of course $\mathbb{R}$ can be replaced here by any uncountable Polish space.)

It also turns out that $E_\infty <_B E_{\mathrm{ctble}}$, so we have

$$\mathbb{N} <_B \mathbb{R} <_B E_0 <_B E_\infty <_B E_{\mathrm{ctble}},$$

and this gives a clear distinction between the case where an equivalence relation is $\leq_B E_\infty$, i.e., one can assign invariants which are equivalence classes of a countable Borel equivalence relation, and the case where it is $\leq_B E_{\mathrm{ctble}}$, where the invariants are arbitrary countable subsets of some Polish space.

To summarize, if $\cong_\sigma$ is potentially $\mathbf{\Sigma}_1^0$ the invariants are integers, if $\cong_\sigma$ is potentially $\mathbf{\Pi}_2^0$ the invariants are reals, if $\cong_\sigma$ is potentially $\mathbf{\Sigma}_3^0$ the invariants are equivalence classes of countable Borel equivalence relations, thus special kinds of countable sets of reals, and if $\cong_\sigma$ is potentially $\mathbf{\Pi}_3^0$ the invariants are arbitrary countable sets of reals. (As shown in Hjorth-Kechris-Louveau [98], this picture can be continued throughout the Borel hierarchy with the invariants climbing up to countable sets of countable sets of reals, countable sets of countable sets of countable sets of reals, etc.)

We next discuss some specific examples and problems in order to illustrate this hierarchy.

(i) If σ is the theory of one equivalence relation or a unary injective function, then $\cong_\sigma$ is concretely classifiable.

(ii) As explained in Hjorth-Kechris [96], if σ is a theory whose models have "finite rank" in some sense, $\cong_\sigma$ is often potentially $\mathbf{\Sigma}_2^0$, so $\leq_B E_\infty$. Examples include $\gamma_{cf} = $ the theory of connected locally finite graphs (*locally finite* means that every vertex has only finitely many neighbors), $\tau_{cf} = $ the theory of locally finite trees (connected acyclic graphs), $\alpha_n = $ the theory of torsion-free abelian groups of rank $\leq n$ (i.e., subgroups of $\mathbb{Q}^n$). As it turns out we actually have:

$$(\cong_{\gamma_{cf}}) \sim_B (\cong_{\tau_{cf}}) \sim_B E_\infty.$$

If on the other hand $\tau_{cf}^* = $ the theory of rigid locally finite trees, then

$$(\cong_{\tau_{cf}^*}) \sim_B E_\infty T \; (<_B E_\infty).$$

Also it turns out that, by using a classical classification theorem,

$$(\cong_{\alpha_1}) \sim_B E_0$$

and it has been conjectured in Hjorth-Kechris [96] that

$$(\cong_{\alpha_n}) \sim_B E_\infty, \quad \text{if } n \geq 2,$$

but this is still open. For a discussion of the relevance of this conjecture to the classical classification problem of torsion-free abelian groups of rank ≥ 2, see again Hjorth-Kechris [96].

(iii) If $\gamma_f = $ the theory of locally finite graphs (not necessarily connected), $\varepsilon_p = $ the theory of two equivalence relations E_1, E_2 with $E_1 \subseteq E_2$, $\rho = $ the theory of infinitely many unary relations, then it turns out that

$$(\cong_{\gamma_f}) \sim_B (\cong_{\varepsilon_p}) \sim_B (\cong_\rho) \sim_B E_{\text{ctble}}.$$

7

Turbulence I: Overview

Let E be an equivalence relation on X. We say that E *admits classification by countable structures* if there is a language L and a Borel map $f : X \to X_L$ which assigns to each $x \in X$ a countable L-structure $f(x)$ (with universe $\mathbb{N}$) such that

$$x E y \Leftrightarrow f(x) \cong f(x).$$

Equivalently, by 6.1, this means that

$$E \leq_B E^Y_{S_\infty},$$

for some Borel S_∞-space Y.

For example, if $E \leq_B E_{\text{ctble}}$, i.e., E can be classified by invariants which are countable sets of reals, then E admits classification by countable structure, but this notion is much more extensive.

Here are some examples:

(i) Let $Aut(I, \lambda)$ be the Polish group of measure preserving automorphisms of the unit interval. A classical problem of ergodic theory is to classify $T \in Aut(I, \lambda)$ up to conjugacy by invariants. The well-known theorem of Ornstein solves this in the case of the Bernoulli automorphisms, where a complete invariant is the entropy, a real number, so we have in this case a concrete classification. Another standard (and earlier) result is the Halmos-von Neumann Theorem, which classifies discrete spectrum ergodic T by the following complete invariant

$$\sigma_p(T) = \{\lambda \in \mathbb{T} : \lambda \text{ is an eigenvalue of } T\}$$

(i.e., the point spectrum of T). Thus conjugacy of discrete spectrum measure preserving automorphisms admits classification by countable structures.

(A reference for all this is, for example, Walters [82].)

(ii) Consider minimal (i.e., having dense orbits) homeomorphisms of the Cantor space, and the following equivalence relation on them

$$fEg \Leftrightarrow f, g \text{ are orbit equivalent}$$
$$\Leftrightarrow \exists \text{ a homeomorphism } h \text{ of the Cantor set}$$
$$\text{mapping the orbits of } f \text{ onto the orbits of } g.$$

In Giordano-Putnam-Skau [95] it is shown how to assign to each such f of a canonical countable abelian (partially) ordered group with distinguished ordered unit, $\mathcal{A}_f$, such that $fEg \Leftrightarrow \mathcal{A}_f \cong \mathcal{A}_g$, and moreover $f \mapsto \mathcal{A}_f$ is of course Borel. Thus E admits classification by countable structures.

(iii) The conjugacy relation on the Polish group of increasing homeomorphisms of the unit interval I admits classification by countable structures (see Hjorth [00]).

(iv) Any orbit equivalence relation induced by a Borel G-space, where G is a product of countably many Polish locally compact groups, admits classification by countable structures (Hjorth [00]).

We now consider the following general question: Given a Polish G-space X, when does E_G^X admit classification by countable structures? The following theory has been recently developed by Hjorth [00] to address this question.

Definition 7.1. Let G be a Polish group and X a Polish G-space. Fix an open set $U \subseteq X$ and a symmetric open nbhd V of the identity 1 of G. The (U, V)-*local graph* is the following symmetric, reflexive relation on U:

$$xR_{U,V}y \Leftrightarrow x, y \in U \ \& \ \exists g \in V(g \cdot x = y).$$

The (U, V)-*local orbit* of $x \in U$, $\mathcal{O}(x, U, V)$, is the connected component of x in this graph. (If $U = X$, $V = G$, $\mathcal{O}(x, U, V) = G \cdot x =$ the orbit of x.)

The Polish G-space X (or the corresponding action) is *turbulent* if every orbit is dense, meager and every local orbit is somewhere dense (i.e., its closure has non-empty interior).

Here are some examples:

(i) (Hjorth) Let $G \subseteq \mathbb{R}^{\mathbb{N}}$ be a proper subgroup containing $\mathbb{R}^{<\mathbb{N}}$. Suppose G is Polishable, i.e., G is a Borel subgroup of $\mathbb{R}^{\mathbb{N}}$ which is Borel isomorphic to a Polish group. Then the translation action of G on $\mathbb{R}^{\mathbb{N}}$ is turbulent. Examples of such G are ℓ^p ($1 \le p < \infty$) and c_0.

(ii) (Hjorth) Every infinite-dimensional separable Banach space X (with addition) has a turbulent action.

(iii) (Hjorth-Kechris) Every closed subgroup of a countable product of S_∞ and Polish locally compact groups does *not* have turbulent actions.

Definition 7.2. Let G be a Polish group and X a Polish G-space. We say that X is *generically turbulent* if its restriction to an invariant dense G_δ set is turbulent.

We now have the following result.

Theorem 7.3 (Hjorth [00]). *Let G be a Polish group, X a Polish G-space and assume that every orbit is meager and some orbit is dense. Then the following are equivalent:*

(i) *X is generically turbulent.*

(ii) *For any Borel S_∞-space Y and any Baire measurable $f : X \to Y$ which is invariant, i.e., $xE_G^X y \Rightarrow f(x)E_{S_\infty}^Y y$, there is a dense G_δ set $C \subseteq X$ which is mapped by f to a single S_∞-orbit.*

Corollary 7.4. *If X is a generically turbulent G-space and $E_G^X \le_B E$, then E does not admit classification by countable structures.*

One has also the following (strong) converse of 7.4, at least for "nice" Polish groups, which shows that turbulence is intrinsically connected with the problem of classification by countable structures. We will omit below the technical definition of a GE group (see Section 13). Suffice to say that countable products of locally compact groups, abelian, nilpotent, admitting an invariant metric, Polish groups are GE.

Theorem 7.5 (Hjorth [00]). *Let G be a GE Polish group. Then for each Polish G-space exactly one of the following holds:*

(i) *E_G^X admits classification by countable structures.*

(ii) *There is a turbulent Polish G-space Y and a continuous G-embedding $\pi : Y \to X$ (so that in particular $E_G^Y \le_B E_G^X$).*

We now present some applications:

(i) (Hjorth [00]) Conjugacy in the homeomorphism group $H(I^2)$ of the unit square does not admit classification by countable structures.

(ii) (Hjorth [97]) Conjugacy of ergodic measure preserving automorphism on the unit interval I (with Lebesgue measure λ) does not admit classification by countable structures.

(iii) (Hjorth-Kechris [00]) Isomorphism (i.e., biholomorphic equivalence) of two-dimensional complex manifolds does not admit classification by countable structures.

(iv) (Kechris-Sofronidis [01]) Unitary equivalence of unitary operators does not admit classification by countable structures. Similarly for equivalence ($\sim$) of Borel probability measures on any uncountable Polish space.

We will sketch a proof of the result about measure equivalence to illustrate the methods used in such proofs. (It then follows for unitary equivalence, since it is $\sim_B$ to measure equivalence; see Section 1.) Since any two uncountable Polish spaces are Borel isomorphic, it is enough to work with measures on the Cantor space $2^{\mathbb{N}}$.

Fix a sequence $1 \geq a_n \geq 0$ with $a_n \to 0$, and $\sum a_n^2 = \infty$. Define the following ideal $\mathcal{J}$ on $\mathbb{N}$:

$$A \in \mathcal{J} \Leftrightarrow A \subseteq \mathbb{N} \ \& \ \sum_{n \in A} a_n^2 < \infty.$$

Clearly $\mathcal{J}$ is Borel. Obviously $\mathcal{J}$ is also a subgroup of the Cantor group $(\mathcal{P}(\mathbb{N}), \Delta)$, when Δ denotes symmetric difference. Moreover $(\mathcal{J}, \Delta)$ is Polishable, i.e., there is a (unique) Polish group topology on $\mathcal{J}$ inducing its Borel structure (as a Borel subset of $\mathcal{P}(\mathbb{N})$). This topology is given by the complete metric

$$d(A, B) = \sum_{n \in A \Delta B} a_n^2.$$

It is not hard to check that the action of $\mathcal{J}$ by translation on $(\mathcal{P}(\mathbb{N}), \Delta)$ is turbulent, thus the corresponding equivalence relation

$$A E_{\mathcal{J}} B \Leftrightarrow (A \Delta B) \in \mathcal{J} \Leftrightarrow \sum_{n \in A \Delta B} a_n^2 < \infty$$

does not admit classification by countable structures. (For more general results about ideals and their actions on $(\mathcal{P}(\mathbb{N}), \Delta)$ by translation, in relation to turbulence, see Kechris [98].)

Now define $f : \mathcal{P}(\mathbb{N}) \to [0,1]^{\mathbb{N}}$ by

$$f(A)(n) = \begin{cases} 0, \text{if } n \notin A; \\ a_n, \text{if } n \in A. \end{cases}$$

Clearly f is continuous and

$$A E_{\mathfrak{I}} B \Leftrightarrow f(A) - f(B) \in \ell^2.$$

For $(\alpha_n) \in (0,1)^{\mathbb{N}}$, let $\mu_{(\alpha_n)}$ be the product measure on $2^{\mathbb{N}}$ for which the nth coordinate is the $(a_n, 1 - a_n)$ measure on $\{0,1\}$. Then by a theorem of Kakutani (see, e.g., Hewitt-Stromberg [69], p. 456) we have that if $\delta \leq \alpha_n, \beta_n \leq 1 - \delta$ for some $\delta > 0$, then

$$\mu_{(\alpha_n)} \sim \mu_{(\beta_n)} \Leftrightarrow \sum_{n=0}^{\infty} (\alpha_n - \beta_n)^2 < \infty$$

$$\Leftrightarrow (\alpha_n) - (\beta_n) \in \ell^2.$$

Now given $x = (x_n) \in [0,1]^{\mathbb{N}}$, let $\bar{x}_n = \frac{1+x_n}{4}$, so that $\frac{1}{4} \leq \bar{x}_n \leq \frac{1}{2}$, and put $g(x) = \mu_{(\bar{x}_n)}$. Then $x - y \in \ell^2 \Leftrightarrow \bar{x} - \bar{y} \in \ell^2 \Leftrightarrow g(x) \sim g(y)$, so we have for $h(A) = g(f(A))$

$$A E_{\mathfrak{I}} B \Leftrightarrow f(A) - f(B) \in \ell^2$$

$$\Leftrightarrow \mu_{h(A)} \sim \mu_{h(B)},$$

so, as $A \mapsto \mu_{h(A)}$ is Borel, we have

$$E_{\mathfrak{I}} \leq_B (\sim)$$

and $\sim$ does not admit classification by countable structures.

As with the argument given for the last example, all these results are obtained by embedding (in the sense of $\leq_B$) the orbit equivalence relation of a turbulent action into the relevant equivalence relation and then using 7.4. In view though of the strong general ergodicity property revealed in Theorem 7.3, it is also of interest to show that the conjugacy action on various Polish groups like $Aut(I, \lambda)$, $U(H)$, $H^*(I^2)$ (the group of homeomorphisms of I^2 fixing the boundary) is itself generically turbulent. This is still open for the groups $Aut(I, \lambda)$, $H^*(I^2)$ but the following has been established recently:

Theorem 7.6 (Kechris-Sofronidis [01]). *The conjugation action on the unitary group $U(H)$ is generically turbulent.*

Thus any reasonable assignment of countable structures to unitary operators, which is conjugacy invariant, is fixed up to isomorphism on a dense G_δ set. (An example of such an invariant is the point spectrum, which is $\emptyset$ on a dense G_δ set.)

A similar result has been proved for measure equivalence.

Theorem 7.7 (Kechris-Sofronidis [01]). *Let X be an uncountable compact metric space, $P(X)$ the compact metric space of probability Borel measures on X with the weak*-topology and let $\sim$ be measure equivalence. If $f : P(X) \to X_L$, L some countable language, is Baire measurable and invariant, i.e., $\mu \sim \nu \Rightarrow f(\mu) \cong f(\nu)$, then f maps an $\sim$-invariant dense G_δ set into a single isomorphism class.*

8
Turbulence II: Basic Facts

The rest of this paper will be devoted to an exposition of the basic theory of turbulence, developed by Hjorth [00]. All the main ideas and results, unless otherwise stated, are due to Hjorth.

Below we fix a Polish group G and a Polish G-space X. Throughout we let U (with various embellishments) vary over nonempty open sets in X and V (with various embellishments) vary over symmetric open nbhds of $1 \in G$.

Definition 8.1. The (U, V)-*local graph* is the following symmetric, reflexive relation on U:

$$x R_{U,V} y \Leftrightarrow x, y \in U \ \& \ \exists g \in V(g \cdot x = y).$$

The (U, V)-*local orbit* of $x \in U$, in symbols

$$\mathcal{O}(x, U, V),$$

is the connected component of x in the (U, V)-local graph. Equivalently, if we define the equivalence relation $\sim_{U,V}$ on U, by

$$x \sim_{U,V} y \Leftrightarrow$$
$$\exists g_0, g_1, \dots, g_k \in V(x_0 = x \ \& x_{i+1} = g_i \cdot x_i \ \& \ x_{k+1} = y \ \& \ x_i \in U),$$

then $\mathcal{O}(x, U, V) = [x]_{\sim_{U,V}}$.

Notice that if $U = X$, $V = G$, then $\mathcal{O}(x, U, V) = G \cdot x =$ the orbit of x.

Definition 8.2. A point $x \in X$ is *turbulent* if for every U with $x \in U$ and every V we have that $\mathcal{O}(x, U, V)$ is somewhere dense, i.e., $\overline{\mathcal{O}(x, U, V)}$ has nonempty interior.

Let

$$T = \{x \in X : x \text{ is turbulent}\}.$$

We will first check that T is a G-invariant set. This follows immediately from the following simple lemma:

Lemma 8.3. $g \cdot \mathcal{O}(x, U, V) = \mathcal{O}(g \cdot x, g \cdot U, gVg^{-1})$.

Thus we can talk about *turbulent orbits*.

Next we provide a couple of equivalent characterizations of turbulent points.

Proposition 8.4. *The following are equivalent:*

(i) x *is turbulent.*

(ii) *For every U containing x and every V, there is $U' \subseteq U$ containing x so that $U' \subseteq \overline{\mathcal{O}(x, U', V)}$.*

(iii) *For every U containing x and every V, $x \in \mathrm{Int}(\overline{\mathcal{O}(x, U, V)})$.*

Proof. Let $U' = \mathrm{Int}(\overline{\mathcal{O}(x, U, V)}) \cap U$. Then U' is $\sim_{U,V}$-invariant, so if $U' \neq \emptyset$, i.e., $\mathrm{Int}(\overline{\mathcal{O}(x, U, V)}) \neq \emptyset$, we have that $\mathcal{O}(x, U, V) \subseteq U'$, $\mathcal{O}(x, U, V)$ is dense in U' and $\mathcal{O}(x, U', V) = \mathcal{O}(x, U, V)$. From this the equivalence of (i)-(iii) is clear. ⊣

The next result shows that if one dense turbulent orbit exists then there are actually comeager many turbulent orbits.

Proposition 8.5. *Assume there is a dense turbulent orbit. Then T is comeager.*

Proof. Put

$$T_{U,V} = \{x \in U : \exists U' \subseteq U, x \in U'(U' \subseteq \overline{\mathcal{O}(x, U', V)})\}.$$

If we fix a countable open basis $\mathcal{B}$ for X and a countable basis $\mathcal{N}$ of symmetric open nbhds of $1 \in G$ we see easily, using 8.4, that

$$x \notin T \Leftrightarrow \exists U \in \mathcal{B} \exists V \in \mathcal{N}(x \in U \ \& \ x \notin T_{U,V}),$$

so it is enough to show that $T_{U,V}$ is comeager on U.

Let

$$S_{U,V} = \{U' \subseteq U : \exists x \in U'(U' \subseteq \overline{\mathcal{O}(x, U', V)})\}.$$

We will show that the union of the members of $S_{U,V}$ is dense (and of course) open in U and that $T_{U,V}$ is comeager in each $U' \in S_{U,V}$. It follows (see, e.g., Kechris [95, 8.29]) that $T_{U,V}$ is comeager.

The first claim is easy, since $T \cap U$ is dense in U (as T is dense) and every element of $T \cap U$ is in $\bigcup S_{U,V}$, by 8.4 (ii).

For the second claim, fix $U' \in S_{U,V}$. Then an easy computation shows that $A = \{x \in U' : U' \subseteq \overline{\mathcal{O}(x, U', V)}\}$ is G_δ. Moreover, if $x \in A$ (which exists as $U' \in S_{U,V}$), then $\mathcal{O}(x, U', V) \subseteq A$, so A is dense in U', thus A is comeager in U'. But clearly $A \subseteq T_{U,V}$, so $T_{U,V}$ is comeager in U'. ⊣

The preceding fact is analogous to the standard observation that if there is a dense orbit, the set of points with dense orbit is comeager. This is simply because the set of points with dense orbit is G_δ. However, we do not know if the set of turbulent points is G_δ.

Definition 8.6. The Polish G-space X (or the action) is called *turbulent* if every orbit is dense, meager and turbulent. It is called *generically turbulent* if its restriction to some invariant dense G_δ set $Y \subseteq X$ is turbulent.

Note the following equivalences:

Proposition 8.7. *The following are equivalent for each Polish G-space X:*

(i) *X is generically turbulent.*

(ii) *There is a dense, turbulent orbit and every orbit is meager.*

(iii) *The set of dense, turbulent and meager orbits is comeager (i.e., the set of points whose orbits have these properties is comeager).*

Proof. (i)⇒(ii) is clear since, in the presence of a dense orbit, an orbit which is non-meager must be comeager (see Kechris [95, 8.46]).

(ii)⇒(iii). By 8.5 and the remark preceding 8.6, the set of points with dense, turbulent orbits must be comeager.

(iii)⇒(i). Let $C \subseteq X$ be the set of points whose orbits are dense, turbulent and meager. Then there is a dense G_δ set $A \subseteq C$. Let A^* be the Vaught transform of A, i.e., $A^* = \{x \in X : \forall^* g(g \cdot x \in A)\}$ (see Kechris [95, 16.B]). Then A^* is a dense G_δ invariant subset of C, and clearly the action restricted to A^* in turbulent. ⊣

9

Turbulence III: Induced Actions

Let G, H be Polish groups and $\pi : H \to G$ a continuous homomorphism of H *onto* G. Then every Polish G-space X gives rise canonically to a Polish H-space on X by defining

$$h \cdot x = \pi(h) \cdot x.$$

It is clear that the orbits of any $x \in X$ in these two actions are the same. Moreover, using subscripts to indicate which action we are considering, we have the following:

$$\mathcal{O}_G(x, U, V) = \mathcal{O}_H(x, U, \pi^{-1}(V))$$

and

$$\mathcal{O}_H(x, U, V) = \mathcal{O}_G(x, U, \pi(V)),$$

where we implicitly use the fact that π is an open mapping to justify this notation (see Becker-Kechris [96, 1.2.6]). Thus the set of local orbits of x in these two actions is exactly the same and so the concepts of turbulence at a point, the whole space, or generically, coincide for the two actions.

Thus turbulence is preserved when we go upwards from a Polish group G to any Polish group for which G is a quotient of H. We will next consider such a preservation in the case when G is a closed subgroup of H.

Let G, H be a Polish groups with $G \subseteq H$ a closed subgroup of H. Let X be a Polish G-space. Then there is a canonical "minimal" way to extend the Polish G-space X to a Polish H-space X due to Mackey and called the *induced action*.

This is defined as follows: Consider G acting by left-multiplication on H and let $X \times H$ be the product G-space: $g \cdot (x, h) = (g \cdot x, gh)$. Let

$(X \times H)/G$ be the orbit space of this action with the quotient topology. This is a Polish space. Let H act on $(X \times H)/G$ by $h \cdot [x, h'] = [x, h'h^{-1}]$, where $[x, h'] =$ the orbit of (x, h') in $X \times H$. Identifying $x \in X$ with $[x, 1]$ makes X a closed subset of $(X \times H)/G$. Moreover $(X \times H)/G$ is a Polish H-space, the G-action on X is the same as the restriction of the H-action to G on $X \subseteq (X \times H)/G$ and every orbit of H on $(X \times H)/G$ contains exactly one orbit of G on X. It is customary to denote the H-space $(X \times H)/G$ by $X \times_G H$ and call it the *induced H-space of the G-space* X. (Mackey originally defined this for Borel G-spaces and the above analog for Polish G-spaces has been worked out by Hjorth; see Becker-Kechris [96, 2.3.5].)

We now prove that turbulence is preserved under induced actions.

Theorem 9.1 (Kechris). *Let G, H be Polish groups with G a closed subgroup of H. Let X a be a Polish G-space and let $X \times_G H$ be the induced H-space. If X is turbulent, so is $X \times_G H$.*

Proof. First we verify that all H-orbits in $X \times_G H$ are dense. Fix U open nonempty in $X \times_G H$. Since every H-orbit contains an element of the form $[x, 1]$, it is enough to find $h \in H$ so that $h \cdot [x, 1] = [x, h^{-1}] \in U$. By considering the projection of $X \times H$ onto $X \times_G H$, this amounts to showing that for any open nonempty G-invariant $U_0 \subseteq X \times H$ and for every x there is $h \in H$ with $(x, h) \in U_0$. Let $\pi_1 : X \times H \to X$ be the first projection. Then $\pi_1(U_0)$ is open nonempty in X so, since every G-orbit in X is dense, we have some $g \in G$ with $g \cdot x \in \pi_1(U_0)$. Let $h' \in H$ be such that $(g \cdot x, h') \in U_0$. Then $g^{-1} \cdot (g \cdot x, h') = (x, g^{-1}h') \in U_0$ and we are done.

Next we check that all H-orbits are meager in $X \times_G H$. If this fails, then $H \cdot [x, 1]$ is not-meager for some $x \in X$, so by Effros' Theorem (see Becker-Kechris [96, 2.2.2]), $H \cdot [x, 1]$ is G_δ in $X \times_G H$ and the map $h/H_{[x,1]} \mapsto h \cdot [x, 1]$ from $H/H_{[x,1]}$ onto $H \cdot [x, 1]$, where $H_{[x,1]}$ is the stabilizer of $[x, 1]$, is a homeomorphism. Now

$$
\begin{aligned}
h \in H_{[x,1]} &\Leftrightarrow h \cdot [x, 1] = [x, 1] \\
&\Leftrightarrow [x, h^{-1}] = [x, 1] \\
&\Leftrightarrow \exists g \in G(g \cdot x = x \;\&\; gh^{-1} = 1) \\
&\Leftrightarrow h \in G_x,
\end{aligned}
$$

where G_x is the stabilizer of x. So the canonical map of H/G_x onto $H \cdot [x, 1]$ is a homeomorphism and thus so is its restriction to G/G_x, which is a closed subset of H/G_x. But the image of this canonical map is $G \cdot x$, so $G \cdot x \subseteq X$ is also G_δ, thus dense G_δ in X, a contradiction.

Finally fix $V \subseteq H$ open symmetric nbhd of 1, $U \subseteq X \times H$ open, $[x, h] \in U$. We will find open nonempty $U' \subseteq U$ so that for each open $W \subseteq X \times_G H$, $W \subseteq U$, we have

$$W \cap U' \neq \emptyset \Rightarrow W \cap \mathcal{O}([x, h], U, V) \neq \emptyset$$

(thus $U' \subseteq \overline{\mathcal{O}([x, h], U, V)}$). By considering the projection map of $X \times H$ onto $X \times_G H$ we view U, U', W as G-invariant open subsets of $X \times H$.

First choose an open symmetric nbhd of 1 in H, say $\tilde{V}$, such that

$$\tilde{V} h^{-1} \tilde{V} h \tilde{V} \subseteq V.$$

Let $\pi_1 : X \times H \to X$ be the first projection. Let $U_1 = \pi_1(U \cap (X \times h\tilde{V}))$. Since $(x, h) \in U$, we have $x \in U_1$. So by applying turbulence in X, we have that $\mathcal{O}(x, U_1, \tilde{V})$ is somewhere dense. So fix open nonempty $U'' \subseteq U_1$, with $U'' \subseteq \mathcal{O}(x, U_1, \tilde{V})$. Let

$$U_0 = \{(y, p) : (y, p) \in U \ \& \ p \in h\tilde{V} \ \& \ y \in U''\},$$

so that $U_0 \subseteq U$ is open nonempty. Let $U' \subseteq U$ be the G-saturation of U_0. We claim that this works.

So fix $W \subseteq X \times H$, $W \subseteq U$, open invariant with $W \cap U' \neq \emptyset$, thus $W \cap U_0 \neq \emptyset$. Let $W_1 = \pi_1(W \cap U_0)$ which is a nonempty open subset of X. Clearly, $W_1 \subseteq U_1$, $W_1 \cap U'' \neq \emptyset$. So $W_1 \cap \mathcal{O}(x, U_1, \tilde{V}) \neq \emptyset$. Let $g_1, \cdots, g_k \in \tilde{V}$ be such that all the points

$$x, g_1 \cdot x, \cdots, g_i g_{i-1} \cdots g_1 \cdot x, \cdots, g_k g_{k-1} \cdots g_1 \cdot x$$

are in U_1 and $g_k g_{k-1} \cdots g_1 \cdot x \in W_1$. So find $h_1, \cdots, h_k \in h\tilde{V}$ so that $(g_1 \cdot x, h_1) \in U, \cdots, (g_i g_{i-1} \cdots g_i \cdot x, h_i) \in U, \cdots, (g_{k-1} \cdots g_1 \cdot x, h_{k-1}) \in U$, $(g_k \cdots g_1 \cdot x, h_k) \in W(\subseteq U)$. Also recall that $(x, h) \in U$. Since both U, W are G-invariant, we also have

$$(x, h), (x, g_1^{-1} h_1), \cdots, (x, g_1^{-1} g_2^{-1} \cdots g_i^{-1} h_i),$$
$$(x, g_1^{-1} \cdots g_{k-1}^{-1} h_{k-1}) \in U,$$
$$(x, g_1^{-1} g_2^{-1} \cdots g_k^{-1} h_k) \in W.$$

Now (putting also $h_0 = h$) $h_i = h\tilde{h}_i$ with $\tilde{h}_i \in \tilde{V}$. Note that if $p_i = g_1^{-1} \cdots g_i^{-1} h_i = g_i^{-1} \cdots g_i^{-1} h\tilde{h}_i$ ($p_0 = h_0 = h$), then

$$p_i \tilde{h}_i^{-1} h^{-1} g_{i+1}^{-1} h\tilde{h}_{i+1} = p_{i+1},$$

and $\tilde{h}_i^{-1}(h^{-1}g_{i+1}^{-1}h)\tilde{h}_{i+1} \in \tilde{V}(h^{-1}\tilde{V}h)\tilde{V} \subseteq V$, so $p_i q_i^{-1} = p_{i+1}$ for some $q_i \in V$. Since

$$q_i \cdot [x, p_i] = [x, p_i q_i^{-1}] = [x, p_{i+1}]$$

and $[x, p_i] \in U$ (by now going to $X \times_G H$) and $[x, p_k] \in W$, we have that $\mathcal{O}([x, h], U, V) \cap W \neq \emptyset$, and the proof is complete. $\dashv$

10

Turbulence IV: Some Examples

We will now discuss some examples of turbulent actions. The first two are due to Hjorth [00].

(i) Let $H =$ the Cantor group $= (\mathcal{P}(\mathbb{N}), \Delta)\, (\cong (\mathbb{Z}_2^{\mathbb{N}}, +))$ and let $G \subseteq H$ be a Polishable subgroup with $\mathrm{FIN} \subseteq G \subsetneq \mathcal{P}(\mathbb{N})$, where $\mathrm{FIN} = \{A \subseteq \mathbb{N} : A$ is finite$\}$. Assume, in the unique Polish topology of G coming from the fact that G is Polishable, that $\{n\} \to \emptyset$ (= the identity of H). Then the translation action of G on H is turbulent.

First since $G \subsetneq \mathcal{P}(\mathbb{N})$, G is meager and so is any orbit, being a translate of G. Also G is dense, as it contains FIN, and so is every orbit. Finally fix $A \in \mathcal{P}(\mathbb{N})$, U an open nbhd of A in $\mathcal{P}(\mathbb{N})$ and V an open nbhd of $\emptyset$ in the Polish topology of G. For some large enough n_0, $x \in U' = \{B \in \mathcal{P}(\mathbb{N}) : B \cap n_0 = A \cap n_0\} \subseteq U$, and for all $n \geq n_0, \{n\} \in V$. We check that $\mathcal{O}(x, U', V)$ is dense in U'. Indeed, fix $B \in U'$. Then for any $m > n_0$ there are $n_0 \leq n_1 < n_2 < \ldots < n_k < m$ with $(A\Delta\{n_1\}\Delta \cdots \Delta\{n_k\}) \cap m = B \cap m$ and $A\Delta\{n_1\}\Delta \cdots \Delta\{n_i\} \in U'$ for all $i \leq m$. So $A\Delta\{n_1\}\Delta \cdots \Delta\{n_k\} \in \mathcal{O}(A, U', V)$ and thus we can approximate as closely as we want B by elements of $\mathcal{O}(A, U', V)$, so we are done.

Remark. In case G is actually an ideal on $\mathbb{N}$ (i.e., is closed under subsets and finite unions) one can actually characterize exactly when the action of G on $\mathcal{P}(\mathbb{N})$ by translation is turbulent; see Kechris [98].

(ii) Let $G \subsetneq \mathbb{R}^{\mathbb{N}}$ be a Polishable subgroup of $\mathbb{R}^{\mathbb{N}}$ which is *strongly dense* in the sense that for every $(x_0, \ldots, x_{n-1}) \in \mathbb{R}^{<\mathbb{N}}$ there is $y \in G$ with $x_i = y_i$ for $i < n$. (Examples of such G include ℓ^p, c_0.) Then it follows that for each n the map $(x_0, x_1, \ldots) \in G \mapsto (x_0, \ldots, x_{n-1}) \in \mathbb{R}^n$ is an onto continuous homomorphism, so it is open. Using this, one can show easily that the action of G on $\mathbb{R}^n$ by translation is turbulent.

This has as a consequence that every infinite-dimensional (say real) separable Banach space, viewed as a Polish group under $+$, has a turbulent action. To see this notice that, by 9.1 and the well-known fact that a separable infinite-dimensional Banach space has an infinite-dimensional closed subspace with a basis, we can assume that X has a basis, say $\{e_n\}$. Then the map

$$x = \sum \alpha_n e_n \mapsto (\alpha_n) \in \mathbb{R}^{\mathbb{N}}$$

is an isomorphism of $(X, +)$ with a Polishable subgroup $G \subseteq \mathbb{R}^{\mathbb{N}}$ and clearly $\mathbb{R}^{<\mathbb{N}} \subseteq G \subsetneqq \mathbb{R}^{\mathbb{N}}$, so we are done.

(iii) Let X be a (real) separable Frechet (i.e., Polish locally convex linear topological) space and let $Y \subsetneqq X$ be a dense linear subspace which is Borel in X and Polishable in the sense that there is a (necessarily unique) Polish topology on Y generating its Borel structure in which Y becomes a topological vector space. (Examples of such pairs include $(\ell^p, \mathbb{R}^{\mathbb{N}})$, $(c_0, \mathbb{R}^{\mathbb{N}})$, $(C([0,1]), L^p([0,1]))$, etc.) Then the action of Y on X by translation is turbulent: First it is clear that the orbits are dense and meager. Now fix $x \in X$, $U \subseteq X$ open nbhd of x and V a symmetric open nbhd of 0 in the topology of Y. Since X is locally convex, let $U_0 \subseteq U$ be a convex open nbhd of 0 such that $U_0 + x \subseteq U$. We will check that $\mathcal{O}(x, U, V)$ is dense in $U_0 + x$. Fix $x' \in U_0 + x$ and an open nbhd W of x' in X. Then find $y' \in Y$ with $y' + x \in W$, $y' \in U_0$. It is enough to show that $y' + x \in \mathcal{O}(x, U, V)$. By convexity $ty' \in U_0$ for $0 \le t \le 1$ and, as $t \mapsto ty'$ is continuous from $[0, 1]$ into Y (with its Polish topology), there is $\delta > 0$ so that $t < \delta \Rightarrow ty' \in V$. So if we choose $0 = t_0 < t_1 < \ldots < t_{n-1} < t_n = 1$ with $t_{i+1} - t_i < \delta$, and $y_i = (t_i - t_{i-1})y'$ $(1 \le i \le n)$, we have that $y_i \in V$, $y_1 + \ldots + y_i + x = t_i y' + x \in U_0 + x \subseteq U$ and $y_1 + \ldots + y_n + x = y' + x$, so $y' + x \in \mathcal{O}(x, U, V)$.

(iv) (Kechris-Sofronidis [01]) The conjugation action of $U(H)$ is generically turbulent.

11

Turbulence V: Calmness

We will see here that certain Polish groups never admit turbulent actions. We are mainly aiming at proving Hjorth's result that a countable product of Polish locally compact groups and closed subgroups of S_∞ has this property, but we will take a little detour which throws some further light into the concept of turbulence.

Proposition 11.1. *Let G be a Polish group and X a Polish G-space with every orbit meager. Then for every U_0, there is $U \subseteq U_0$ and V such that $R_{U,V}$ is nowhere dense.*

Proof. By Kuratowski-Ulam (the category analog of Fubini) $E_G^X \subseteq X^2$ is meager. Let $E_G^X \subseteq \bigcup_n F_n$, with $F_n \subseteq X^2$ closed nowhere dense. Then the proposition follows from the following general lemma. (A direct proof can also be given.)

Lemma 11.2. *Let X be a Polish G-space. Let $\{g_n\}$ be dense in G, containing $1 \in G$, and let $E_G^X \subseteq \bigcup_n F_n$, with $F_n \subseteq X^2$ closed. Put*

$$F_{n,g} = \{(x,y) : (x, g \cdot y) \in F_n\}.$$

Then for every U_0 there is $U \subseteq U_0, V, n, m$, such that $R_{U,V} \subseteq F_{n,g_m}$.

Proof. Let $G_n = X^2 \setminus F_n$. Fix U_0 and suppose that for every $U \subseteq U_0, V, n, m$ we have that

$$R_{U,V} \cap G_{n,g_m} \neq \emptyset \tag{$*$}$$

where $G_{n,g} = X^2 \setminus F_{n,g}$. Fix a compatible metric d_G for G and a complete compatible metric d_X for X. Using $(*)$ we will inductively construct two decreasing sequences $\{U_i'\}, \{U_i''\}$ with $\overline{U_{i+1}'} \subseteq U_i', \overline{U_{i+1}''} \subseteq U_i'', U_0' = U_0'' = U_0, d_X(U_i'), d_X(U_i'') \to 0, U_{i+1}' \times U_{i+1}'' \subseteq G_i$ and a sequence $\{h_i\} \subseteq \{g_n\}$ such that $h_0 = 1, h_i \cdot U_i' = U_i''$ and $d_G(h_i, h_{i+1}) < 2^{-i}$. Then if

152

$x_i' \in U_i', x_i' \to x'$ for some x', and if $h_i \cdot x_i' = x_i''$, then $x_i'' \to x''$ for some x''. Also $h_i \to h$ for some h, thus $h \cdot x' = x''$ and so $(x', x'') \in E_G^X$. On the other hand, $(x', x'') \in \bigcap_i (U_i' \times U_i'') \subseteq \bigcap_i G_i$, a contradiction.

To see how the construction proceeds, assume U_i', U_i'', h_i are given. Let V be such that $h \in V \Rightarrow d_G(h_i, h_i h) < 2^{-(i+1)}$. By $(*)$ there are $(x_0, x_1) \in R_{U_i', V}$ such that $(x_0, h_i \cdot x_1) \in G_i$. Say $h \cdot x_0 = x_1$ for $h \in V$. Let $g = h_i h$. Then $d_G(h_i, g) < 2^{-(i+1)}, g \cdot x_0 = y_0$, where $(x_0, y_0) \in G_i, y_0 \in U_i''$. It is now clear that we can find $h_{i+1} \in \{g_n\}$ such that

$$d_G(h_i, h_{i+1}) < 2^{-(i+1)}, (x_0, h_{i+1} \cdot x_0) \in G_i, h_{i+1} \cdot x_0 \in U_i''.$$

Then take U_{i+1}', U_{i+1}'' to be small enough nbhds of $x_0, h_{i+1} \cdot x_0$ resp. $\dashv$

Thus for any Polish G-space X, and some U, V, for a comeager set of $x \in U$, the set of neighbors $R_{U,V}(x)$ of x in the (U, V)-graph is nowhere dense. On the other hand, turbulence requires $\mathcal{O}(x, U, V)$, the connected component of x in the (U, V)-graph to be somewhere dense.

The following condition gives an easy criterion for non-turbulence.

Proposition 11.3. *Let X be a Polish G-space. Assume for every U, V and $x \in U$ there is $U', x \in U' \subseteq U, V' \subseteq V$ such that $\mathcal{O}(x, U', V') \subseteq R_{U', V'}(x)$. Then the action is not turbulent.*

Proof. By Proposition 11.1, choose U, V with $R_{U,V}$ nowhere dense and then $x \in U$ with $R_{U,V}(x)$ nowhere dense. Then let U', V' be as above, so that $\mathcal{O}(x, U', V') \subseteq R_{U', V'}(x) \subseteq R_{U,V}(x)$ is nowhere dense, contradicting turbulence. $\dashv$

Definition 11.4. Call an action that satisfies the hypothesis of Proposition 11.3 *calm*.

Proposition 11.5. *Any Polish G-space, where G is of the form $G = G_1 \times G_2$, with G_1 a closed subgroup of S_∞ and G_2 is locally compact, is calm.*

Proof. First we check that any Polish G-space with G locally compact is calm. Indeed given $U, V, x \in U$, let $V_1 \subseteq V$ be such that $\overline{V}_1$ is compact and let $(V')^2 \subseteq V_1$. We claim that there is $U', x \in U' \subseteq U$ such that $V_1 \cdot x \cap U' \subseteq V' \cdot x$. Indeed, otherwise we can find $g_n \in V_1, g_n \cdot x \to x$ with $g_n \cdot x \notin V' \cdot x$. By compactness, we can assume that $g_n \to g$, so $g \cdot x = x$. Since $g_n g^{-1} \to 1$ we can also assume that $g_n g^{-1} \in V'$, so $g_n \in V'g$ and $g_n \cdot x \in V'g \cdot x = V' \cdot x$, a contradiction. It is now easy to see that $\mathcal{O}(x, U', V') \subseteq R_{U', V'}(x)$.

Now consider a $G_1 \times G_2$-space, X, where G_1 is a closed subgroup of S_∞ and G_2 is locally compact. It is clear that we can view, identifying G_1 with $G_1 \times \{1\}$, X as a G_1-space and similarly as a G_2-space. Moreover, $(g_1, g_2) \cdot x = g_1 \cdot g_2 \cdot x = g_2 \cdot g_1 \cdot x$.

Given U, V and $x \in U$ we can assume that $V = V_1 \times V_2$, where V_1 is open in G_1 and V_2 is open in G_2. Now, considering the G_2-action, we can find, by the preceding, $U_2' \subseteq U$, $x \in U_2'$ and $V_2' \subseteq V_2$ with $\mathcal{O}(x, U_2', V_2') \subseteq R_{U_2', V_2'}(x)$, where everything refers here to the G_2-action. Next let $V_1' \subseteq V_1$ be an open subgroup containing $1 \in G$, and $U' \subseteq U_2'$, $x \in U'$ be such that $V_1' \cdot U' \subseteq U_2'$ (for the G_1-action). We claim then that if $V' = V_1' \times V_2'$, we have $\mathcal{O}(x, U', V') \subseteq R_{U', V'}(x)$ for the G-action and the proof is complete. Indeed let $y \in \mathcal{O}(x, U', V')$, so that $y = g_k \cdot h_k \cdot g_{k-1} \cdot h_{k-1} \cdots g_0 \cdot h_0 \cdot x$, where $g_i \in V_1', h_i \in V_2'$ and for any $i \leq k$, $g_i \cdot h_i \cdots g_0 \cdot h_0 \cdot x \in U'$. Now $y_i = g_i \cdot h_i \cdots g_0 \cdot h_0 \cdot x = g_i \cdots g_0 \cdot h_i \cdots h_0 \cdot x$, and since $g^i = g_i \cdots g_0 \in V_1', y_i \in U'$ we have that $h_i \cdots h_0 \cdot x \in U_2'$, so $h_k h_{k-1} \cdots h_0 \cdot x \in \mathcal{O}(x, U_2', V_2')$, thus $h_k \cdots h_0 \cdot x = h \cdot x$ for some $h \in V_2'$, so $y = g_k \cdots g_0 \cdot h \cdot x = g \cdot h \cdot x = (g, h) \cdot x$ for some $(g, h) \in V'$, i.e., $y \in R_{U', V'}(x)$ and we are done. $\dashv$

Theorem 11.6 (Hjorth [00]). *Let $G = G_0 \times G_1 \times \cdots$, where each G_i is a closed subgroup of S_∞ or else locally compact. Then no Polish G-space is turbulent.*

Proof. We can assume of course that $G = G_0 \times G_1 \times \cdots$, where G_0 is a closed subgroup of S_∞ and $G_1, G_2, \cdots$ are locally compact.

We recall first some general facts about universal spaces.

Let H be a Polish group and $d < 1$ a left-invariant compatible metric. Let $\mathcal{L}(H) = \{f : H \to [0,1] : \forall h_1, h_2 \in H(|f(h_1) - f(h_2)| \leq d(h_1, h_2))\}$. Put on $\mathcal{L}(H)$ the pointwise convergence topology, so it becomes compact metrizable. H acts continuously on $\mathcal{L}(H)$ by $(h_1 \cdot f)(h_2) = f(h_1^{-1} h_2)$. $\mathcal{L}(H)^{\mathbb{N}}$ is a universal Polish H-space in the following strong sense: Let X be a Polish H-space. Let $\{U_n\}$ be an open basis for X. Let

$$f_n^x(g) = d(g, \{h : h \cdot x \in U_n\}^{-1})$$

(with $d(g_i, \emptyset) = 1$). Let $F(x) = (f_n^x) \in \mathcal{L}(H)^{\mathbb{N}}$. Then F is an injective H-map of X into $\mathcal{L}(H)^{\mathbb{N}}$, F is Baire class 1 and open as a map from X onto $F(X)$ and $F(X)$ is G_δ (see Hjorth [00], Kechris [00]).

Now suppose H_1, H_2 are two Polish groups, and $H = H_1 \times H_2$. Let $\pi_1 : H \to H_1$ be the first projection. Given a Polish H-space X, define

$F : X \to \mathcal{L}(H_1)^{\mathbb{N}}$ by $F(x) = (f_n^x)$, where $f_n^x \in \mathcal{L}(H_1)$ is given by

$$f_n^x(g) = d_1(g, \{h_1 : \exists h \in H(\pi_1(h) = h_1 \ \& \ h \cdot x \in U_n)\}^{-1}),$$

with d_1 the metric for H_1. First, it is easy to check that if we view $\mathcal{L}(H_1)^{\mathbb{N}}$ as an H-space via $(h_1, h_2) \cdot y = h_1 \cdot y$, then F is an H-map, i.e., $F((h_1, h_2) \cdot x) = h_1 \cdot F(x)$. Also it is easy to check that F is Baire class 1.

Let $G_n^* = G_0 \times G_1 \times \cdots \times G_n, G^n = G_{n+1} \times G_{n+2} \times \cdots (n \geq 1)$, so that $G = G_n^* \times G^n$. Let X be a Polish G-space and assume it is turbulent, towards a contradiction. By Proposition 11.1, fix U, V_0 so that R_{U,V_0} is nowhere dense. Consider, for each $n \geq 1$, the map $F_n : X \to \mathcal{L}(G_n^*)^{\mathbb{N}}$ corresponding to the product $G = G_n^* \times G^n$. It is of Baire class 1, thus it has a dense G_δ set of continuity points. So we can find $n, x \in U, V$ such that V is of the form $V = V_n \times G^n$, for some V_n open nbhd of $1 \in G_n^*, V^2 \cdot x \subseteq U, F_n$ is continuous at x, and $R_{U,V^2}(x)$ is nowhere dense.

We will show that for an appropriate $U' \subseteq U, x \in U', V' \subseteq V$,

$$\overline{\mathcal{O}(x, U', V')} \subseteq \overline{R_{U,V^2}(x)},$$

which violates turbulence.

Since G_n^* is a product of a closed subgroup of S_∞ and a locally compact group, the G_n^*-space $\mathcal{L}(G_n^*)^{\mathbb{N}}$ is calm by Proposition 11.5. So we can find an open nbhd U'' of $F_n(x)$ and $V'' \subseteq V_n$ such that $\mathcal{O}(F_n(x), U'', V'') \subseteq R_{U'',V''}(F_n(x))$. Put $V' = V'' \times G^n \subseteq V$. Let $U' \subseteq U$ be an open nbhd of x such that $F_n(U') \subseteq U''$ (by the continuity of F_n at x). We will show that this works. Indeed let W be basic open with $W \cap \mathcal{O}(x, U', V') \neq \emptyset$, and pick $y \in W \cap \mathcal{O}(x, U', V')$. Then $F_n(y) \in \mathcal{O}(F_n(x), U'', V'')$, so $F_n(y) \in R_{U'',V''}(F_n(x))$, i.e., for some $g \in V'', F_n(y) = g \cdot F_n(x) = F_n(g' \cdot x)$, where $g' \in V'$. Since, if m is such that $W = U_m$, we have $f_m^y = f_m^{g' \cdot x}$, and $y \in U_m$, so that $f_m^y(1_{G_n^*}) = 0$, it follows that there is a sequence $\epsilon_i \in G_n^*, \varepsilon_i \to 1$ in G_n^*, with $(\epsilon_i, h_i) \cdot g' \cdot x \in U_m$ for some $h_i \in G^n$. For large $i, \varepsilon_i \in V''$, so $(\epsilon_i, h_i) \in V'$ and thus, since $g' \in V'$, we have that $(\epsilon_i, h_i) \cdot g' \cdot x \in (V')^2 \cdot x \subseteq U$, so $W \cap R_{U,(V')^2}(x) \neq \emptyset$, and the proof is complete. ⊣

From 11.6 and 9.1 it follows that no closed subgroup of a Polish group of the form $G = G_0 \times G_1 \times \cdots$, where G_0 is a closed subgroup of S_∞ and $G_1, G_2, \cdots$ are locally compact, can admit turbulent actions. Equivalently, no one of the groups (like infinite-dimensional Banach spaces

$(X, +))$ that have turbulent actions can be a closed subgroup of such a product.

12

Turbulence VI: The First Main Theorem

We will first need a few definitions.

Definition 12.1. Let X, Y be Polish spaces and E, F equivalence relations on X, Y resp. We say that E is *generically F-ergodic* if for every Baire measurable $f : X \to Y$ which is (E, F)-invariant, i.e., $xEy \Rightarrow f(x)Ff(y)$, there is a comeager set in X which f maps into a single F-equivalence class.

Trivially, if E is generically F-ergodic and every E-equivalence class is meager, then $E \not\leq_{BM} F$, where

$$E \leq_{BM} F \Leftrightarrow \exists \text{ Baire measurable } f : X \to Y \text{ with } xEy \Leftrightarrow f(x)Ff(y).$$

Definition 12.2. Let E_{ctble} be the following equivalence relation on $(2^{\mathbb{N}})^{\mathbb{N}}$:

$$(x_n)E_{\text{ctble}}(y_n) \Leftrightarrow \{x_n : n \in \mathbb{N}\} = \{y_n : n \in \mathbb{N}\}.$$

Thus, the quotient space $(2^{\mathbb{N}})^{\mathbb{N}}/E_{\text{ctble}}$ can be canonically identified with the set of countable (nonempty) subsets of $2^{\mathbb{N}}$.

Theorem 12.3 (Hjorth [00]). *Let E be an equivalence relation on a Polish space X. Then the following are equivalent:*

(i) *E is generically E_{ctble}-ergodic.*

(ii) *E is generically $E_{S_\infty}^Y$-ergodic, for any Borel S_∞-space Y.*

Proof. (ii)$\Rightarrow$(i). It is easy to see that $E_{\text{ctble}} \sim_B E_{S_\infty}^Y$ for some S_∞-space Y.

(i)$\Rightarrow$(ii). Fix some canonical coding system of hereditarily countable sets by elements of $2^{\mathbb{N}}$. This coding provides a $\mathbf{\Pi}_1^1$ set $\widehat{HC} \subseteq 2^{\mathbb{N}}$ and a surjection $\pi : \widehat{HC} \to HC$ such that the relations $\pi(x) = \pi(y)$ and $\mathrm{rank}(\pi(x)) \leq \mathrm{rank}(\pi(y))$ are $\mathbf{\Pi}_1^1$.

If $f : X \to Y$ is $(E, E_{S_\infty}^Y)$-invariant, then using the fact that every Borel S_∞-space is Borel isomorphic to the logic action of S_∞ on the countable models of an $L_{\omega_1\omega}$ sentence, and then making use of canonical Scott sentences, we see that there is a C-measurable map $g : Y \to 2^{\mathbb{N}}$ such that $g(Y) \subseteq \widehat{HC}$ and

$$yE_{S_\infty}^Y z \Leftrightarrow \pi(g(y)) = \pi(g(z)).$$

Let $h = g \circ f$. Then

$$xEy \Rightarrow \pi(h(x)) = \pi(h(y)).$$

For $\alpha < \omega_1$, let $HC_\alpha = V_\alpha \cap HC$. Let $A_\alpha = \{x \in X : \pi(h(x)) \in HC_\alpha\}$. We have that $X = \bigcup_{\alpha<\omega_1} A_\alpha$ and by standard facts it follows that for some $\alpha_0 < \omega_1, \bigcup_{\alpha<\alpha_0} A_\alpha$ is comeager, so on a comeager set $\rho(x) = \pi(h(x)) \in V_{\alpha_0} \cap HC$.

We will then prove by induction on $\alpha \leq \alpha_0$ that there is a comeager set C_α such that

$$TC(\rho(x)) \cap V_\alpha$$

is constant on C_α. For $\alpha = \alpha_0$ this shows that $TC(\rho(x))$, and thus $\rho(x)$, is constant on C_{α_0}, so on a comeager set f maps into a single $E_{S_\infty}^Y$-equivalence class.

For $\alpha = 0$ and through limit ordinals this assertion is clear. So assume $\alpha < \alpha_0$ and $TC(\rho(x)) \cap V_\alpha$ is constant on C_α, say $TC(\rho(x)) \cap V_\alpha = A_\alpha \in HC \cap V_{\alpha+1}$. Let $A_a = \{a_n\}$. If $a \in TC(\rho(x)) \cap V_{\alpha+1}$, so that $a \subseteq A_\alpha$, let $x_a \in 2^{\mathbb{N}}$ be defined by $x_a(n) = 1 \Leftrightarrow a_n \in a$. Since we can assume that $h|C_\alpha$ is Borel, there is a Borel function $p : C_\alpha \to (2^{\mathbb{N}})^{\mathbb{N}}$ so that for $x \in C_\alpha, (p(x)_n)$ enumerates the $x_a \in 2^{\mathbb{N}}$ for which $a \in TC(\rho(x)) \cap V_{\alpha+1}$. Then for $x, y \in C_\alpha$,

$$\begin{aligned} &TC(\rho(x)) \cap V_{\alpha+1} = TC(\rho(y)) \cap V_{\alpha+1} \\ \Leftrightarrow\ &\{x_a : a \in TC(\rho(x)) \cap V_{\alpha+1}\} \\ =\ &\{y_a : a \in TC(\rho(y)) \cap V_{\alpha+1}\} \\ \Leftrightarrow\ &\{p(x)_n\} = \{p(y)_n\} \\ \Leftrightarrow\ &p(x)E_{\mathrm{ctble}}p(y). \end{aligned}$$

Since $xEy \Rightarrow \rho(x) = \rho(y) \Rightarrow p(x)E_{\mathrm{ctble}}p(y)$, for $x, y \in C_\alpha$, it follows using our hypothesis (i), that there is a comeager set $C_{\alpha+1} \subseteq C_\alpha$ such that $\{p(x)_n\}$ is constant on $C_{\alpha+1}$, so $TC(\rho(x)) \cap V_{\alpha+1}$ is constant on $C_{\alpha+1}$, and the proof is complete. ⊣

Before we state the main result in this section we will also formulate an ostensibly weaker notion of turbulence.

Definition 12.4. Let G be a Polish group and X a Polish G-space. We say that the action is *weakly generically turbulent* if

(i) Every orbit is meager;

(ii) $\forall^* x \in X \forall^* y \in X \forall U \forall V (x \in U \Rightarrow \overline{\mathcal{O}(x,U,V)} \cap G \cdot y \neq \emptyset)$.

Note that (ii) implies that there is a dense orbit. Indeed, let $x \in X$ be such that $\forall^* y \in X \forall U, V(x \in U \Rightarrow \overline{\mathcal{O}(x,U,V)} \cap G \cdot y \neq \emptyset)$. Taking $U = X, V = G$, we see that $\overline{G \cdot x}$ meets $G \cdot y$ for a comeager set of y. But if $G \cdot x$ is not dense, then $\overline{G \cdot x}$ is disjoint from an invariant nonempty open set, a contradiction.

We now have the main

Theorem 12.5 (Hjorth [00]). *Let G be a Polish group and X a Polish G-space with every orbit meager and some orbit dense. Then the following are equivalent:*

(i) X *is weakly generically turbulent;*

(ii) E_G^X *is generically E_{ctble}-ergodic;*

(iii) *For any Borel S_∞-space Y, E_G^X is generically $E_{S_\infty}^Y$-ergodic;*

(iv) X *is generically turbulent.*

Proof. (i)⇒(ii): Let $f : X \to Y = (2^{\mathbb{N}})^{\mathbb{N}}$ be Baire measurable. Put

$$A = \{a \in 2^{\mathbb{N}} : \forall^* x (a \in \{f(x)_n\})\},$$

where for $y \in Y, \{y_n\} = \{y_n : n \in \mathbb{N}\}$.

Claim 1. A is countable.

Proof of Claim 1. Since f is continuous on a dense G_δ set, A is Borel. So if it is uncountable, it contains a Cantor set C. Then

$$\forall^*_C a \forall^* x (a \in \{f(x)_n\}),$$

so, by Kuratowski-Ulam,

$$\forall^* x \forall^*_C a (a \in \{f(x)_n\}),$$

thus for some $x, \{f(x)_n\}$ is uncountable, which is obviously absurd.

We will show that $\forall^* x(A = \{f(x)_n\})$ which will complete the proof that (i)$\Rightarrow$(ii).

Assume not, towards a contradiction. Since $\{x : A = \{f(x)_n\}\}$ has the Baire property and is invariant, and since there is a dense orbit, it follows that $C_1 = \{x : \{f(x)_n\} \not\supseteq A\}$ is comeager (since $\{x : A \subseteq \{f(x)_n\}\}$ is comeager).

Let also

$$C_2 = \{x : \forall^* y \forall U \forall V (x \in U \Rightarrow G \cdot y \cap \overline{\mathcal{O}(x, U, V)} \neq \emptyset)\},$$

so that C_2 is comeager as well, by assumption.

Next fix a comeager set $C_0 \subseteq X$ with $f|C_0$ continuous. For $B \subseteq X$ let

$$C_B = \{x : x \in B \Leftrightarrow \exists \text{ open hbhd } U \text{ of } x \text{ with } \forall^* y \in U(y \in B)\}.$$

Then if B has the Baire property, C_B is comeager (see Kechris [95, 8G]). Finally fix a countable dense subgroup $G_0 \subseteq G$ and find a countable collection $\mathcal{C}$ of comeager sets in X with the following properties:

(i) $C_0, C_1, C_2 \in \mathcal{C}$;

(ii) $C \in \mathcal{C}, g \in G_0 \Rightarrow g \cdot C \in \mathcal{C}$;

(iii) $C \in \mathcal{C} \Rightarrow C^* = \{x : \forall^* g(g \cdot x \in C)\} \in \mathcal{C}$.

(iv) If $\{V_n\}$ enumerates a local basis of open symmetric nbhds of 1 in G, then, letting

$$A_{\ell,n} = \{x : \forall^* g \in V_n(f(x)_\ell = f(g \cdot x)_\ell)\},$$

we have that $C_{A_{\ell,n}} \in \mathcal{C}$.

(v) If $\{U_n\}$ enumerates a basis for X, and

$$C_{m,n,\ell} = \{x : x \notin U_m \text{ or } \forall^* g \in V_n(f(x)_\ell = f(g \cdot x)_\ell)\},$$

then $\mathcal{C}$ contains all $C_{m,n,\ell}$ which are comeager.

For simplicity, if $x \in \bigcap \mathcal{C}$ (and there are comeager many such x), we call x "generic".

So fix a generic x. Then there is $a \notin A$ so that $a \in \{f(x)_n\} = \{f(g \cdot x)_n\}$ for all g. So $\forall g \exists \ell(a = f(g \cdot x)_\ell)$, thus there is $\ell \in \mathbb{N}$ and open

$W \subseteq G$ so that $\forall^* g \in W(f(g \cdot x)_\ell = a)$. Fix $p_0 \in G_0$ and V a basic symmetric nbhd of 1 so that $Vp_0 \subseteq W$. Let $p_0 \cdot x = x_0$, so that x_0 is generic too, and $\forall^* g \in V(f(g \cdot x_0)_\ell = a)$. Now $\forall^* g \in V(g \cdot x_0 \in C_0)$, so we can find $g_i \in V, g_i \to 1$ with $g_i \cdot x_0 \in C_0$ and $f(g_i \cdot x_0)_\ell = a$, so as $g_i \cdot x_0 \to x_0 \in C_0$, by continuity we have $f(x_0)_\ell = a$. Also since $\forall^* g \in V(f(x_0)_\ell = f(g \cdot x_0)_\ell)$ and x_0 is generic, there is basic open U, with $x_0 \in U$ such that

$$\forall^* z \in U \forall^* g \in V(f(z)_\ell = f(g \cdot z)_\ell),$$

i.e., if $U = U_m$, $V = V_n$, then $C_{m,n,\ell}$ is comeager, so is in $\mathcal{C}$. Since $a \notin A$, $\forall^* y(a \notin \{f(y)_n\})$, so choose y to be generic, with $a \notin \{f(y)_n\}$, and also

$$\forall \tilde{U}, \tilde{V}(x_0 \in \tilde{U} \Rightarrow G \cdot y \cap \overline{\mathcal{O}(x_0, \tilde{U}, \tilde{V})} \neq \emptyset).$$

Thus we have $G \cdot y \cap \overline{\mathcal{O}(x_0, U, V)} \neq \emptyset$. So choose $g_0, g_1, \cdots \in V$ so that if $g_i \cdot x_i = x_{i+1}$, then $x_i \in U$ and some subsequence of (x_i) converges to some $y_1 \in G \cdot y$. Fix a compatible metric d for X.

Since $\forall^* h(h \cdot x_0$ is generic$)$ and $\forall^* g \in V(f(g \cdot x_0)_\ell = a)$, we can find h_1 so that $h_1 g_0 \in V, g_1 h_1^{-1} \in V, \bar{x}_1 = h_1 \cdot x_1 \in U, d(x_1, \bar{x}_1) < \frac{1}{2}, \bar{x}_1 = h_1 \cdot g_0 \cdot x_0$ is generic, and so $\forall^* g \in V(f(\bar{x}_1)_\ell = f(g \cdot \bar{x}_1)_\ell)$ (as $\bar{x}_1 \in C_{m,n,\ell}$), and $f(\bar{x}_1)_\ell = a$, so also $\forall^* g \in V(f(g \cdot \bar{x}_1)_\ell = a)$. Note that $g_1 h_1^{-1} \cdot \bar{x}_1 = x_2$ and $g_1 h_1^{-1} \in V$, so since $\forall^* h(h \cdot \bar{x}_1$ is generic$)$ and $\forall^* g \in V(f(g \cdot \bar{x}_1)_\ell = a)$, we can find h_2 so that $h_2 g_1 h_1^{-1} \in V, g_2 h_2^{-1} \in V, \bar{x}_2 = h_2 \cdot x_2 \in U, d(x_2, \bar{x}_2) < \frac{1}{4}, \bar{x}_2 = h_2 g_1 h_1^{-1} \cdot \bar{x}_1$ is generic, and so $\forall^* g \in V(f(\bar{x}_2)_\ell = f(g \cdot \bar{x}_2)_\ell)$, and $f(\bar{x}_2)_\ell = a$, so $\forall^* g \in V(f(g \cdot \bar{x}_2) = a)$, etc.

Repeating this process, we get $x_0, \bar{x}_1, \bar{x}_2, \cdots$ generic and belonging in the (V, U)-local orbit of x_0, so that some subsequence of $\{\bar{x}_i\}$ converges to y_1 and $\forall^* g \in V(f(g \cdot \bar{x}_i)_\ell = a)$. Now

$$\forall^* g(g \cdot \bar{x}_i \in C_0), \quad \forall^* g(g \cdot y_1 \in C_0), \quad \forall^* g \in V(f(g \cdot \bar{x}_i)_\ell = a),$$

so fix g satisfying all these conditions. Then for some subsequence $\{n_i\}$ we have $\bar{x}_{n_i} \to y_1$, so $g \cdot \bar{x}_{n_i} \to g \cdot y_1$ and $g \cdot \bar{x}_{n_i}, g \cdot y_1 \in C_0$, so, by continuity, $f(g \cdot \bar{x}_{n_i})_\ell = a \to f(g \cdot y_1)_\ell$, so $f(g \cdot y_1)_\ell = a$, i.e., $a \in \{f(g \cdot y_1)_n\} = \{f(y_1)_n\} = \{f(y)_n\}$, a contradiction.

(ii)$\Rightarrow$(iii): By 12.3.

(iii)$\Rightarrow$(iv): Assume actually only that E_G^X is generically E_{ctble}-ergodic (i.e., (ii) holds). We will prove the following claim.

Claim 2. $\forall U, V \forall^* x \in U (\mathcal{O}(x, U, V)$ is somewhere dense).

Assuming this we will complete the proof as follows: By restricting ourselves to a countable basis in X and a countable local basis of $1 \in G$ we see that $\forall^* x \forall U, V (x \in U \Rightarrow \mathcal{O}(x, U, V)$ is somewhere dense). So let

$$C = \{x : \forall U, V (x \in U \Rightarrow \mathcal{O}(x, U, V) \text{ is somewhere dense})\}$$
$$\cap \{x : G \cdot x \text{ is dense}\}.$$

Then C is comeager and invariant, since if $x \in U$, then $g \cdot x \in g \cdot U$ and $g \cdot \mathcal{O}(x, U, V) = \mathcal{O}(g \cdot x, g \cdot U, gVg^{-1})$. So C contains a dense G_δ set C_0 and then $X_0 = C_0^*$ is an invariant dense G_δ set with $X_0 \subseteq C$. It is then routine to verify that the G-action on X_0 is turbulent.

Proof of Claim 2. Fix U, V. First notice that there are only countably many local orbits $\mathcal{O}(y, U, V)$, when $y \in U$ is restricted to some fixed orbit $G \cdot x$. This is because if $g_1 \cdot x = y, g_2 \cdot x = z$ and $\mathcal{O}(y, U, V) \neq \mathcal{O}(z, U, V)$, then $g_2^{-1} g_1 \notin V$. Thus for each x, $\{\mathcal{O}(y, U, V) : y \in G \cdot x \cap U\}$ is countable. Fix a countable basis $\{U_n\}$ for X and for each $x \in U$, let $a_x \in 2^{\mathbb{N}}$ be defined by $a_x(n) = 1 \Leftrightarrow U_n \cap \mathcal{O}(x, U, V) \neq \emptyset$. So a_x encodes $\overline{\mathcal{O}(x, U, V)}$. It is easy to find then a Baire measurable $f : X \to (2^{\mathbb{N}})^{\mathbb{N}}$ such that $\{f(x)_n\} = \{a_y : y \in G \cdot x \cap U\}$, if $G \cdot x \cap U \neq \emptyset$. So if $x, y \in U$, $x E_G^X y \Rightarrow f(x) E_{\text{ctble}} f(y)$. It follows that there is a comeager in U set $C \subseteq U$ such that $\{f(x)_n\}$ is constant for $x \in C$, so for $x \in C, y \in C$, $\{\overline{\mathcal{O}(x', U, V)} : x' \in G \cdot x, x' \in U\} = \{\overline{\mathcal{O}(y', U, V)} : y' \in G \cdot y, y' \in U\}$.

We will prove that for $x \in C$, $\mathcal{O}(x, U, V)$ is somewhere dense. Otherwise, $F = \overline{\mathcal{O}(x, U, V)}$ is meager, so for some $x_0 \in C, \forall^* g(g \cdot x_0 \notin F)$. Now as $x, x_0 \in C$, there is some g_0 with $g_0 \cdot x_0 \in U$ and $\overline{\mathcal{O}(g_0 \cdot x_0, U, V)} = \overline{\mathcal{O}(x, U, V)} = F$. Thus $hg_0 \cdot x_0 \in F$ for all $h \in V$ with $hg_0 \cdot x_0 \in U$, which is an open set of h's, so $\exists^* g(g \cdot x_0 \in F)$, a contradiction.

(iv)$\Rightarrow$(i). Let $X_0 \subseteq X$ be invariant dense G_δ on which the action is turbulent. Fix $x \in X_0, y \in X_0, U, V$ such that $x \in U$. Now, working in X_0, we see that $\overline{\mathcal{O}(x, U \cap X_0, V)}^{X_0}$ has nonempty interior. Since $G \cdot y$ is dense, $G \cdot y \cap \overline{\mathcal{O}(x, U \cap X_0, V)}^{X_0} \neq \emptyset$, so $G \cdot y \cap \overline{\mathcal{O}(x, U, V)} \neq \emptyset$.

$\dashv$

Corollary 12.6. *If G is a Polish group and X a generically turbulent G-space, then $E_G \not\leq_{BM} E_{S_\infty}^Y$ for any Polish S_∞-space Y.*

13

Turbulence VII: The Second Main Theorem

Let G be a Polish group and X a Polish G-space. Fix from now on a countable dense subgroup $G_0 \subseteq G$, a countable open basis (of nonempty sets) $\mathcal{B}$ for X, closed under $U \mapsto g_0 \cdot U$ for $g_0 \in G_0$, and containing X. Also fix a countable basis $\mathcal{N}$ of symmetric open nbhds of $1 \in G$ closed under $V \mapsto g_0 V g_0^{-1}$ for $g_0 \in G_0$, and containing G. Below U, V will vary over $\mathcal{B}, \mathcal{N}$ resp. If needed, we will identify $\mathcal{B}, \mathcal{N}$ with $\mathbb{N}$ by fixing a 1-1 enumeration for each of them.

Definition 13.1. For x, U, V and any ordinal $\alpha \geq 1$, define by induction

$$\varphi_\alpha(x, U, V) \in \mathcal{P}^\alpha(\mathbb{N})$$

(where $\mathcal{P}^\alpha(\mathbb{N}) = \mathbb{N}, \mathcal{P}^\alpha(\mathbb{N}) =$ the set of countable subsets of $\mathcal{P}^{<\alpha}(\mathbb{N}) \cup \mathbb{N}$, with $\mathcal{P}^{<\alpha}(\mathbb{N}) = \bigcup_{\beta < \alpha} \mathcal{P}^\beta(\mathbb{N})$) as follows:

$$\varphi_1(x, U, V) = \{U' \in \mathcal{B} : U' \cap \mathcal{O}(x, U, V) \neq \emptyset\};$$

(we identify here $\mathcal{B}$ with $\mathbb{N}$);

$$\varphi_{\alpha+1}(x, U, V) = \{\langle U', V', \varphi_\alpha(y, U', V') \rangle :$$
$$U' \subseteq U, V' \subseteq V, y \in \mathcal{O}(x, U, V)\}$$

(where we identify again $\mathcal{B}, \mathcal{N}$ with $\mathbb{N}$ here and use some canonical coding function $\langle \rangle$ for tuples in $\mathbb{N} \cup \mathcal{P}^\alpha(\mathbb{N})$ by elements of $\mathcal{P}^\alpha(\mathbb{N})$);

$$\varphi_\lambda(x, U, V) = \{\varphi_\alpha(x, U, V) : \alpha < \lambda\}.$$

In order to justify that $\varphi_\alpha \in \mathcal{P}^\alpha(\mathbb{N})$, we have to verify that $\varphi_\alpha(x, U, V)$ is countable. This follows immediately from the following facts:

Proposition 13.2. $\varphi_\alpha(x, U, V)$ depends only on $\mathcal{O}(x, U, V)$.

Proof. Trivial by induction on α. ⊣

Proposition 13.3. *Fix $x \in U, V$ and $U' \subseteq U, V' \subseteq V$. Then for $y \in \mathcal{O}(x, U, V)$, $\mathcal{O}(y, U', V') \subseteq \mathcal{O}(x, U, V)$. Moreover, $\mathcal{O}(x, U, V)$ contains only countably many $\mathcal{O}(y, U', V')$.*

Proof. Note that if $G \cdot x$ is the orbit of x, and we give $G \cdot x$ the quotient topology induced by the map $x \mapsto g \cdot x$ (i.e., $A \subseteq G \cdot x$ is open iff $\{x : g \cdot x \in A\}$ is open in G), then $\mathcal{O}(x, U, V)$ is open and so is each $\mathcal{O}(y, U', V')$ contained in it. Since distinct $\mathcal{O}(y, U', V')$ are disjoint and this topology is separable, we are done. ⊣

Note that $\varphi_\alpha(x, U, V) \in \mathcal{P}^\alpha(\mathbb{N}) \setminus \mathcal{P}^{<\alpha}(\mathbb{N})$ (we are assuming here that $\mathbb{N} \cap \mathcal{P}^\alpha(\mathbb{N}) = \emptyset$), so $\varphi_\alpha(x, U, V)$ completely determines α and in particular $\varphi_\alpha(x, U, V)$ completely determines all $\varphi_\beta(x, U, V)$ for $\beta \leq \alpha$.

Occasionally we may want to refer to a local orbit $\mathcal{O}(x, U, V)$ of the (U, V)-graph without explicitly mentioning x. We will simply write then $\mathcal{O}(U, V)$. Since $\varphi_\alpha(x, U, V)$ only depends on $\mathcal{O}(x, U, V)$ we could consider the preceding definition as actually defining $\varphi_\alpha(\mathcal{O}(U, V))$ for each local orbit $\mathcal{O}(U, V)$, as follows:

$$\varphi_1(\mathcal{O}(U, V)) = \{U' \in \mathcal{B} : U' \cap \mathcal{O}(U, V) \neq \emptyset\};$$
$$\varphi_{\alpha+1}(\mathcal{O}(U, V)) = \{\langle U', V', \varphi_\alpha(\mathcal{O}(U', V'))\rangle :$$
$$U' \subseteq U, V' \subseteq V, \mathcal{O}(U', V') \subseteq \mathcal{O}(U, V)\};$$
$$\varphi_\lambda(\mathcal{O}(U, V)) = \{\varphi_\alpha(\mathcal{O}(U, V)) : \alpha < \lambda\}.$$

We will next see how $\varphi_\alpha(x, U, V)$ is transformed under application of any $g_0 \in G_0$. Inductively define $g_0 \cdot \varphi_\alpha(x, U, V)$ as follows:

$$g_0 \cdot \varphi_1(x, U, V) = \{g_0 \cdot U' : U' \cap \mathcal{O}(x, U, V) \neq \emptyset\};$$
$$g_0 \cdot \varphi_{\alpha+1}(x, U, V) = \{\langle g_0 \cdot U', g_0 V' g_0^{-1}, g_0 \cdot \varphi_\alpha(y, U', V')\rangle :$$
$$U' \subseteq U, V' \subseteq V, y \in \mathcal{O}(x, U, V)\};$$
$$g_0 \cdot \varphi_\lambda(x, U, V) = \{g \cdot \varphi_\alpha(x, U, V) : \alpha < \lambda\}.$$

Then it is easy to check by induction the following:

Proposition 13.4. $g_0 \cdot \varphi_\alpha(x, U, V) = \varphi_\alpha(g_0 \cdot x, g_0 \cdot U, g_0 V g_0^{-1})$.

So we also have (with a mild abuse of notation)

$$g_0 \cdot \varphi_\alpha(\mathcal{O}(U, V)) = \varphi_\alpha(\mathcal{O}(g_0 \cdot U, g_0 V g_0^{-1})).$$

The next fact shows that $\varphi_\alpha(\mathcal{O}(U', V'))$ determines $\varphi_\alpha(\mathcal{O}(U, V))$ as long as $U' \subseteq U, V' \subseteq V$ and $\mathcal{O}(U', V') \subseteq \mathcal{O}(U, V)$.

Proposition 13.5. *There is a function f_α such that if $U' \subseteq U, V' \subseteq V, \mathcal{O}(U', V') \subseteq \mathcal{O}(U, V)$, then $f_\alpha(U, V, \varphi_\alpha(\mathcal{O}(U', V'))) = \varphi_\alpha(\mathcal{O}(U, V))$.*

Proof. By induction on α.

For $\alpha = 1$, note that for $U^1 \in \mathcal{B}$:

$$U^1 \in \varphi_1(\mathcal{O}(U, V)) \Leftrightarrow U^1 \cap \mathcal{O}(U, V) \neq \emptyset$$
$$\Leftrightarrow \exists U^2 \in \varphi_1(\mathcal{O}(U', V')) \exists g_0, \cdots, g_n \in V \text{ such}$$
$$\text{that if } U_0 = U^2, g_i \cdot U_i = U_{i+1}, \text{ then}$$
$$U_{n+1} \subseteq U^1 \text{ and } \forall i \leq n + 1(U_i \subseteq U).$$

It is obvious for λ limit. Finally consider the successor case $\alpha + 1$.

Suppose $U_1 \subseteq U, V_1 \subseteq V$ and $\mathcal{O}(U_1, V_1) \subseteq \mathcal{O}(U, V)$. Let

$$x \in \mathcal{O}(U', V') \subseteq \mathcal{O}(U, V).$$

Then for some $g_0, \cdots, g_n$, if $x_0 = x, g_i \cdot x_i = x_{i+1}$, we have $x_i \in U$ and $x_{n+1} \in \mathcal{O}(U_1, V_1)$. Since $h \cdot x_{n+1} \in \mathcal{O}(U_1, V_1)$ for all h close enough to $1 \in G$, it follows that there is $g_0 \in G_0$, with $g_0 \cdot x \in \mathcal{O}(U_1, V_1)$. Choose then $U'' \subseteq U', V'' \subseteq V'$ so that $g_0 \cdot U'' \subseteq U_1, g_0 V'' g_0^{-1} \subseteq V_1$. Thus $\mathcal{O}(g_0 \cdot x, g_0 \cdot U'', g_0 V'' g_0^{-1}) \subseteq \mathcal{O}(U_1, V_1)$, so $f_\alpha(U_1, V_1, \varphi_\alpha(\mathcal{O}(g_0 \cdot x, g_0 \cdot U'', g_0 V'' g_0^{-1}))) = \varphi_\alpha(\mathcal{O}(U_1, V_1))$. Recall also that

$$\varphi_\alpha(\mathcal{O}(g_0 \cdot x, g_0 \cdot U'', g_0 V'' g_0^{-1})) = g_0 \cdot \varphi_\alpha(\mathcal{O}(x, U'', V'')).$$

Thus

$$\varphi_{\alpha+1}(\mathcal{O}(U, V))$$

consists of all

$$\langle U_1, V_1, A \rangle$$

where $U_1 \subseteq U, V_1 \subseteq V$, and A is of the form

$$f_a(U_1, V_1, g_0 \cdot \varphi_\alpha(\mathcal{O}(x, U'', V''))),$$

where $x \in \mathcal{O}(U', V'), g_0 \in G_0$,

$$\langle U'', V'', \varphi_\alpha(\mathcal{O}(x, U'', V'')) \rangle \in \varphi_{\alpha+1}(\mathcal{O}(U', V')),$$
$$g_0 \cdot x \in \mathcal{O}(U_1, V_1), g_0 \cdot U'' \subseteq U_1, g_0 V'' g_0^{-1} \subseteq V_1.$$

$\dashv$

Definition 13.6. Fix an orbit $[x] = G \cdot x$. If $\mathcal{O}(U, V), \mathcal{O}'(U, V)$ are two local orbits contained in $[x]$, let $\alpha(\mathcal{O}, \mathcal{O}')$ be the least ordinal $1 \leq \alpha < \omega_1$ such that $\varphi_\alpha(\mathcal{O}) \neq \varphi_\alpha(\mathcal{O}')$, if such exists; else $\alpha(\mathcal{O}, \mathcal{O}') = 1$.

Let $\alpha(x, U, V) = \sup\{\alpha(\mathcal{O}, \mathcal{O}') : \mathcal{O}(U, V), \mathcal{O}'(U, V) \subseteq [x]\}$. Note that $\alpha(x, U, V)$ depends only on $[x], U, V$, so we can also write

$$\alpha([x], U, V) = \alpha(x, U, V).$$

Let

$$\alpha_x = \alpha_{[x]} = \sup\{\alpha(x, U, V) : U \in \mathcal{B}, V \in \mathcal{N}\}.$$

Then $\alpha_x < \omega_1$. Notice that for any $\mathcal{O}(U, V), \mathcal{O}'(U, V)$ contained in $[x]$:

$$\varphi_{\alpha(x)}(\mathcal{O}(U, V)) = \varphi_{\alpha(x)}(\mathcal{O}'(U, V)) \Rightarrow$$
$$\forall \alpha < \omega_1 (\varphi_\alpha(\mathcal{O}(U, V)) = \varphi_\alpha(\mathcal{O}'(U, V))).$$

Moreover $\alpha(x)$ is the least ordinal with this property. Put

$$\varphi_x = \varphi_{\alpha(x)+2}(x, X, G).$$

Again φ_x depends on $[x]$ only, so we can write

$$\varphi_{[x]} = \varphi_x = \varphi_{\alpha([x])+2}([x]),$$

as $\mathcal{O}(x, X, G) = [x]$.

We have the following fact.

Proposition 13.7. *If* $\varphi_{\alpha(x)+2}([x]) = \varphi_{\alpha(x)+2}([y])$, *then* $\alpha(x) = \alpha(y)$ *and* $\varphi_\alpha([x]) = \varphi_\alpha([y])$ *for all* $\alpha < \omega_1$ *(so in particular* $\varphi_x = \varphi_y$).

Proof. By hypothesis, for each U, V

$$\{\varphi_{\alpha(x)+1}(\mathcal{O}(U, V)) : \mathcal{O}(U, V) \subseteq [x]\} =$$
$$\{\varphi_{\alpha(x)+1}(\mathcal{O}(U, V)) : \mathcal{O}(U, V) \subseteq [y]\}.$$

From this it easily follows that for $\mathcal{O}(U, V), \mathcal{O}'(U, V) \subseteq [y]$:

$$\varphi_{\alpha(x)}(\mathcal{O}(U, V)) = \varphi_{\alpha(x)}(\mathcal{O}'(U, V)) \Rightarrow$$
$$\varphi_{\alpha(x)+1}(\mathcal{O}(U, V)) = \varphi_{\alpha(x)+1}(\mathcal{O}'(U, V)),$$

and so by an easy induction on α, for all U, V

$$\varphi_{\alpha(x)}(\mathcal{O}(U, V)) = \varphi_{\alpha(x)}(\mathcal{O}'(U, V)) \Rightarrow$$
$$\varphi_\alpha(\mathcal{O}(U, V)) = \varphi_\alpha(\mathcal{O}'(U, V)),$$

thus $\alpha(x) \geq \alpha(y)$ and by symmetry (as $\varphi_{\alpha(y)+2}([x]) = \varphi_{\alpha(y)+2}([y])$), we have that $\alpha(x) = \alpha(y)$.

To see that $\forall \alpha < \omega_1 (\varphi_\alpha([x])) = (\varphi_\alpha([y]))$ it is enough to check that for any $\alpha < \omega_1$ and any $\mathcal{O}_x(U,V) \subseteq [x], \mathcal{O}_y(U,V) \subseteq [y]$ we have

$$\varphi_{\alpha(x)}(\mathcal{O}_x(U,V)) = \varphi_{\alpha(x)}(\mathcal{O}_y(U,V)) \Rightarrow$$
$$\varphi_\alpha(\mathcal{O}_x(U,V)) = \varphi_\alpha(\mathcal{O}_y(U,V)).$$

This is easily proved by induction. Assume it is true for α. To prove it for $\alpha + 1$ consider $U' \subseteq U, V' \subseteq V$ and $\mathcal{O}_x(U',V') \subseteq \mathcal{O}_x(U,V)$. We have to find $\mathcal{O}_y(U',V') \subseteq \mathcal{O}_y(U,V)$ with $\varphi_\alpha(\mathcal{O}_y(U',V')) = \varphi_\alpha(\mathcal{O}_x(U',V'))$ (and vice versa; but the argument is clearly symmetric). Let $\tilde{\mathcal{O}}_y(U,V) \subseteq [y]$ be such that $\varphi_{\alpha(x)+1}(\tilde{\mathcal{O}}_y(U,V)) = \varphi_{\alpha(x)+1}(\mathcal{O}_x(U,V))$, and find

$$\tilde{\mathcal{O}}_y(U',V') \subseteq \tilde{\mathcal{O}}_y(U',V')$$

so that $\varphi_{\alpha(x)}(\tilde{\mathcal{O}}_y(U',V')) = \varphi_{\alpha(x)}(\mathcal{O}_x(U',V'))$ and so

$$\varphi_\alpha(\tilde{\mathcal{O}}_y(U',V')) = \varphi_\alpha(\mathcal{O}_x(U',V')).$$

Now $\varphi_{\alpha(x)}(\mathcal{O}_y(U,V)) = \varphi_{\alpha(x)}(\tilde{\mathcal{O}}_y(U,V))$, so

$$\varphi_{\alpha(x)+1}(\mathcal{O}_y(U,V)) = \varphi_{\alpha(x)+1}(\tilde{\mathcal{O}}_y(U,V)),$$

thus there is $\mathcal{O}_y(U',V') \subseteq \mathcal{O}_y(U,V)$ with

$$\varphi_{\alpha(x)}(\mathcal{O}_y(U',V')) = \varphi_{\alpha(x)}(\tilde{\mathcal{O}}_y(U',V'))$$

and thus (as $\alpha(x) = \alpha(y)$),

$$\varphi_\alpha(\tilde{\mathcal{O}}_y(U',V')) = \varphi_\alpha(\mathcal{O}_y(U',V')) = \varphi_\alpha(\mathcal{O}_x(U',V')).$$

$\dashv$

We will next associate to each orbit $[x]$ a canonical countable structure

$$\mathcal{M}^0(x) = \mathcal{M}^0([x]),$$

encoding the process of defining $\varphi_\alpha(\mathcal{O}(U,V))$ for all $\mathcal{O}(U,V) \subseteq [x]$.

First fix a 1-1 enumeration $\{U_\ell\} = \mathcal{B}$. Our structure will be a typed structure with types indexed by the pairs $(U,V) \in \mathcal{B} \times \mathbb{N}$. It will also have a binary relation $\leq$ and unary relations $R_\ell, \ell \in \mathbb{N}$. By the usual procedure this can be converted into a standard (untyped) structure: One simply replaces throughout an element $\mathcal{O}$ of type (U,V) by the triple $\langle U,V,\mathcal{O}\rangle$ and introduces a unary relation $S_{U,V}$ satisfied exactly by these triples.

Fix U,V. The elements of type (U,V) in $\mathcal{M}^0(x)$ are simply the $\mathcal{O}(U,V)$

contained in $[x]$. (So for some (U, V) this may be $\emptyset$, i.e., when $U \cap [x] = \emptyset$.) We define

$$R_\ell(\mathcal{O}(U, V)) \Leftrightarrow U_\ell \cap \mathcal{O}(U, V) \neq \emptyset$$

(So $\{R_\ell\}$ encode the closures of the local orbits.) Finally we let

$$\mathcal{O}(U', V') \leq \mathcal{O}(U, V) \Leftrightarrow U' \subseteq U \ \& \ V' \subseteq V \ \& \ \mathcal{O}(U', V') \subseteq \mathcal{O}(U, V).$$

It is clear that if $\mathcal{M}^0(x)$ is viewed as an untyped structure, then $\leq$ is a partial ordering. We can immediately reconstruct $\varphi_\alpha(\mathcal{O}(U, V))$ from $\mathcal{M}^0(x)$ by the following procedure:

$$\varphi_1(\mathcal{O}(U, V)) = \{U_\ell : R_\ell(\mathcal{O}(U, V))\},$$
$$\varphi_{\alpha+1}(\mathcal{O}(U, V)) = \{\langle U', V', \varphi_\alpha(\mathcal{O}(U', V'))\rangle :$$
$$\mathcal{O}(U', V') \leq \mathcal{O}(U, V)\},$$
$$\varphi_\lambda(\mathcal{O}(U, V)) = \{\varphi_\alpha(\mathcal{O}(U, V)) : \alpha < \lambda\}.$$

Thus it is clear that

$$x E_G^X y \Rightarrow \mathcal{M}^0(x) = \mathcal{M}^0(y) \Rightarrow \varphi_x = \varphi_y.$$

Next we would like for each $x \in X$ to encode $\mathcal{M}^0(x)$, up to isomorphism, as a structure on $\mathbb{N}$ (which will now depend on x and not on $[x]$). Toward that goal, first define an auxiliary structure $\mathcal{M}^1(x) \cong \mathcal{M}^0(x)$ (depending on x and not on $[x]$) by replacing each $\mathcal{O}(U, V)$ in $\mathcal{M}^0(x)$ by $\{g \in G : g \cdot x \in \mathcal{O}(U, V)\} = \mathcal{O}^x(U, V)$. (Note that $\mathcal{O}^x(U, V) \cdot x = \mathcal{O}(U, V)$.) So the elements of type (U, V) in $\mathcal{M}^1(x)$ are simply the equivalence classes of the following open equivalence relation on $U^x = \{g : g \cdot x \in U\}$:

$$g \sim_{U,V}^x g' \Leftrightarrow g \cdot x \sim_{U,V} g' \cdot x.$$

Define

$$R_\ell(\mathcal{O}^x(U, V)) \Leftrightarrow R_\ell(\mathcal{O}(U, V)),$$
$$\mathcal{O}^x(U', V') \leq \mathcal{O}^x(U, V) \Leftrightarrow \mathcal{O}(U', V') \leq \mathcal{O}(U, V)$$
$$\Leftrightarrow U' \subseteq U \ \& \ V' \subseteq V \ \& \ \mathcal{O}^x(U', V') \subseteq \mathcal{O}^x(U', V').$$

Finally, we define a structure $\mathcal{M}^2(x) \cong \mathcal{M}^1(x)$ by replacing each $\mathcal{O}^x(U, V)$ by $\mathcal{O}_0^x(U, V) = G_0 \cap \mathcal{O}^x(U, V)$ (where G_0 is the fixed countable dense subgroup of G). This is nonempty as $\mathcal{O}^x(U, V)$ is open in G. So the elements of type (U, V) in $\mathcal{M}^2(x)$ are simply the equivalence classes of the restriction of $\sim_{U,V}^x$ to G_0. We again define:

$$R_\ell(\mathcal{O}_0^x(U,V)) \Leftrightarrow R_\ell(\mathcal{O}^x(U,V)),$$
$$\mathcal{O}_0^x(U',V') \leq \mathcal{O}_0^x(U,V) \Leftrightarrow \mathcal{O}^x(U',V') \subseteq \mathcal{O}^x(U,V)$$
$$\Leftrightarrow U' \subseteq U' \;\&\; V' \subseteq V \;\&\; \mathcal{O}_0^x(U',V') \subseteq \mathcal{O}_0^x(U,V).$$

(The last equivalence is easily checked by using the fact that $\mathcal{O}^x(U,V) = [\mathcal{O}_0^x(U,V)]_{\sim_{U,V}^x}$, and $U' \subseteq U, V' \subseteq V \Rightarrow \sim_{U',V'}^x \subseteq \sim_{U,V}^x$).

Since the elements of reach type in $\mathcal{M}^2(x)$ are equivalence classes on the fixed countable set G_0, we can easily encode $\mathcal{M}^2(x)$ by an isomorphic copy with universe $\mathbb{N}$, call it $\mathcal{M}^3(x)$. Thus we have

$$xE_G^X y \quad \Rightarrow \quad \mathcal{M}^3(x) \cong \mathcal{M}^3(y)$$
$$\Leftrightarrow \quad \mathcal{M}^0(x) \cong \mathcal{M}^0(y)$$
$$\Rightarrow \quad \varphi_x = \varphi_y.$$

The last implication easily follows from the fact that if $\pi : \mathcal{M}^0(x) \to \mathcal{M}^0(y)$ is an isomorphism, then $\varphi_\alpha(\mathcal{O}(U,V)) = \varphi_\alpha(\pi(\mathcal{O}(U,V))$.

The only problem with the construction of $\mathcal{M}^3(x)$ is that the function $x \mapsto \mathcal{M}^3(x)$ is not necessarily Borel. This is because the equivalence relation $\sim_{U,V}^x |G_0$ is analytic but not necessarily Borel (uniformly in x), unless the equivalence relation E_G^X, and thus, by Becker-Kechris [96, 7.1.2], the function $(x,y) \mapsto G_{x,y} = \{g \in G : g \cdot x = y\}$, is Borel.

We will modify this construction in order to achieve this Borelness condition. Instead of using $\sim_{U,V}^x$, define the equivalence relation

$$g \approx_{U,V}^x g' \Leftrightarrow g \cdot x, g' \cdot x \in U \;\&\; \exists g_0, \cdots, g_k \in V \text{ such}$$
$$\text{that } g' = g_k g_{k-1} \cdots g_0 g \text{ and if}$$
$$x_0 = g \cdot x, x_{i+1} = g_i \cdot x_i, \text{ then } x_i \in U.$$

Thus $\approx_{U,V}^x \subseteq \sim_{U,V}^x$ and again $\approx_{U,V}^x$ is an open equivalence relation. Denote by $\tilde{\mathcal{O}}^x(U,V)$ a typical equivalence class of $\approx_{U,V}^x$. Now note the key fact that if $\tilde{\mathcal{O}}^x(U,V) \subseteq \mathcal{O}^x(U,V)$, then $\tilde{\mathcal{O}}^x(U,V) \cdot x = \mathcal{O}^x(U,V) \cdot x$. Because if $g \in \tilde{\mathcal{O}}^x(U,V)$, so that $g \in \mathcal{O}^x(U,V)$, and $g' \in \mathcal{O}^x(U,V)$, then $g \sim_{U,V}^x g'$, so there are $g_0, g_1, \cdots, g_k \in V$ such that if $x_0 = g \cdot x, x_{i+1} = g_i \cdot x_i$, we have $x_{k+1} = g' \cdot x$ and $x_i \in U$. But then if $g'' = g_k g_{k-1} \cdots g_0 g$, we have that $g'' \cdot x = g' \cdot x$ and $g'' \approx_{U,V}^x g$.

Now define the structure $\tilde{\mathcal{M}}^1(x)$ as follows: The elements of $\tilde{\mathcal{M}}^1(x)$ of type (U, V) are now the $\tilde{\mathbb{O}}^x(U, V)$. We define

$$R_\ell(\tilde{\mathbb{O}}^x(U, V)) \quad \Leftrightarrow \quad R_\ell(\mathbb{O}^x(U, V)), \text{ for the unique}$$
$$\mathbb{O}^x(U, V) \supseteq \tilde{\mathbb{O}}^x(U, V)$$
$$\Leftrightarrow \quad R_\ell(\tilde{\mathbb{O}}^x(U, V) \cdot x);$$
$$\tilde{\mathbb{O}}^x(U', V') \leq \tilde{\mathbb{O}}^x(U, V) \quad \Leftrightarrow \quad U' \subseteq U, V' \subseteq V \text{ and}$$
$$\tilde{\mathbb{O}}^x(U', V') \subseteq \tilde{\mathbb{O}}^x(U, V).$$

Thus the effect of passing from $\mathcal{M}^1(x)$ to $\tilde{\mathcal{M}}^1(x)$ is to replace each $\mathbb{O}^x(U, V)$ by the countably many $\tilde{\mathbb{O}}^x(U, V)$ contained in it, and define $\leq$ on these as before.

Next we define $\tilde{\mathcal{M}}^2(x)$ by replacing each $\tilde{\mathbb{O}}^x(U, V)$ by $\tilde{\mathbb{O}}^x(U, V) \cap G_0$, i.e., the elements of type (U, V) in $\tilde{\mathcal{M}}^2(x)$ are simply the equivalence classes of the restriction of $\approx_{U,V}^x$ to G_0. Clearly $\tilde{\mathcal{M}}^2(x) \cong \tilde{\mathcal{M}}^1(x)$. However $\approx_{U,V}^x |G_0$ is now Borel, uniformly in x, because for $g, g' \in G_0$:

$$g \approx_{U,V}^x g' \Leftrightarrow \quad g \cdot x, g' \cdot x \in U \ \& \ \exists g_0, \cdots, g_k \in V \text{ such}$$
$$\text{that } g' = g_k g_{k-1} \cdots g_0 g \text{ and if}$$
$$x_0 = g \cdot x, x_{i+1} = g_i \cdot x_i, \text{ then } x_i \in U$$
$$\Leftrightarrow \quad g \cdot x, g' \cdot x \in U \ \& \ \exists g_0', \cdots, g_k' \in V \cap G_0 \text{ such}$$
$$\text{that } g' = g_k' g_k' \cdots g_0' g \text{ and if}$$
$$y_0 = g \cdot x, y_{i+1} = g_i' \cdot y_i, \text{ then } y_i \in U.$$

To see the last equivalence, inductively define $h_0, h_1, \cdots, h_{k-1}$ so that if $g_0' = h_0^{-1} g_0, g_1' = h_1^{-1} g_1 h_0, \cdots, g_{k-1}' = h_{k-1}^{-1} g_{k-1} h_{k-1}$ and $y_0 = g \cdot x, y_{i+1} = g_i' \cdot y_i$ for $i < k$, then $y_i \in U$ and $g_i' \in G_0$. Let $g_k' = g_k h_{k-1}$. Then $g_k' \cdots g_0' = g_k \cdots g_0 = g'g^{-1} \in G_0$, so $g_k' \in G_0$ and $y_{k+1} = g_k' \cdot y_{k-1} = g' \cdot x$, so we are done.

It follows that we can replace $\tilde{\mathcal{M}}^2(x)$ by an isomorphic structure $\tilde{\mathcal{M}}^3(x)$ with universe $\mathbb{N}$, so that $x \mapsto \tilde{\mathcal{M}}^3(x)$ is now Borel.

We will next verify that

$$x E_G^X y \Rightarrow \tilde{\mathcal{M}}^1(x) \cong \tilde{\mathcal{M}}^1(y) (\Leftrightarrow \tilde{\mathcal{M}}^3(x) \cong \tilde{\mathcal{M}}^3(y)) \tag{1}$$

and

$$\tilde{\mathcal{M}}^1(x) \cong \tilde{\mathcal{M}}^1(y) \Rightarrow \varphi_x = \varphi_y. \tag{2}$$

For (1): Assume $g \cdot x = y$. Then it is trivial to check that the map

$$\pi_g(\tilde{\mathbb{O}}^x(U, V)) = \tilde{\mathbb{O}}^x(U, V)g^{-1}$$

is an isomorphism of $\tilde{M}^1(x)$ and $\tilde{M}^1(y)$. (Notice here that if $\tilde{O}^x(U,V) = [h]_{\approx^x_{U,V}}$, then $\tilde{O}^x(U,V)g^{-1} = [hg^{-1}]_{\approx^y_{U,V}}$.)

For (2): First for $\tilde{O}^x(U,V) \in \tilde{M}^1(x)$, we define $\varphi_\alpha(\tilde{O}^x(U,V))$ as follows:

$$\varphi_1(\tilde{O}^x(U,V)) = \varphi_1(\tilde{O}^x(U,V) \cdot x);$$
$$\varphi_{a+1}(\tilde{O}^x(U,V)) = \{\langle U',V',\varphi_\alpha(\tilde{O}^x(U',V'))\rangle : U' \subseteq U, V' \subseteq V \text{ and }$$
$$\tilde{O}^x(U',V') \subseteq \tilde{O}^x(U,V)\}$$
$$= \{\langle U',V',\varphi_\alpha(\tilde{O}^x(U',V'))\rangle : \tilde{O}^x(U',V') \leq \tilde{O}^x(U,V)\};$$
$$\varphi_\lambda(\tilde{O}^x(U,V)) = \{\varphi_\alpha(\tilde{O}^x(U,V)) : \alpha < \lambda\}.$$

Then it is clear that $\varphi_\alpha(\tilde{O}^x(U,V))$ depends only on $\tilde{O}^x(U,V)$ and the structure $\tilde{M}^1(x)$.

Next we claim that

$$\varphi_\alpha(\tilde{O}^x(U,V)) = \varphi_\alpha(\tilde{O}^x(U,V) \cdot x)$$

(recall here that $\tilde{O}^x(U,V) \cdot x = O^x(U,V) \cdot x = O(U,V)$ – an appropriate (U,V)-local orbit). This is easily proved by induction, noticing that if $U' \subseteq U$, $V' \subseteq V$ and $O(U',V') \subseteq O(U,V)$, then $\tilde{O}^x(U,V) \cdot x = O(U,V) \supseteq O(U',V')$, so there is $\tilde{O}^x(U',V') \subseteq \tilde{O}^x(U,V)$ with $\tilde{O}^x(U',V') \cdot x = O(U',V')$.

Now assume that $\tilde{M}^1(x) \cong \tilde{M}^1(y)$. To show that $\varphi_x = \varphi_y$, it is enough to show that for each $O(U,V) \in M^0(x)$, there is $O'(U,V) \in M^0(y)$ with $\varphi_\alpha(O(U,V)) = \varphi_\alpha(O'(U,V))$ for all $\alpha < \omega_1$ (and of course vice versa, which is clear by symmetry). Because then $\alpha(x) = \alpha(y)$, and since the type (X,G) in both $M^0(x), M^0(y)$ contains a unique element, i.e., $[x], [y]$ resp., we have that $\varphi_\alpha([x]) = \varphi_\alpha([y])$ for all α and we are done.

So fix $O(U,V) \in M^0(x)$. Consider $\tilde{O}^x(U,V)$ with $\tilde{O}^x(U,V) \cdot x = O(U,V)$. Let $\pi : \tilde{M}^1(x) \cong \tilde{M}^1(y)$ and put $\pi(\tilde{O}^x(U,V)) = \tilde{O}^y(U,V)$. Then if $O'(U,V) = \tilde{O}^y(U,V) \cdot y$, we have for each α,

$$\varphi_\alpha(O(U,V)) = \varphi_\alpha(\tilde{O}^x(U,V)) = \varphi_\alpha(\tilde{O}^y(U,V))$$
$$= \varphi_\alpha(O'(U,V)).$$

So we have proved the following, by putting $\tilde{M}(x) = \tilde{M}^3(x)$.

Proposition 13.8. *There is a countable language $\tilde{L}$ and a Borel map $x \in X \mapsto \tilde{M}(x) \in X_{\tilde{L}}$ such that*

$$xE_G^X y \Rightarrow \tilde{M}(x) \cong \tilde{M}(y) \Rightarrow \varphi_x = \varphi_y.$$

For each structure $\tilde{\mathcal{M}}(x) = \tilde{\mathcal{M}} \in X_{\tilde{L}}$ and each element $a = a(U,V) \in \tilde{\mathcal{M}}$ of type (U,V), we define for $1 \leq \alpha < \omega_1$ the set $\varphi_\alpha(a)$ as follows:

$$\varphi_1(a) = \{\ell : R_\ell(a)\},$$
$$\varphi_{\alpha+1}(a) = \{\langle U',V',\varphi_\alpha(a'(U',V')))\rangle : U' \subseteq U, V' \subseteq V, a' \leq a\},$$
$$\varphi_\lambda(a) = \{\varphi_\alpha(a) : \alpha < \lambda\}.$$

Also define an equivalence relation $\equiv_{U,V}^\alpha$ on the elements of $\tilde{\mathcal{M}}$ of type (U,V) by

$$a \equiv_{U,V}^1 b \Leftrightarrow \forall\ell(R_\ell(a) \Leftrightarrow R_\ell(b)),$$
$$a \equiv_{U,V}^{\alpha+1} b \Leftrightarrow \forall U' \subseteq U \forall V' \subseteq V \forall a'(U',V') \leq a \exists b'(U',V') \leq b(a' \equiv_{U',V'}^\alpha b')$$

and vice versa

$$a \equiv_{U,V}^\lambda b \Leftrightarrow \forall \alpha < \lambda(a \equiv_{U,V}^\alpha b).$$

Clearly,

$$a \equiv_{U,V}^\alpha b \Leftrightarrow \varphi_\alpha(a) = \varphi_\alpha(b).$$

Then by standard facts on positive arithmetical inductive definitions it follows that for some ordinal $\alpha_{\tilde{\mathcal{M}}} \leq \omega_1^{\tilde{\mathcal{M}}}$ we have

$$a \equiv_{U,V}^{\alpha_{\tilde{\mathcal{M}}}} b \Rightarrow \forall \alpha < \omega_1 (a \equiv_{U,V}^\alpha b)$$

for all (U,V) and a,b of type (U,V). Equivalently,

$$\varphi_{\alpha_{\tilde{\mathcal{M}}}}(a) = \varphi_{\alpha_{\tilde{\mathcal{M}}}}(b) \Rightarrow \forall \alpha < \omega_1(\varphi_\alpha(a) = \varphi_\alpha(b)).$$

Thus we have

Proposition 13.9. *In the notation of Proposition 13.8, $\alpha_x \leq \omega_1^{\tilde{\mathcal{M}}(x)}$.*

We will now modify the structure $\tilde{\mathcal{M}}(x)$ to another structure, $\mathcal{M}(x)$, in a countable language L, in order to get a further desired property, namely that the set

$$M = \{\mathcal{M} \in X_L : \mathcal{M} \cong \mathcal{M}(x) \text{ for some } x \in X\}$$

is Borel and there is a Borel map

$$\mathcal{M} \in M \mapsto x(\mathcal{M}) \in X,$$

such that

$$\mathcal{M}(x(\mathcal{M})) \cong \mathcal{M}.$$

We again start with the structure $\tilde{\mathcal{M}}^1(x)$ defined earlier. For each $g_0 \in G_0$ we can define a function F_{g_0} on the universe of this structure, as follows: Consider a type (U, V) and let $U' = g_0 \cdot U, V' = g_0 V g_0^{-1}$. For each element $\tilde{O}^x(U, V)$ of type (U, V) in $\tilde{M}^1(x)$ define the element

$$F_{g_0}(\tilde{O}^x(U, V))$$

of type (U', V') in $\tilde{M}^1(x)$ by

$$F_g(\tilde{O}^x(U, V)) = g_0 \tilde{O}^x(U, V).$$

(It is easy to check that this is well-defined, i.e., $g_0 \tilde{O}(U, V)$ is an equivalence class of $\approx_{U', V'}^x$: If $\tilde{O}^x(U, V) = [h]_{\approx_{U,V}^x}$, then $g_0 \tilde{O}^x(U, V) = [g_0 h]_{\approx_{U', V'}^x}$.) Denote by

$$\tilde{\tilde{M}}^1(x) = \langle \tilde{M}^1(x), F_{g_0} \rangle_{g_0 \in G_0},$$

the expansion of $\tilde{M}^1(x)$ by the addition of these functions F_{g_0}. Since also

$$g_0(\tilde{O}^x(U, V) \cap G_0) = (g_0 \tilde{O}^x(U, V)) \cap G_0,$$

F_{g_0} is naturally defined also on the universe of $\tilde{M}^2(x)$, and is denoted again by F_{g_0}. Let

$$\tilde{\tilde{M}}^2(x) = \langle \tilde{M}^2(x), F_{g_0} \rangle_{g_0 \in G_0}$$

be the corresponding expansion and $\tilde{\tilde{M}}^3(x)$ the isomorphic copy of $\tilde{\tilde{M}}^2(x)$ with universe $\mathbb{N}$, so that $\tilde{\tilde{M}}^3(x)$ is of the form

$$\tilde{\tilde{M}}^3(x) = \langle \tilde{M}^3(x), F_{g_0} \rangle_{g_0 \in G_0}.$$

i.e., again an expansion of $\tilde{M}^3(x)$. It is clear that $\tilde{\tilde{M}}^3(x) \cong \tilde{\tilde{M}}^2(x) \cong \tilde{\tilde{M}}^1(x)$ and that $x \mapsto \tilde{\tilde{M}}^3(x)$ is again Borel. We next verify that we still have the following properties:

$$x E_G^X y \Rightarrow \tilde{\tilde{M}}^1(x) \cong \tilde{\tilde{M}}^1(y) (\Leftrightarrow \tilde{\tilde{M}}^3(x) \cong \tilde{\tilde{M}}^3(y)) \tag{3}$$

and

$$\tilde{\tilde{M}}^1(x) \cong \tilde{\tilde{M}}^1(y) \Rightarrow \varphi_x = \varphi_y. \tag{4}$$

Since $\tilde{\tilde{M}}^1(x) \cong \tilde{\tilde{M}}^1(y) \Rightarrow \tilde{M}^1(x) \cong \tilde{M}^1(y)$, it is enough to verify (3).

For (3): If $x E_G^X y$, say $g \cdot x = y$, note that $\pi_g(\tilde{O}^x(U, V)) = \tilde{O}^x(U, V) g^{-1}$ is an isomorphism of $\tilde{M}^1(x)$ and $\tilde{M}^1(y)$. So it is enough to check that it preserves F_{g_0}, i.e., $\pi_g(F_{g_0}(\tilde{O}^x(U, V))) = F_{g_0}(\pi_g(\tilde{O}^x(U, V)))$, which is clear as both sides are equal to

$$g_0 \tilde{O}^x(U, V) g^{-1}.$$

Let

$$\mathcal{M}(x) = \tilde{\mathcal{M}}^3(x).$$

The following result summarizes the basic properties of $\mathcal{M}(x)$ proved so far, and states the aforementioned additional property.

Theorem 13.10 (Hjorth [00]). *Let G be a Polish group and X a Polish G-space. Then there is a countable language L and a Borel map $x \in X \mapsto \mathcal{M}(x) \in X_L$ such that*

(i) $xE_G^X y \Rightarrow \mathcal{M}(x) \cong \mathcal{M}(y) \Rightarrow \varphi_x = \varphi_y.$

(ii) $M = \{\mathcal{M} \in X_L : \exists x \in X(\mathcal{M} \cong \mathcal{M}(x))\}$ *is Borel and there is a Borel map $\mathcal{M} \in M \mapsto x(\mathcal{M}) \in X$ such that*

$$\mathcal{M}(x(\mathcal{M})) \cong \mathcal{M}.$$

Proof. We have already noted (i). We will omit the proof of (ii) which can be found in Hjorth [00], 6.2. ⊣

Remark. It is clear that the analog of Proposition 13.9 goes through as well for the structure $\mathcal{M}(x)$.

Before we state the next result we will need the following fact from topology.

Proposition 13.11. *Let (X, τ) be a Polish space and $A \subseteq X$ a set which is the intersection of a closed and an open set. Then the topology generated by $\tau \cup \{A\}$ is Polish. Similarly, if $A_n \subseteq X, n \in \mathbb{N}$, and each A_n is the intersection of a closed and an open set, the topology generated by $\tau \cup \{A_n : n \in \mathbb{N}\}$ is Polish.*

Proof. Let $A = W \cap F$ with W open, F closed. Put $H = X \setminus W$. Let $d \leq 1$ be a compatible metric for X and assume that $X \neq W$ (otherwise $A = F$ is closed and the result is clear). Define

$$f : X \to [0, 1]$$

by

$$f(x) = \begin{cases} d(x, H), & \text{if } x \in A \\ 0, & \text{if } x \notin A. \end{cases}$$

Easily graph$(f) = X' \subseteq X \times [0, 1]$ is G_δ, so Polish in the relative topology. Using the bijection $x \mapsto (x, f(x))$ between X and X', transfer this topology to X and notice that it is generated by $\tau \cup \{A\}$.

The second statement follows from what we just proved and Kechris [95, 13.3]. ⊣

Proposition 13.12. *For each $\alpha < \omega_1, x \in X$,, there is a Polish topology $\tau_\alpha = \tau_\alpha^x$ on X such that*

(i) $\tau_0 = $ *the topology of X.*

(ii) $\alpha \le \beta \Rightarrow \tau_\alpha \subseteq \tau_\beta$.

(iii) *For $\alpha \ge 1$, τ_α is generated by the following sets (where U, U' vary over $\mathcal{B}$ and V, V' over $\mathcal{N}$):*

   (a) U;

   (b) $\{y \in U : U' \cap \mathcal{O}(y, U, V) = \emptyset\}$;

   (c) $\{y : \varphi_\beta(y, U, V) = \varphi_\beta(\mathcal{O}(U, V))\}, \mathcal{O}(U, V) \subseteq [x], \beta \le \alpha$;

   (d) *For $\alpha > 1$,* $\{y : \varphi_{\beta+1}(y, U, V) \subseteq \varphi_{\beta+1}(\mathcal{O}(U, V))\}, \mathcal{O}(U, V) \subseteq [x], \beta + 1 \le \alpha$;

   (e) *For $\alpha > 1$,* $\{y : \langle U', V', \varphi_\beta(\mathcal{O}(U', V'))\rangle \notin \varphi_{\beta+1}(y, U, V)\}$, $\mathcal{O}(U', V') \subseteq [x], U' \subseteq U, V' \subseteq V, \beta + 1 \le \alpha$;

   (f) *For $\alpha > 1$,* $\{y : \forall \mathcal{O}(U, V) \subseteq [x](\varphi_\beta(y, U, V) \ne \varphi_\beta(\mathcal{O}(U, V)))\}$, $\beta + 1 \le \alpha$.

Proof. Notice that each set of the form $C_{U',U,V} = \{y \in U : U' \cap \mathcal{O}(y, U, V) = \emptyset\}$ is the intersection of an open and a closed set in τ_0, so, by 13.11, the topology generated by τ_0 and these sets is Polish. Call it τ_0'. Now

$$\varphi_1(y, U, V) = \varphi_1(\mathcal{O}(U, V)) \Leftrightarrow$$
$$y \in U \ \& \ \forall U'(y \in C_{U',U,V} \Rightarrow U' \cap \mathcal{O}(U, V) = \emptyset) \ \&$$
$$\forall U'(U' \cap \mathcal{O}(U, V) = \emptyset \Rightarrow U' \cap \mathcal{O}(y, U, V) = \emptyset),$$

so $\{y : \varphi_1(y, U, V) = \varphi_1(\mathcal{O}(U, V))\}$ is the intersection of a closed and an open set in τ_0', so the topology generated by τ_0' and these sets is Polish. Call it τ_1.

Assume now τ_α is defined and consider $\alpha + 1$. First notice that the sets

$$\{y : \forall \mathcal{O}(U, V) \subseteq [x](\varphi_\alpha(y, U, V) \ne \varphi_\alpha(\mathcal{O}(U, V)))\}$$

are closed in τ_α, so let τ_α' be the Polish topology generated by τ_α and these sets.

We next claim that the sets

$$\{y : \varphi_{\alpha+1}(y, U, V) \subseteq \varphi_{\alpha+1}(\mathcal{O}(U, V))\}, \mathcal{O}(U, V) \subseteq [x] \qquad (5)$$

$$\{y : \langle U', V', \varphi_\alpha(\mathcal{O}(U', V')) \rangle \notin \varphi_{\alpha+1}(y, U, V)\}, \mathcal{O}(U', V') \subseteq [x], \qquad (6)$$

are the intersection of U and a closed set in τ'_α. Since for $\mathcal{O}(U, V) \subseteq [x]$,

$$\varphi_{\alpha+1}(y, U, V) = \varphi_{\alpha+1}(\mathcal{O}(U, V)) \Leftrightarrow y \in U \ \&$$
$$\varphi_{\alpha+1}(y, U, V) \subseteq \varphi_{\alpha+1}(\mathcal{O}(U, V)) \ \&$$
$$\forall U' \subseteq U \forall V' \subseteq V \forall \mathcal{O}(U', V') \subseteq \mathcal{O}(U, V)$$
$$(\langle U', V', \varphi_\alpha(\mathcal{O}(U', V')) \rangle \in \varphi_{\alpha+1}(y, U, V)),$$

it follows that

$$\{y : \varphi_{\alpha+1}(y, U, V) = \varphi_{\alpha+1}(\mathcal{O}(U, V))\}, \qquad (7)$$

for $\mathcal{O}(U, V) \subseteq [x]$, is the intersection of an open set and closed set in the Polish topology τ''_α generated by the τ'_α and the sets (5), (6) above, so we put $\tau_{\alpha+1}$ = the topology generated by τ''_α and the sets of the form (7).

Proof of the Claim.

For (6). Fix $y \in U$ such that

$$\langle U', V', \varphi_\alpha(\mathcal{O}(U', V')) \rangle \in \varphi_{\alpha+1}(y, U, V),$$

so for some $\mathcal{O}_y(U', V') \subseteq \mathcal{O}(y, U, V)$ we have

$$\varphi_\alpha(\mathcal{O}(U', V')) = \varphi_\alpha(\mathcal{O}_y(U', V')).$$

Let $g_0, \cdots, g_k \in V$ be such that if $y = y_0, g_i \cdot y_i = y_{i+1}$, then $y_i \in U$ and $y_{k+1} \in \mathcal{O}_y(U', V')$. Choose $h \in V$ so that $hg_k \cdots g_0 = h_0 \in G_0$ and $h \cdot y_{k+1} \in \mathcal{O}_y(U', V')$, so $\varphi_\alpha(h_0 \cdot y, U', V') = \varphi_\alpha(\mathcal{O}(U', V'))$. Then $h_0^{-1} \cdot \varphi_\alpha(h_0 \cdot y, U', V') = h_0^{-1} \cdot \varphi_\alpha(\mathcal{O}(U', V'))$. Put $h_0^{-1} \cdot U' = U'', h_0^{-1} V' h_0 = V''$, so that

$$\varphi_\alpha(y, U'', V'') = \varphi_\alpha(\mathcal{O}(U'', V'')),$$

for $\mathcal{O}(U'', V'') = h_0^{-1} \cdot \mathcal{O}(U', V') \subseteq [x]$. Now the set

$$\{\bar{y} : \bar{y} \in U \ \& \ g_0 \cdot \bar{y} \in U \ \& \ g_1 g_0 \cdot \bar{y} \in U \ \& \ \cdots \ \& \ g_k \cdots g_0 \cdot \bar{y} \in U \ \&$$
$$hg_k \cdots g_0 \cdot \bar{y} \in U\}$$

is open in τ_0 and contains y. Also the set

$$\{\bar{y} : \varphi_\alpha(\bar{y}, U'', V'') = \varphi_\alpha(\mathcal{O}(U'', V''))\}$$

is open in τ_α and contains y. If $\bar{y} \in U$ belongs to both of these sets then, by reversing the above steps, we get that

$$\varphi_\alpha(h_0 \cdot \bar{y}, U', V') = \varphi_\alpha(\mathcal{O}(U', V'))$$

and $h_0 \cdot \bar{y} \in \mathcal{O}(\bar{y}, U, V)$, so

$$\langle U', V', \varphi_\alpha(\mathcal{O}(U', V')) \rangle \in \varphi_{\alpha+1}(\bar{y}, U, V).$$

For (5): Fix $y \in U$ with

$$\varphi_{\alpha+1}(y, U, V) \not\subseteq \varphi_{\alpha+1}(\mathcal{O}(U, V)).$$

Fix $U' \subseteq U$, $V' \subseteq V$, $\mathcal{O}_y(U', V') \subseteq \mathcal{O}(y, U, V)$ with

$$\langle U', V', \varphi_\alpha(\mathcal{O}_y(U', V')) \rangle \not\subseteq \varphi_{\alpha+1}(\mathcal{O}(U, V))$$

Fix $g_0, \cdots, g_k, h \in V, h_0 \in G_0, y_i \in U, y_{k+1} \in \mathcal{O}_y(U', V')$ as in the case of (6) before, so that $\varphi_\alpha(h_0 \cdot y, U', V') \not\subseteq \varphi_{\alpha+1}(\mathcal{O}(U, V))$. For U'', V'' as before again, we get

$$\varphi_\alpha(y, U'', V'') \not\subseteq \varphi_{\alpha+1}(\mathcal{O}(U^*, V^*)),$$

where $h_0^{-1} \cdot U = U^*$, $h_0^{-1} V h_0 = V^*$, for some $\mathcal{O}(U^*, V^*) \subseteq [x]$.

If now $\forall \mathcal{O}(U'', V'') \subseteq [x](\varphi_\alpha(y, U'', V'') \neq \varphi_\alpha(\mathcal{O}(U'', V'')))$, then any $\bar{y}$ satisfying this condition (which is open in τ_α) and satisfying ($\bar{y} \in U$ & $g_0 \cdot \bar{y} \in U$ & $\cdots$ & $g_k \cdots g_0 \cdot \bar{y} \in U$ & $h g_k \cdots g_0 \in U$) (which is open in τ_0), must satisfy that $\varphi_\alpha(h_0 \cdot \bar{y}, U', V') \neq \varphi_\alpha(\mathcal{O}(U', V'))$ for any $\mathcal{O}(U', V') \subseteq [x]$, so clearly $\varphi_{\alpha+1}(\bar{y}, U, V) \not\subseteq \varphi_{\alpha+1}(\mathcal{O}, U, V))$.

Otherwise, let $\mathcal{O}(U'', V'') \subseteq [x]$ be such that

$$\varphi_\alpha(y, U'', V'') = \varphi_\alpha(\mathcal{O}(U'', V'')).$$

Let $\bar{y}$ satisfy this condition, which is open in τ_α, and ($\bar{y} \in U$ & $g_0 \cdot \bar{y} \in U$ & $\cdots$ & $g_k \cdots g_0 \cdot \bar{y} \in U$ & $h g_k \cdots g_0 \in U$). Then, as before, $\varphi_\alpha(h_0 \cdot \bar{y}, U', V') = \varphi_\alpha(h_0 \cdot y, U', V') \not\subseteq \varphi_{\alpha+1}(\mathcal{O}(U, V))$, so

$$\varphi_{\alpha+1}(\bar{y}, U, V) \not\subseteq \varphi_{\alpha+1}(\mathcal{O}(U, V)).$$

This finishes the proof of the claim and the successor case.

For λ limit, it is enough to show that each set of the form

$$\{y : \varphi_\lambda(y, U, V) = \varphi_\lambda(\mathcal{O}(U, V))\}, \text{ for } \mathcal{O}(U, V) \subseteq [x],$$

is the intersection of U and a closed set in the topology generated by $\bigcup_{\alpha < \lambda} \tau_\lambda$, because then we can take $\tau_\lambda = $ the topology generated by

$\bigcup_{\alpha<\lambda} \tau_\alpha$ and these sets. To prove this, it is enough to show that for each $\alpha < \lambda$,

$$\{y : \varphi_{\alpha+1}(y, U, V) \neq \varphi_{\alpha+1}(\mathcal{O}(U,V))\}$$

is open in $\tau_{\alpha+1}$. Now

$$\varphi_{\alpha+1}(y, U, V) \neq \varphi_{\alpha+1}(\mathcal{O}(U,V)) \Leftrightarrow y \in U \ \&$$
$$[(\varphi_{\alpha+1}(y, U, V) \nsubseteq \varphi_{\alpha+1}(\mathcal{O}(U,V))) \text{ or}$$
$$\exists U' \subseteq U \exists V' \subseteq V \exists \mathcal{O}(U', V') \subseteq \mathcal{O}(U,V)$$
$$(\langle U', V', \varphi_\alpha(\mathcal{O}(U', V'))\rangle \notin \varphi_{\alpha+1}(y, U, V))].$$

Now the first condition is open in $\tau'_\alpha \subseteq \tau_{\alpha+1}$ and the second is also open in $\tau_{\alpha+1}$, so we are done. $\dashv$

Put

$$X_{(\varphi_x)} = \{y \in X : \varphi_x = \varphi_y\}.$$

Let τ_x be the topology $\tau^x_{\alpha(x)+2}$ defined as before. Notice that τ_x depends only on φ_x, so we can write

$$\tau_{(\varphi_x)} = \tau_x.$$

Clearly, $X_{(\varphi_x)}$ is an invariant Borel subset of X. Since

$$\varphi_y = \varphi_x \Leftrightarrow \varphi_{\alpha(x)+2}(y, X, G) = \varphi_{\alpha(x)+2}(x, X, G),$$

it follows that $X_{(\varphi_x)}$ is open in $\tau_{(\varphi_x)}$, so $(X_{(\varphi_x)}, \tau_{(\varphi_x)})$ is a Polish space. We next claim that the topology $\tau_{(\varphi_x)}$ on $X_{(\varphi_x)}$ is generated only by the open sets of type (c) in Proposition 13.11.

Proposition 13.13. *The topology $\tau_{(\varphi_x)}$ on $X_{(\varphi_x)}$ is generated by the sets of the form*

$$\{y \in X_{(\varphi_x)} : \varphi_\beta(y, U, V) = \varphi_\beta(\mathcal{O}(U,V))\}$$

for $\mathcal{O}(U,V) \subseteq [x]$, $\beta \leq \alpha(x)$, and these sets also form a basis. Moreover for each $\alpha < \omega_1$, $\mathcal{O}(U,V) \subseteq [x]$, there is $\beta \leq \alpha(x)$, with

$$\{y \in X_{(\varphi_x)} : \varphi_\alpha(y, U, V) = \varphi_\alpha(\mathcal{O}(U,V))\} =$$
$$\{y \in X_{(\varphi_x)} : \varphi_\beta(y, U, V) = \varphi_\beta(\mathcal{O}(U,V))\}.$$

Proof. We prove the second assertion first. Notice that for $y \in X_{(\varphi_x)}$, $\alpha < \omega_1$,

$$\{\varphi_\alpha(\mathcal{O}_y(U,V)) : \mathcal{O}_y(U,V) \subseteq [y]\}$$
$$= \{\varphi_\alpha(\mathcal{O}_x(U,V)) : \mathcal{O}_x(U,V) \subseteq [x]\}.$$

Because if $\mathcal{O}_y(U,V) \subseteq [y]$, find $\mathcal{O}_x(U,V) \subseteq [x]$ such that

$$\varphi_{\alpha(x)+1}(\mathcal{O}_y(U,V)) = \varphi_{\alpha(x)+1}(\mathcal{O}_x(U,V)),$$

and so

$$\varphi_{\alpha(x)}(\mathcal{O}_y(U,V)) = \varphi_{\alpha(x)}(\mathcal{O}_x(U,V)).$$

Then by the proof of Proposition 13.7, we have that

$$\varphi_\alpha(\mathcal{O}_y(U,V)) = \varphi_\alpha(\mathcal{O}_x(U,V)).$$

So fix $\mathcal{O}(U,V) \subseteq [x]$ and $\alpha < \omega_1$ and consider

$$\{y \in X_{(\varphi_x)} : \varphi_\alpha(y,U,V) = \varphi_\alpha(\mathcal{O}(U,V)\}.$$

If $\alpha \le \alpha(x)$, there is nothing to prove. So assume $\alpha > \alpha(x)$. Then

$$\{y \in X_{(\varphi_x)} : \varphi_\alpha(y,U,V) = \varphi_\alpha(\mathcal{O}(U,V))\}$$
$$= \{y \in X_{(\varphi_x)} : \varphi_{\alpha(x)}(y,U,V) = \varphi_{\alpha(x)}(\mathcal{O}(U,V)))\},$$

by preceding remarks, so we are done.

We will now prove the first assertion.

First note that every open set of type (f) in Proposition 13.12 (iii) has empty intersection with $X_{(\varphi_x)}$, so these sets can be neglected. Concerning sets of type (d) in Proposition 13.12 (iii) notice that

$$\varphi_{\beta+1}(y,U,V) \subseteq \varphi_{\beta+1}(\mathcal{O}(U,V)) \Leftrightarrow$$
$$\exists \mathcal{O}'(U,V) \subseteq [x](\varphi_{\beta+1}(y,U,V) = \varphi_{\beta+1}(\mathcal{O}'(U,V)) \ \&$$
$$\varphi_{\beta+1}(\mathcal{O}'(U,V)) \subseteq \varphi_{\beta+1}(\mathcal{O}(U,V))),$$

and for those of type (e) in Proposition 13.12 (iii) notice that

$$\langle U',V',\varphi_\beta(\mathcal{O}(U',V'))\rangle \notin \varphi_{\beta+1}(y,U,V) \Leftrightarrow$$
$$\exists \mathcal{O}'(U,V) \subseteq [x](\varphi_{\beta+1}(\mathcal{O}'(U,V)) = \varphi_{\beta+1}(y,U,V) \ \&$$
$$\langle U',V',\varphi_\beta(\mathcal{O}(U',V'))\rangle \notin \varphi_{\beta+1}(\mathcal{O}'(U,V))).$$

For sets of type (a), (b) in Proposition 13.12 (iii) notice that

$$U \cap X_{(\varphi_x)} = \bigcup_{\mathcal{O}(U,V) \subseteq [x]} \{y \in X_{(\varphi_x)} : \varphi_1(y,U,V) = \varphi_1(\mathcal{O}(U,V))\}$$

and

$$\{y \in U : U' \cap \mathcal{O}(y,U,V) = \emptyset\} \cap X_{(\varphi_x)} =$$
$$\bigcup_{\mathcal{O}(U,V) \subseteq [x], U' \cap \mathcal{O}(U,V) = \emptyset} \{y \in X_{(\varphi_x)} : \varphi_1(y,U,V) = \varphi_1(\mathcal{O}(U,V))\}.$$

Finally, we check that the sets of the form

$$\{y \in X_{(\varphi_x)} : \varphi_\alpha(y, U, V) = \varphi_\alpha(\mathcal{O}(U, V))\}$$

form a basis. Let $y \in X_{(\varphi_x)}$ satisfy the conditions

$$\varphi_{\alpha_1}(y, U_1, V_1) = \varphi_{\alpha_1}(\mathcal{O}(U_1, V_1)), \cdots,$$
$$\varphi_{(\alpha_k)}(y, U_k, V_k) = \varphi_{\alpha_k}(\mathcal{O}(U_k, V_k)).$$

Let $\alpha = \max\{\alpha_1, \cdots, \alpha_k\}$, let $y \in U \subseteq U_i$ and $V \subseteq V_i$, for $i \leq k$. Then

$$y \in \{\bar{y} : \varphi_\alpha(\bar{y}, U, V) = \varphi_\alpha(y, U, V)\} = N,$$

N is of the form

$$\{\bar{y} : \varphi_\alpha(\bar{y}, U, V) = \varphi_\alpha(\mathcal{O}(U, V))\},$$

and every $\bar{y} \in N$, satisfies, by Proposition 13.5,

$$\varphi_\alpha(\bar{y}, U_i, V_i) = \varphi_\alpha(y, U_i, V_i),$$

so, as $\alpha_i \leq \alpha$,

$$\varphi_{\alpha_i}(\bar{y}, U_i, V_i) = \varphi_{\alpha_i}(y, U_i, V_i) = \varphi_{\alpha_i}(\mathcal{O}(U_i, V_i)),$$

and the proof is complete.   ⊣

Proposition 13.14. *The action of G on $(X_{(\varphi_x)}, \tau_{(\varphi_x)})$ is continuous.*

Proof. It is sufficient to show that it is separately continuous.

$g \mapsto g \cdot y$ *is continuous:* Note that the condition

$$\varphi_\alpha(g \cdot y, U, V) = \varphi_\alpha(\mathcal{O}(U, V))$$

depends only on $\mathcal{O}(g \cdot y, U, V)$, so since $\mathcal{O}(g' \cdot y, U, V) = \mathcal{O}(g \cdot y, U, V)$ for g' in a sufficiently small nbhd of g, it is clear that $\{g : \varphi_\alpha(g \cdot y, U, V) = \varphi_\alpha(\mathcal{O}(U, V))\}$ is open.

$y \mapsto g \cdot y$ *is continuous:* Suppose that $\varphi_\alpha(g \cdot y, U, V) = \varphi_\alpha(\mathcal{O}(U, V))$. Let $h \in V$ be such that $hg = h_0 \in G_0$ and $h_0 \cdot y \in U$. Then $\varphi_\alpha(g \cdot y, U, V) = \varphi_\alpha(h_0 \cdot y, U, V) = \varphi_\alpha(\mathcal{O}(U, V))$, so, by applying h_0^{-1}, we get that $\varphi_\alpha(y, U', V') = \varphi_\alpha(\mathcal{O}(U', V'))$, for $U' = h_0^{-1} \cdot U, V' = h_0^{-1}Vh_0$ and an appropriate $\mathcal{O}(U', V') \subseteq [x]$. If now $\bar{y}$ satisfies the open condition

$$\varphi_\alpha(\bar{y}, U', V') = \varphi_\alpha(\mathcal{O}(U', V')) \ \& \ g \cdot \bar{y} \in U,$$

which is satisfied by y, then we get, by applying h_0, that

$$\varphi_\alpha(h_0 \cdot \bar{y}, U, V) = \varphi_\alpha(\mathcal{O}(U, V))$$

and $g \cdot \bar{y} \in \mathcal{O}(h_0 \cdot \bar{y}, U, V)$, so

$$\varphi_\alpha(g \cdot \bar{y}, U, V) = \varphi_\alpha(\mathcal{O}(U, V))$$

and we are done. ⊣

Proposition 13.15. *Consider the Polish G-space* $(X_{(\varphi_x)}, \tau_{(\varphi_x)})$. *Every orbit is dense and every local orbit is somewhere dense. So if every orbit is meager, then this Polish G-space is turbulent.*

Proof. Consider a basic nbhd

$$N = \{y \in X_{(\varphi_x)} : \varphi_\alpha(y, U, V) = \varphi_\alpha(\mathcal{O}(U, V))\},$$

where $\mathcal{O}(U, V)$ is a local orbit of the Polish G-space X contained in $[x]$. Given $y \in X_{(\varphi_x)}$ there is $\mathcal{O}'(U, V) \subseteq [y]$ with $\varphi_\alpha(\mathcal{O}'(U, V)) = \varphi_\alpha(\mathcal{O}(U, V))$. Let $g \cdot y \in \mathcal{O}'(U, V)$. Then $\varphi_\alpha(g \cdot y, U, V) = \varphi_\alpha(\mathcal{O}'(U, V)) = \varphi_\alpha(\mathcal{O}(U, V))$, so $g \cdot y \in N$. Thus every orbit of the Polish G-space $(X_{(\varphi_x)}, \tau_{(\varphi_x)})$ is dense.

Next we check that every local orbit of the action of G on $(X_{(\varphi_x)}, \tau_{(\varphi_x)})$ is somewhere dense in $(X_{(\varphi_x)}, \tau_{(\varphi_x)})$. Let N vary over the basic nbhds of $(X_{(\varphi_x)}, \tau_{(\varphi_x)})$, i.e., the sets of the form

$$N = \{y \in X_{(\varphi_x)} : \varphi_\alpha(\mathcal{O}(y, U, V)) = \varphi_\alpha(\mathcal{O}(U, V))\},$$

for $\mathcal{O}(U, V) \subseteq [x]$. Denoting by $\mathcal{O}^*$ local orbits in the space $(X_{(\varphi_x)}, \tau_{(\varphi_x)})$, we note that for $y \in N$ as above and $V' \subseteq V$ we have that $\mathcal{O}^*(y, N, V') = \mathcal{O}(y, U, V')$, so it is clearly enough to show that every local orbit $\mathcal{O}(U, V)$ contained in $X_{(\varphi_x)}$ is somewhere dense in the space $(X_{(\varphi_x)}, \tau_{(\varphi_x)})$.

Fix such a $\mathcal{O}(U, V)$. Let

$$W = \{y \in X_{(\varphi_x)} : \exists U'' \subseteq U, V'' \subseteq V, \mathcal{O}(U'', V'') \subseteq \mathcal{O}(U, V)$$
$$[\varphi_{\alpha(x)+1}(\mathcal{O}(y, (U'', V''))) = \varphi_{\alpha(x)+1}(\mathcal{O}(U'', V''))]\}$$

Clearly W is a nonempty open set in $(X_{(\varphi_x)}, \tau_{(\varphi_x)})$. We will show that $\mathcal{O}(U, V)$ is dense in it (in the topology $\tau_{(\varphi_x)}$). So fix a basic nbhd

$$N' = \{y \in X_{(\varphi_x)} : \varphi_\alpha(\mathcal{O}(y, U', V')) = \varphi_\alpha(\mathcal{O}(U', V'))\},$$

with $\mathcal{O}(U', V') \subseteq [x]$, which intersects W, in order to show that it intersects $\mathcal{O}(U, V)$.

Let $y \in N' \cap W$, so that $\varphi_\alpha(\mathcal{O}(y, U', V')) = \varphi_\alpha(\mathcal{O}(U', V'))$ and for some $U'' \subseteq U, V'' \subseteq V, \mathcal{O}(U'', V'') \subseteq \mathcal{O}(U, V)$ we have

$$\varphi_{\alpha(x)+1}(\mathcal{O}(y, U'', V'')) = \varphi_{\alpha(x)+1}(\mathcal{O}(U'', V'')).$$

Since $y \in U' \cap U''$, let $U''' \subseteq U' \cap U''$ be such that $y \in U'''$, and let $V''' \subseteq V' \cap V''$. Since

$$\langle U''', V''', \varphi_{\alpha(x)}(y, U''', V''') \rangle \in \varphi_{\alpha(x)+1}(y, U'', V'')$$
$$= \varphi_{\alpha(x)+1}(\mathcal{O}(U'', V'')),$$

let $z \in \mathcal{O}(U'', V''') \subseteq \mathcal{O}(U, V)$ be such that

$$\varphi_{\alpha(x)}(y, U''', V''') = \varphi_{\alpha(x)}(z, U''', V''').$$

Thus $\varphi_\alpha(y, U''', V''') = \varphi_\alpha(z, U''', V''')$, so by Proposition 13.5,

$$\varphi_\alpha(z, U', V') = \varphi_\alpha(y, U', V') = \varphi_\alpha(\mathcal{O}(U', V')),$$

i.e., $z \in N'$. Since $z \in \mathcal{O}(U'', V'') \subseteq \mathcal{O}(U, V), N' \cap \mathcal{O}(U, V) \neq \emptyset$ and we are done.   ⊣

Definition 13.16. A Polish group G is called *a GE (Glimm-Effros) group* if for any minimal Polish G-space X (i.e., one for which every orbit is dense), if X contains a G_δ orbit, then X is transitive (i.e., has only one orbit).

It is known that the following Polish groups are *GE* (see Hjorth [00]):

 (i) nilpotent (Hjorth-Solecki);

 (ii) admitting an invariant compatible metric (Hjorth-Solecki);

(iii) countable products of locally compact groups (Hjorth).

Corollary 13.17. *If G is a GE group, X a Polish G-space and for some $x, X_{(\varphi_x)}$ has more than one orbit, then the Polish space $(X_{(\varphi_x)}, \tau_{(\varphi_x)})$ is turbulent.*

Proof. It is enough to show that every orbit in $X_{(\varphi_x)}$ is meager in the space $(X_{(\varphi_x)}, \tau_{(\varphi_x)})$. Otherwise one of these orbits is dense G_δ, a contradiction.   ⊣

We can now state the final main theorem:

Theorem 13.18 (Hjorth [00]). *Let G be a GE Polish group and X a Polish G-space. Then exactly one of the following holds:*

 (I) *There is a turbulent G-space Y with $E_G^Y \leq_B E_G^X$.*

or

 (II) *There is a Polish S_∞-space Z with $E_G^X \leq_B E_{S_\infty}^Z$.*

Moreover we have the following equivalences:

(I) *is equivalent to:*

(I)′ *There is an invariant Borel set $X_0 \subseteq X$ and a Polish topology τ_0 on X_0, extending its relative topology, so that (X_0, τ_0) is a turbulent Polish G-space.*

(I)″ *Same as (I) but with $\leq_B$ replaced by $\leq_{BM}$.*

(II) *is equivalent to:*

(II)′ $xE_G^X y \Leftrightarrow \varphi_x = \varphi_y$.

(II)″ *Same as (II) but with $\leq_B$ replaced by $\leq_{UB}$ (where $\leq_{UB}$ means reducible by a map f which has the property that $f \circ g$ is Baire measurable for any Borel g).*

(II)‴ *There is a Polish S_∞-space Z and Borel $f : X \to Z$ such that*

$$xE_G^X y \Leftrightarrow f(x)E_{S_\infty}^Z f(y),$$

$S_\infty \cdot f(X) = Z$ *(i.e., the saturation of $f(X)$ is Z) and there is a Borel map $g : Z \to X$ such that*

$$f(g(z))E_{S_\infty}^Z z.$$

Proof. (I) $\Rightarrow \neg$ (II)″ by Corollary 12.6. $\neg$ (II)″ $\Rightarrow \neg$ (II) is obvious. $\neg$ (II) $\Rightarrow \neg$ (II)′ by Theorem 13.10. $\neg$ (II)′ $\Rightarrow$ (I)′ by Corollary 13.17. (I)′ $\Rightarrow$ (I) is obvious. So we have proved the equivalence of (I), (I)′, $\neg$ (II), $\neg$ (II)′, $\neg$ (II)″. Also (I) $\Rightarrow$ (I)″ is obvious and (I)″ $\Rightarrow \neg$ (II) follows from Corollary 12.6. So (I) $\Leftrightarrow$ (I)″. Finally, (II)‴ $\Rightarrow$ (II) is obvious and (II)′ $\Rightarrow$ (II)‴ by Theorem 13.10, so (II)‴ $\Leftrightarrow$ (II). ⊣

It is open whether (I) or (II) always holds for a general Polish group G.

References

Adams, S. [88]. Indecomposability of treed equivalence relations, *Israel J. Math.*, **64(3)** (1988), 362–380.

Adams, S. and Kechris, A.S. [00]. Linear algebraic groups and countable Borel equivalence relations, *J. Amer. Math. Society*, **13 (4)** (2000), 909–943.

Ambrose, W. [41]. Representation of ergodic flows, *Ann. of Math.*, **42** (1941), 723–739.

Becker, H. and Kechris, A.S. [96]. *The Descriptive Set Theory of Polish Group Actions*, London Math. Society Lecture Note Series, **232** (1996), Cambridge University Press.

Becker, J., Henson, C.W., and Rubel, L. [80]. First order conformal invariants, *Ann. of Math.*, **102** (1980), 124–178.

Burgess, J. [79]. A selection theorem for group actions, *Pac. J. Math.*, **80** (1979), 333–336.

Connes, A., Feldman, T., and Weiss, B. [81]. An amenable equivalence relation is generated by a single transformation, *Ergodic Theory and Dynamical Systems*, **1** (1981), 431–450.

Dougherty, R., Jackson, S., and Kechris, A.S. [94]. The structure of hyperfinite Borel equivalence relations, *Trans. Amer. Math. Soc.*, **341(1)** (1994), 193–225.

Feldman, J., Hahn, P., and Moore, C.C. [79]. Orbit structure and countable sections for actions of continuous groups, *Adv. in Math.*, **26** (1979), 186–230.

Feldman, J. and Moore, C.C. [77]. Ergodic equivalence relations, cohomology and von Neumann algebras I, *Trans. Amer. Math. Soc.*, **234** (1977), 289–324.

Giordano, T., Putnam, I.F., and Skau, C.F. [95]. Topological orbit equivalence and C^*-crossed products, *J. Reine Angew. Math.*, **469** (1995), 51–111.

Harrington, L., Kechris, A.S., and Louveau, A. [90]. A Glimm-Effros dichotomy for Borel equivalence relations, *J. Amer. Math. Soc.*, **3 (4)** (1990), 903–928.

Hewitt, E. and Stromberg, K. [69]. *Real and Abstract Analysis,* (1969), Springer-Verlag, New York.

Hjorth, G. [97]. On the isomorphism problem for measure preserving transformations, (April 1997), preprint.

Hjorth, G. [00]. *Classification and orbit equivalence relations*, (2000), Math. Surveys and Monographs, **75**, Amer. Math. Society, Providence, RI.

Hjorth, G. and Kechris, A.S. [95]. Analytic equivalence relations and Ulm-type classifications, *J. Symb. Logic*, **60(4)** (1995), 1273–1300.

Hjorth, G. and Kechris, A.S. [96]. Borel equivalence relations and classifications of countable models, *Ann. Pure and Applied Logic*, **82** (1996), 221–272 .

Hjorth, G. and Kechris, A.S. [00]. The complexity of the classification of Riemann surfaces and complex manifolds, *Ill. J. Math.*, **(44)1** (2000), 104–137.

Hjorth, G., Kechris, A.S., and Louveau A. [98]. Borel equivalence relations induced by actions of the symmetric group, *Ann. Pure and Applied Logic,* **92** (1998), 63–112.

Jackson, S., Kechris, A.S., and Louveau, A. [00]. Countable Borel equivalence relations, (2000), preprint.

Kechris, A.S. [91]. Amenable equivalence relations and Turing degrees, *J. Symb. Logic*, **56** (1991), 182–194.

Kechris, A.S. [92]. Countable sections for locally compact actions, *Ergodic Theory and Dynamical Systems*, **12** (1992), 283-295.

Kechris, A.S. [95]. *Classical Descriptive Set Theory*, Graduate Texts in Mathematics, **156** (1995), Springer-Verlag, New York.

Kechris, A.S. [98]. Rigidity properties of Borel ideals on the integers, *Topology and Its Appl.*, **85** (1998), 195–205.

Kechris, A.S. [00]. Descriptive dynamics, *Descriptive Set Theory and Dynamical Systems*; ed. M. Foreman et al., London Math. Society Lecture Notes Series, **277** (2000), 231–258, Cambridge Univ. Press.

Kechris, A.S. and Sofronidis, N.E. [01]. A strong generic ergodicity property of unitary and self-adjoint operators, *Ergod. Th. & Dynam. Sys.*, **21** (2001), 1459–1479.

Mackey, G.W. [78]. *Unitary Group Representations in Physics, Probability and Number Theory*, (1978), Addison-Wesley, New York.

Mauldin, D. and Ulam, S. [87]. Mathematical problems and games, *Adv. in Appl. Math.*, **8** (1987), 281–344.

Nadkarni, M.G. [91]. On the existence of a finite invariant measure, *Proc. Indian Acad. Sci. Math. Sci.*, **100** (1991), 203–220.

Ornstein, D. and Weiss, B. [80]. Ergodic theory of amenable group actions I: The Rohlin theorem, *Bull. Amer. Math. Soc.*, **2** (1980), 161–164.

Schmidt, K. [90]. *Algebraic ideas in ergodic theory*, CBMS Regional Conf. Ser. in Math., **76** (1990), Amer. Math. Soc., Providence, RI.

Slaman, T. and Steel, J. [88]. Definable functions on degrees, *Cabal Seminar 81-85*, Lecture Notes in Math. **1333**, (1988), 37–55 , Springer-Verlag, New York.

Sullivan, D., Weiss, B., and Wright, J.D.M. [86]. Generic dynamics and monotone complete C^*-algebras, *Trans. Amer. Math. Soc.*, **295** (1986), 795–809.

Wagh, V.M. [88]. A descriptive version of Ambrose's representation theorem for flows, *Proc. Ind. Acad. Sci. Math. Sci.*, **98** (1988), 101–108.

Walters, P. [82]. *An Introduction to Ergodic Theory*, Grad. Texts in Math., **79** (1982), Springer-Verlag, New York.

Weiss, B. [84]. Measurable dynamics, *Conf. Modern Analysis and Probability* (New Haven, CT, 1982); ed. R. Beals et al., Contemp. Math. **26** (1984), 395–421, Amer. Math. Soc., Providence, RI.

Index

$X \times_G H$, 147
X_L, 133
$X_{(\varphi_x)}$, 178
$\approx_{U,V}^x$, 169
$\cong_B$, 128
$\cong_D$, 132
$\cong_R$, 132
$\cong_\gamma$, 134
$\cong_\sigma$, 134
$\sqsubseteq_c$, 122
$\equiv_{U,V}^\alpha$, 172
$<_B$, 118
$\leq_B$, 118
$\leq_{BM}$, 157
$\varphi_\alpha(x, U, V)$, 163
$\varphi_\alpha(a)$, 172
φ_x, 166
$\sim$, 119
$\sim_B$, 118
$\sim_{U,V}^x$, 168

admits classification by countable
    structures, 137
α_n, 136
$\alpha(\mathcal{O}, \mathcal{O}')$, 165
$\alpha(x, U, V)$, 165
α_x, 166
amenable group, 129
aperiodic equivalence relation, 128
$Aut(\mathcal{A})$, 133
$Aut(I, \lambda)$, 123, 137

Borel cardinality, 118
Borel G-space, 123
Borel isomorphism, 127
Borel reducible, 118
Borel selector, 120

calm action, 153
comeager, 119
complete discrete section, 131
concretely classifiable, 120
conformal equivalence, 131
countable equivalence relation, 126

Δ, 140, 150

ε_p, 136
$\mathcal{E}(E)$, 128
E-ergodic, 121
E-nonatomic, 121
E_0, 122
E_{ctble}, 135, 157
$E(F_2, 2)$, 130
$E^*(F_2, 2)$, 130
E_H, 121
$E_{\mathfrak{g}}$, 140
E_R, 121
E_S, 121
E_S^*, 128
E_t, 127
E_V, 119
E_G^X, 123
E_∞, 129

F_2, 121
free action, 129

γ_{cf}, 136
γ_f, 136
GE group, 182
General Glimm-Effros Dichotomy, 122
generically F-ergodic, 157
generically turbulent, 139
generically turbulent action, 145

$H^*(I^2)$, 141

hyperfinite equivalence relation, 126

I_∞, 128
I_n, 128
II_∞, 128
II_n, 128
III_λ, 128
induced action, 146
induced space, 147

local graph, 138, 143
local orbit, 138, 143
locally finite, 136
logic action, 133

$\mathcal{M}^0(x)$, 167
$\mathcal{M}^1(x)$, 168
$\mathcal{M}^2(x)$, 168
$\mathcal{M}^3(x)$, 169
$\tilde{\mathcal{M}}^1(x)$, 169
$\tilde{\mathcal{M}}^2(x)$, 170
$\tilde{\mathcal{M}}^3(x)$, 170
$\tilde{\tilde{\mathcal{M}}}^1(x)$, 173
$\tilde{\tilde{\mathcal{M}}}^2(x)$, 173
$\tilde{\tilde{\mathcal{M}}}^3(x)$, 173
$\tilde{\mathcal{M}}(x)$, 171
$\mathcal{M}(x)$, 172
meager, 119
$Mod(\sigma)$, 134

orbit equivalence, 117
orbit equivalence relation, 123
$\mathcal{O}(x, U, V)$, 138, 143

$\mathcal{P}(\mathbb{N})$, 140, 150
$\mathcal{P}^\alpha(\mathbb{N})$, 163
Poincaré flow, 128
Polish G-space, 123
Polish group, 117
Polishable, 140
potential class, 135

quasi-invariant, 127

ρ, 136
$R_{U,V}$, 138, 143

$\sigma_p(T)$, 137
smooth, 120
Spectral Theorem, 119
strongly dense, 150
S_∞, 133

tail equivalence, 127

τ_α, 175
τ_α^x, 175
τ_{cf}, 136
τ_{cf}^*, 136
$\tau_{(\varphi_x)}$, 178
τ_x, 178
tree, 130
treeable equivalence relation, 130
turbulent, 138
turbulent action, 145
turbulent orbit, 144
turbulent point, 143

$U(H)$, 124
Ulm-classifiable, 124

Vitali equivalence relation, 119

weakly generically turbulent, 159

On Subspaces, Asymptotic Structure, and Distortion of Banach Spaces; Connections with Logic

Edward Odell

Department of Mathematics
The University of Texas at Austin
Austin, Texas 78712 USA
odell@math.utexas.edu

1

Introduction

The most outstanding problems in the theory of infinite dimensional Banach spaces, those that were central to the study of the general structure of a Banach space, finally yielded their secrets in the 1990's. In this survey we shall discuss these problems and their solutions and more. For many years researchers have been aware of deep connections between both the theorems and ideas of logic and set theory and Banach space theory. We shall try to illuminate these connections as well.

For example the ideas of Ramsey theory played a key role in H. Rosenthal's magnificent ℓ_1-theorem in 1974 [R1]. But there is also a less direct connection with the Banach space question as to whether or not separable infinite dimensional Hilbert space, ℓ_2, is distortable. This is equivalent to the following approximate Ramsey problem. Let $S_{\ell_2} = \{x \in \ell_2 : \|x\| = 1\}$ be the unit sphere of ℓ_2. Finitely color the sphere by colors $C_1, \ldots, C_k$ and let $\varepsilon > 0$. Does there exist an i_0 and an infinite dimensional closed linear subspace X of ℓ_2 so that the unit sphere of X, S_X, is a subset of $(C_{i_0})_\varepsilon \equiv \{y \in S_{\ell_2} : \|y - x\| < \varepsilon$ for some $x \in C_{i_0}\}$? It suffices to let (e_i) be an orthonormal basis for ℓ_2 and confine the search to block subspaces — those spanned by block bases of (e_i) (these terms are defined precisely below). Thus we are compelled to ask if there is a known or undiscovered Ramsey type theorem which will allow us to assert the existence of such a block subspace X and color C_{i_0}? Ramsey theorems certainly exist for colorings of subsequences of $\mathbb{N}$. Also block type theorems (e.g., Hindman's theorem [H], the Carlson-Simpson theorem [CS]) exist. Is there an approximate block type Ramsey theorem, one that considers all block bases of (e_i)?

Generally speaking Ramsey theorems are of the type that a function

Acknowledgment. Research supported by NSF and TARP.

191

into a finite set can be restricted to some sort of substructure on which it is constant. In an approximate Ramsey theorem we seek a substructure where the function is nearly constant in some appropriate sense; a joining of analysis and Ramsey theory.

The answer turns out to be yes and no. There is such a theorem and it was discovered by the 1998 Fields medalist W.T. Gowers [G1]. But it cannot be used to solve the distortion problem for ℓ_2 which we posed. Rather it leads to Gowers' spectacular Dichotomy Theorem for Banach spaces. And previously known Ramsey theorems do lead to a number of other results in Banach space theory.

These and more are discussed in this survey. Chapter 2 introduces a number of Banach space problems, provides background information and relevant definitions. In Chapter 3 we discuss the famous unconditional basic sequence problem and its connections with distortion. Chapter 4 concerns Gowers' block Ramsey theorem for Banach spaces. In Chapter 5 we discuss distortion. Chapter 6 involves asymptotic structure. There are two main notions here, both a type of logical connection between the finite dimensional (or local) structure of a space X and the infinite dimensional structure of X. Chapter 7 involves certain ordinal indices that have been constructed (e.g., by Szlenk and Bourgain) to study the structure of infinite dimensional Banach spaces. Chapter 8 is a brief discussion of the homogeneous Banach space problem: If X is isomorphic to all of its infinite dimensional subspaces is X isomorphic to Hilbert space?

Our focal point is problems that have been recently solved but we also mention some open problems as well. This survey is not all inclusive. The selection of material is based on the interests of the author. There are many other results, especially in the theory of nonseparable Banach spaces, that have also been strongly influenced by set theory and logic. To help the reader keep track of what is known we have adopted the notation $(P1), (P2), \ldots$ for solved problems and $(Q1), (Q2), \ldots$ for open questions. We use standard notation as may be found in [LT1]. Thus X^* is the dual space of a Banach space X, x^* is an element of X^*, an *isomorphism* $T : X \to Y$ is a bounded linear invertible operator from X onto Y, Z is a *quotient* of X if $Z = X/Y$ for some closed subspace Y of X (equivalently, there exists a bounded linear operator T from X onto Z), a closed linear subspace Y of X is *complemented* in X if Y is the range of a *projection* (bounded linear idempotent operator) on X, etc.

Other source books for our survey are [D], [Gu] and the books [HHZ]. We assume that the reader is well versed in basic functional analysis.

Many results are given without proof and others are mostly sketched and thus we have included a large bibliography.

I want to thank all of the researchers upon whose work this survey is based for making this subject both exciting and beautiful. In particular this includes my advisor W.B. Johnson and my unofficial advisor, H. Rosenthal. In addition many thanks are due to a number of people who read all or part of earlier versions of this manuscript and pointed out errors, made suggestions, allowed preprints to be cited, etc.. I wish to especially thank George Androulakis, Spyros Argyros, Steve Dilworth, Ioannis Gasparis, Petr Habala, Rob Judd, Antonio Martínez-Abejón, Nicole Tomczak-Jaegermann and extend my deepest appreciation to my co-author on a number of adventures, Thomas Schlumprecht. Special kudos go to Margaret Combs who did her usual magnificent job composing this paper.

2
Background material: The 60's and 70's

In this chapter we introduce most of the problems with which we are concerned. We also present a number of definitions, terminology and some background material to give context to the problems. Our problems are often of the kind, given a (separable infinite dimensional) Banach space X does there exist a subspace (closed linear infinite dimensional) Y with a given "nice" structure. For example as discussed shortly, every X contains a Y with a basis. This relatively mild "nice" structure means that Y can be regarded as a sequence space; a space given by a complete norm $\| \cdot \|$ on a certain linear subspace of the space of null sequences of reals with the unit vector basis (e_i) being normalized and with the natural projections $P_n(\sum a_i e_i) = \sum_1^n a_i e_i$ being uniformly bounded. Moreover every $y \in Y$ can be uniquely expressed as $y = \sum_{i=1}^{\infty} a_i e_i$ for some $(a_i) \subseteq \mathbb{R}$.

Thus our generic search for a nice space Y can be restricted to spaces X having a basis. For the properties with which we are concerned one need only then consider block subspaces, subspaces having a basis which is a block basis (defined below) of the given basis. This follows from standard perturbation arguments (see any of the source books cited above) and we do not give the proofs.

Of course when attacking problems it is necessary to have a good selection of examples in addition to the classical Banach spaces such as c_0, ℓ_p, L_p and $C(K)$. In this chapter we also recall Tsirelson's famous space T. This was the first Banach space whose norm was not given by an explicit formula. Its construction was a pivotal point in the development of the theory. The construction of T solved (P1), below. Not every X contains c_0 or ℓ_p for some $1 \le p < \infty$. Although T was discovered in 1974 its importance in relation to the solution of the *unconditional basic sequence problem* and the distortion problem (problems (P2) and

(P4) below remained hidden until 1991. At that time Schlumprecht constructed S, his famous second generation Tsirelson type space. This led to the breakthrough solutions of these problems. But first let us return to the basics.

We use $X, Y, Z, \ldots$ to denote separable infinite dimensional real Banach spaces. Perhaps the most fundamental structure a Banach space can possess is a basis. $(x_i)_{i=1}^{\infty}$ is a *basis* for X if for all $x \in X$ there exists a unique sequence of scalars $(a_i)_{i=1}^{\infty}$ so that $x = \sum_{i=1}^{\infty} a_i x_i$. A sequence $(x_i)_{i=1}^{\infty} \subseteq X$ is *basic* if it is a basis for its *closed linear span* $[(x_i)]$. Equivalently, $x_i \neq 0$ for all i and for some $K < \infty$ and all $(a_i)_1^m \subseteq \mathbb{R}$ and $n < m$, $\|\sum_{i=1}^{n} a_i x_i\| \leq K \|\sum_{i=1}^{m} a_i x_i\|$. (See [LT1], [D] or [HHZ] for a proof of this equivalence along with those of many of the things we are about to say.) The smallest such K is called the *basis constant* of (x_i) and (x_i) is said to be K-*basic*.

If (x_i) is a basis for X, the biorthogonal functionals (x_i^*), given by $x_i^*(\sum a_j x_j) = a_i$, are contained in X^*. In his famous book [Ba] Banach raised the problem as to whether every X has a basis. Banach's problem was solved in the negative in 1972 by P.Enflo [E1] who produced a subspace of c_0 not having a basis (in fact not having a weaker property called the approximation property). Enflo's work subsequently was extended to produce subspaces of ℓ_p ($p \neq 2$) without a basis (nor the approximation property). The best result of this type is due to Szankowski [S] who proved that if X is not very very close to being a Hilbert space (technically, not of type $2 - \varepsilon$ and cotype $2 + \varepsilon$ for all $\varepsilon > 0$) then X contains a subspace without a basis. Of course every subspace of ℓ_2 is again ℓ_2 and hence has a basis. Every finite dimensional Banach space has a basis but the best basis constant of a basis can grow as the dimension grows [Gl], [Sza2]. Szarek [Sza1] proved that a space X need not even possess the following weaker structure: one cannot always write $X = \overline{\cup E_n}$ where $E_1 \subseteq E_2 \subseteq \cdots$ are all finite dimensional spaces having bases of uniformly bounded basis constant.

For nonseparable Banach spaces the following problem remains open.

(Q1) Let B be a *nonseparable Banach space*. Is there a sequence of closed subspaces $B_1 \subsetneq B_2 \subsetneq \cdots$ with $\overline{\cup B_n} = B$?

This is a reformulation (see [JR]) of the well known separable quotient problem: For every nonseparable Banach space B does there exist a (separable infinite dimensional) space X and a bounded linear onto map $T : B \to X$?

However it is relatively easy to see (sketched below) that for $\varepsilon > 0$

every X contains a basic sequence with basis constant less than $1 + \varepsilon$. The classical Banach spaces $L_p[0,1]$ $(1 \leq p < \infty)$, $C(K)$ (K compact metric), ℓ_p $(1 \leq p < \infty)$ and c_0 all have *monotone* bases, i.e., bases of basis constant 1.

Two other structural problems arose even before Enflo's beautiful construction. It is difficult to say much about the structure of an arbitrary space X. So one naturally searches for nice subspaces.

(P1) Does every X contain an isomorph of ℓ_p for some $1 \leq p < \infty$ or c_0?

(P2) Does every X contain an unconditional basic sequence?

(x_i) is an *unconditional basic sequence* if each $x_i \neq 0$ and there exists $K < \infty$ so that for all $(a_i)_1^n \subseteq \mathbb{R}$ and all $\varepsilon_i = \pm 1$,

$$\left\| \sum_1^n \varepsilon_i a_i x_i \right\| \leq K \left\| \sum_1^n a_i x_i \right\| .$$

The smallest such K is called the *unconditional basis constant* of (x_i) which is then said to be K-*unconditional*. It can be shown that a basis (x_i) is unconditional iff whenever $\sum_{i=1}^\infty a_i x_i = x$, the series converges unconditionally (in any rearrangement) to x. The unit vector basis of ℓ_p $(1 \leq p < \infty)$ or c_0 is 1-unconditional. $L_p[0,1]$ has an unconditional basis (the Haar basis) for $1 < p < \infty$. $L_1[0,1]$ and $C[0,1]$ do not even embed into spaces with unconditional bases.

Of course both L_1 and $C[0,1]$ contain nice subspaces in the sense of (P1). In fact L_1 contains isometric copies of ℓ_p for all $1 \leq p \leq 2$ while $C[0,1]$ contains isometric copies of all separable Banach spaces.

In order to gain some insight into the difficulties involved in the *unconditional basic sequence problem* (P2), we briefly recall the proof that X contains a $1 + \varepsilon$-basic sequence (x_i). Choose $x_1 \in S_X = \{x \in X : \|x\| = 1\}$. If $(x_1, \ldots, x_n)$ have been chosen, using the compactness of $S_{\langle x_i \rangle_1^n}$ ($\langle \cdot \rangle$ denotes linear span), we may find a finite set $F_n \subseteq S_{X^*}$ which is $\frac{1}{1+\varepsilon}$ *norming* for $S_{\langle x_i \rangle_1^n}$ and $F_n \supseteq F_{n-1}$. This means for $x \in S_{\langle x_i \rangle_1^n}$,

$$1 = \|x\| \geq \sup\{f(x) : f \in F_n\} \geq \frac{1}{1+\varepsilon} .$$

We then choose $x_{n+1} \in \left(\bigcap_{f \in F_n} \ker f \right) \cap S_X$.

Suppose $(a_i)_1^m \subseteq \mathbb{R}$, $n < m$ and $\| \sum_1^n a_i x_i \| = 1$. Choose $f \in F_n$ with

$$f\left(\sum_1^n a_i x_i \right) \geq \frac{1}{1+\varepsilon} .$$

Thus

$$\left\|\sum_1^m a_i x_i\right\| \geq f\left(\sum_1^m a_i x_i\right) = f\left(\sum_1^n a_i x_i\right) \geq \frac{1}{1+\varepsilon}$$

so

$$\left\|\sum_1^n a_i x_i\right\| \leq (1+\varepsilon)\left\|\sum_1^m a_i x_i\right\| .$$

Obtaining an unconditional basic sequence (x_n) is equivalent to finding nonzero (x_n) and $K < \infty$ so that for all $(a_i)_1^m \subseteq \mathbb{R}$ and all $F \subseteq (1,\ldots,m)$, $\|\sum_F a_i x_i\| \leq K\|\sum_1^m a_i x_i\|$. To generalize the argument above would require choosing $(x_i)_1^\infty$ so that for all $F \subseteq \mathbb{N}$ there exists a set G_F which norms $\langle x_i \rangle_F$ and for which $x_i \in \bigcap_{f \in G_F} \ker f$ for $i \notin F$. This is infinitely more difficult and, as we shall see, not always possible.

Another way to view the difference between being basic versus unconditionally basic is that having (x_i) basic requires making a sequence of finite rank projections (from $[\langle x_i \rangle]$ onto $\langle x_i \rangle_1^n$ with kernel $[\langle x_i \rangle_{n+1}^\infty]$ where $[\cdot]$ denotes closed linear span) uniformly bounded. If (x_i) is unconditional then $[(x_i)]$ admits (uncountably many) uniformly bounded projections of infinite rank. One can project onto $[(x_i)]_{i \in M}$ for any infinite $M \subseteq \mathbb{N}$. Thus $[(x_i)]$ must admit nontrivial operators.

Problems (P1) and (P2) gained prominence in the 60's. Lindenstrauss and Tzafriri, in some deep work, proved that every Orlicz space contains an isomorph of ℓ_p for some $1 < p < \infty$ [LT2]. V.D. Milman [M1] found a sufficient condition for insuring that the answer to (P1) is positive. Given $(X, \|\cdot\|)$, $S_X = \{x \in X : \|x\| = 1\}$ is the unit sphere of X. A function $f : S_X \to \mathbb{R}$ *stabilizes* if for all $\varepsilon > 0$ and all $Y \subseteq X$ (i.e., Y is an infinite dimensional closed subspace of X) there exists $Z \subseteq Y$ so that

$$\text{osc}(f, S_Z) \equiv \sup\{f(z_1) - f(z_2) : z_1, z_2 \in S_Z\} < \varepsilon$$

We then say that f is *oscillation stable*.

(P3) Let $f : S_X \to \mathbb{R}$ be Lipschitz. Is f oscillation stable?

Milman showed that a "yes" answer to (P3) yields a "yes" for (P1). In fact he showed that one need only show that every equivalent norm on X is oscillation stable, i.e., X does not contain a *distortable* subspace (see below). Norms $|\cdot|$ and $\|\cdot\|$ are *equivalent* on X if there exists positive constants A, B with

$$A|x| \leq \|x\| \leq B|x| \quad \text{for all} \quad x \in X .$$

This just says that the identity map from $(X, \| \cdot \|)$ to $(X, | \cdot |)$ is an isomorphism or that the unit ball in either norm is contained in a multiple of the other unit ball. Milman's work fore-shadowed the theory of types developed by Krivine and Maurey in the mid 70's [KM]. A Banach space X is *stable* if for all bounded sequences $(x_n), (y_n) \subseteq X$, $\lim_{n \to \infty} \lim_{m \to \infty} \|x_n + y_m\| = \lim_{m \to \infty} \lim_{n \to \infty} \|x_n + y_m\|$ provided both limits exist (equivalently one could take the limits over ultrafilters). X is *weakly stable* if the above condition holds whenever (x_n) and (y_n) are both weakly convergent.

If X and Y are isomorphic the *Banach Mazur distance* between them is

$$d(X, Y) = \inf\{\|T\| \, \|T^{-1}\| : T : X \to Y \text{ is an onto isomorphism}\} \,.$$

A word of caution is needed before using this "distance." Note that $\|T\| \, \|T^{-1}\| \geq \|TT^{-1}\| = 1$. Thus $d(X, Y) \geq 1$. Also $d(X, Z) \leq d(X, Y) d(Y, Z)$. In order to get a metric on the space of all spaces isomorphic to a fixed space one must take $\log d(\cdot, \cdot)$. Even in this it is possible to have if X is infinite dimensional, $d(X, Y) = 1$ yet X and Y are not isometric, i.e., X and Y are different Banach spaces. To say X *contains almost isometric copies of* ℓ_p means that for all $\varepsilon > 0$ there exists $Y \subseteq X$ with $d(Y, \ell_p) \leq 1 + \varepsilon$.

2.1 Theorem. [KM] *If X is stable then X contains almost isometric copies of ℓ_p for some $1 \leq p < \infty$.* [ANZ] *If X is weakly stable then X contains almost isometric copies of ℓ_p for some $1 \leq p < \infty$ or c_0.*

This extended Aldous' theorem [Ald] that every subspace of $L_1[0, 1]$ contains almost isometric copies of ℓ_p for some $1 \leq p \leq 2$.

A *trivial type* on X is a function $\mathcal{T}_b(x) = \|b + x\|$ where $b \in X$. (For more on types see [KM], [HM], [O2], [Gu].) Milman proved the following result:

2.2 Theorem. [M1] *Suppose that all trivial types on X stabilize. Then X contains almost isometric copies of ℓ_p for some $1 \leq p < \infty$ or c_0.*

From this theorem one obtains that if X does not contain almost isometric copies of some ℓ_p or c_0 then X contains a *distortable* subspace Y (i.e., some equivalent norm on X does not stabilize). Indeed one can show that for $x \in X$, $| \cdot |_x$ is an equivalent norm on X where

$$|y|_x = \left\| x\|y\| + y \right\| + \left\| x\|y\| - y \right\| \,.$$

Furthermore it can be shown that $\mathcal{T}_x$ stabilizes on X iff $| \cdot |_x$ stabilizes on X. A proof of Theorem 2.2 along these lines appears in [OS9] based upon arguments in [HORS].

To date there are few effective conditions for proving that a space X contains an isomorph of some ℓ_p (or c_0) that do not yield almost isometric copies. One of these criteria is due to Deville [De]. He proves that if X admits a nonzero real valued C^∞-smooth function with bounded support then X contains an isomorph of either c_0 or ℓ_p for some even integer p. It remains open to find an isomorphic variant. We record this necessarily vague question as

(Q2) Find a (nontrivial) isomorphically invariant condition (C) so that if X has (C) then X contains an isomorph of c_0 or ℓ_p for some $1 \leq p < \infty$.

Dvoretsky's famous theorem [Dv] proved that every X contains almost isometric copies of ℓ_2^n (finite dimensional Hilbert space) for all $n \in \mathbb{N}$. Later extensions of this work [M3] are based on the fact that (P3) has a positive local answer (see [FLM], [M2, M4, M5], [MS]).

2.3 Theorem. [MS], [M3] *For all K, $n \in \mathbb{N}$ and $\varepsilon > 0$ there exists m so that if $\dim F = m$ and $f : S_F \to \mathbb{R}$ is K-Lipschitz then there exists an n dimensional $G \subseteq F$ with $\mathrm{osc}(f, S_G) < \varepsilon$.*

Despite these positive indications (P1) was answered in the negative in 1974 by B.S. Tsirelson [T] and thus also a distortable space was found. Tsirelson's space T (actually the dual of the original space) had a forerunner, an example of Schreier [Sc] (see also [CaS]) from 1930.

To describe both spaces we require some notation which we shall employ throughout this paper. c_{00} is the linear space of finitely supported real sequences. $(e_i)_{i=1}^\infty$ is the unit vector basis for c_{00}. Both Tsirelson's space and Schreier's space (and many other spaces to come) will be the completion of c_{00} under certain norms defined on c_{00}. If $E, F \subseteq \mathbb{N}$ we write $E < F$ if $\max E < \min F$ or if either is empty. $S_0 = \{\{n\} : n \in \mathbb{N}\} \cup \emptyset$. The *first Schreier class* is

$$S_1 = \{E \subseteq \mathbb{N} : \min E \geq |E|\} \cup \{\emptyset\}$$

($|\cdot|$ denotes cardinality).

Sets $(E_i)_1^n$ are *1-admissible* if $n \leq E_1 < \cdots < E_n$ (equivalently, $E_1 < \cdots < E_n$ and $(\min E_i)_1^n \in S_1$). Schreier's norm is

$$\|x\| = \sup\left\{ \left| \sum_{i \in E} x(i) \right| : E \in S_1 \right\} .$$

He constructed this norm to give an example of a normalized weakly null sequence, namely (e_i), such that no subsequence converged Cesáro to 0. Schreier's space is the completion of c_{00} under this norm. The norm on c_{00} yielding T (as described in [FJ]) is the one that solves the following equation. Thus the norm is given implicitly rather than explicitly.

$$\|x\| = \|x\|_\infty \vee \sup\left\{ \frac{1}{2} \sum_{i=1}^{n} \|E_i x\| : (E_i)_1^n \text{ is 1-admissible} \right\}. \quad (2.1)$$

In (2.1) the notation $E_i x$ refers to the restriction of x to E_i. Thus $E_i x(j) = x(j)$ if $j \in E_i$ and 0 otherwise. Of course it must be shown that such a norm exists.

Tsirelson type norms have become commonplace by now. These norms are fixed points of certain mappings on $\mathcal{N}$, the class of all norms $\|\cdot\|$ on c_{00} for which (e_i) is a normalized monotone basis for $\overline{(c_{00}, \|\cdot\|)}$ and such that $\|\sum a_i e_i\| \geq \max_i |a_i|$. The following simple principle [OS1] insures the existence of the norm for T and many others. $P : \mathcal{N} \to \mathcal{N}$ is *order preserving* if $|\cdot| \leq \|\cdot\| \Rightarrow P(|\cdot|) \leq P(\|\cdot\|)$. (By $|\cdot| \leq \|\cdot\|$ we mean $\forall\, x \in c_{00},\ |x| \leq \|x\|$).

2.4 Proposition. *Let $P : \mathcal{N} \to \mathcal{N}$ be order preserving. Then P admits a smallest fixed point. Thus there exists $\|\cdot\| \in \mathcal{N}$ with $\|\cdot\| = P(\|\cdot\|)$ and $\|\cdot\| \leq |\cdot|$ whenever $P(|\cdot|) = |\cdot|$.*

The proof is by transfinite induction. The existence of the norm in (2.1) follows by defining

$$P(\|\cdot\|)(x) = \|x\|_\infty \vee \sup\left\{ \frac{1}{2} \sum_{i=1}^{n} \|E_i x\| : (E_i)_1^n \text{ is 1-admissible} \right\}.$$

The information given by (2.1) is enough to prove that X is reflexive, the unit vector basis (e_i) is 1-unconditional and yet X does not contain an isomorph of ℓ_p or c_0 [FJ]. The only such space it could contain is ℓ_1, but it doesn't.

T is distortable, in fact it is $2 - \varepsilon$ *distortable* for all $\varepsilon > 0$. More generally we say that Y is λ-*distortable* for some $\lambda > 1$ if there exists an equivalent norm $|\cdot|$ on Y so that for all $X \subseteq Y$

$$\sup\left\{ \frac{|x|}{|y|} : x, y \in (S_X, \|\cdot\|) \right\} \geq \lambda. \quad (2.2)$$

The fact that T is $2 - \varepsilon$ distortable is shown in some detail in Chapter 3.

The current state of knowledge is that if one can write down an explicit equation for the norm of a Banach space then one can prove (sometimes with difficulty) that X contains c_0 or ℓ_p (and in fact almost isometrically). All spaces thus far constructed which do not contain c_0 or ℓ_p involve an implicit definition of the norm like that of T. Is there a possible theorem here? We leave it to the logicians to precisely formulate the problem.

(Q3) Formulate a condition which interprets the statement "the norm $\|\cdot\|$ on c_{00} is explicit." Then prove that if X is the completion of c_{00} under an explicit norm, then X contains (almost isometric) copies of c_0 or ℓ_p for some $1 \le p < \infty$.

The unconditional basic sequence problem (P2) remained unsolved throughout the 70's and 80's. The ideas involved in the construction of T and their impact upon the solution of (P2) would remain unrealized until the 90's. But we get ahead of ourselves.

Problem (P3) evolved into the question as to which spaces are distortable? James [J1], in the 1960's, proved that c_0 and ℓ_1 are not distortable. This naturally left

(P4) Is ℓ_2 distortable? Is ℓ_p distortable for $1 < p < \infty$?

James' proof is relatively easy and reveals some of the difficulties of (P4). We sketch it for the ℓ_1 case. Let (x_i) be a normalized basis satisfying for some K and all (a_i),

$$\frac{1}{K}\sum|a_i| \le \left\|\sum a_i x_i\right\| \le \sum |a_i| .$$

By passing to a subsequence of (x_i) and redefining K if necessary we may assume that the constant "$\frac{1}{K}$" cannot be improved by more than an ε. Thus one can find a normalized *block basis* (y_i) of x_i so that for some $p_1 < p_2 < \cdots$ and $(b_j) \subseteq \mathbb{R}$, each $y_i = \sum_{j=p_{i-1}+1}^{p_i} b_j x_j$ where $\sum_{j=p_{i-1}+1}^{p_i} |b_j| \approx K$. Thus

$$\left\|\sum a_i y_i\right\| = \left\|\sum_i \sum_{j=p_{i-1}+1}^{p_i} a_i b_j x_j\right\|$$

$$\ge \frac{1}{K}\sum_i \sum_{j=p_{i-1}+1}^{p_i} |a_i||b_j| \approx \sum_i |a_i| .$$

In words the y_i's each sucked up the worst part of the lower ℓ_1 estimate and thus as a whole had to have a very good lower ℓ_1 estimate. From this one obtains that for all $\varepsilon > 0$ there exists a normalized block basis

(y_i) of (x_i) satisfying

$$\left\| \sum a_i y_i \right\| \geq (1 - \varepsilon) \sum |a_i| \qquad (2.3)$$

for all $(a_i) \subseteq \mathbb{R}$. Hence ℓ_1 is not distortable. The same sort of argument used in an isomorph of ℓ_p would yield a block basis with a very good lower ℓ_p estimate or a similar argument would yield a block basis with a very good upper ℓ_p estimate but not both simultaneously. Since ℓ_1 (and also c_0) are "extremal" in the ℓ_p's only a one sided estimate is required.

More recently Dowling, Johnson, Lennard and Turett [DJLT] have considered sequences satisfying a stronger condition than (2.3). A normalized sequence X contains an *asymptotically isometric* copy of ℓ_1 if there exists a normalized $(y_i) \subseteq X$ satisfying

$$\left\| \sum a_i y_i \right\| \geq \sum (1 - \varepsilon_i)|a_i|$$

for all $(a_i) \subseteq \mathbb{R}$ and some $\varepsilon_i \downarrow 0$. In [DJLT] it is shown that ℓ_1 can be renormed so as to not contain an asymptotically isometric copy of ℓ_1 so the sequence obtained in (2.3) is optimal. They also give a c_0 version of this result.

Let us mention another famous problem involving Hilbert space, again unsolved until the 90's.

(P5) If X is isomorphic to all of its infinite dimensional subspaces is X isomorphic to ℓ_2?

Such a space X is called *homogeneous*. This problem is raised in Banach's book [Ba]. A homogeneous space has the property that all of its subspaces have a basis. This is not sufficient to insure that X is isomorphic to ℓ_2. Indeed W.B. Johnson [Jo1] showed that *convexified Tsirelson space* T_2 has such a property (even that every subspace of a quotient has a basis) but does not even contain ℓ_2. The norm in T_2 is given by

$$\left\| \sum a_i e_i \right\|_{T_2} = \left(\left\| \sum a_i^2 e_i \right\|_T \right)^{1/2}.$$

T_2 was the first nontrivial example of what is called a weak Hilbert space (see [Pi2], also, [CaS]). Many of the basic properties about T, T_2 and certain other variations of T can be found in [CaS].

In 1994 P. Mankiewicz and N. Tomczak-Jaegermann obtained the following [MTJ2]: If every quotient of every subspace of $(X \oplus X \oplus \ldots)_{\ell_2}$ has a basis then X must be isomorphic to ℓ_2. The following problem remains open.

(Q4) Suppose that all subspaces of X have an unconditional basis. Is X isomorphic to ℓ_2?

Of course if X is isomorphic to ℓ_2 then every subspace Y of X is *complemented* in X, i.e., Y is the range of a projection on X. Lindenstrauss and Tzafriri [LT3] proved that this condition in turn forces X to be isomorphic to ℓ_2.

But back to the 70's. When searching for an unconditional basic sequence in X it suffices to assume X has a basis (x_i) and to confine the search to block bases. (y_i) is a *block basis* of (x_i) if $y_i \neq 0$ for all i and for some $(a_j) \subseteq \mathbb{R}$ and $p_1 < p_2 < \cdots$, $y_i = \sum_{j=p_{j-1}+1}^{p_i} a_j x_j$ for all i. (y_i) is automatically basic with basis constant not exceeding that of (x_i). $Y = [(y_i)]$ is called a *block subspace* of X. To see if (x_i) admits an unconditional block basis was beyond current techniques. However one could look at subsequences of (x_i) via Ramsey theory. First some notation. $[\mathbb{N}]^\omega$ denotes all infinite subsequences of $\mathbb{N}$. If $M \in [\mathbb{N}]^\omega$, $[M]^k$ is the set of all subsequences of M of length k and $[M]^{<\omega} = \bigcup_{k=1}^\infty [M]^k$. Ramsey proved

2.5 Theorem. [Ra] *Let $k \in \mathbb{N}$ and $\mathcal{A} \subseteq [\mathbb{N}]^k$. For all $M \in [\mathbb{N}]^\omega$ there exists $N \in [M]^\omega$ so that either $[N]^k \subseteq \mathcal{A}$ or $[N]^k \subseteq \sim \mathcal{A}$.*

This was used by Brunel and Sucheston [BS] (see also [BL]) to construct spreading models. Given a normalized basic sequence $(x_i) \subseteq X$ and $\varepsilon_n \downarrow 0$ one can find a subsequence (y_i) of (x_i) so that for all n and $n \leq i_1 < \cdots < i_n$, $n \leq j_1 < \cdots < j_n$ and all $(a_i)_1^n \subseteq [-1,1]^n$,

$$\left| \left\| \sum_{k=1}^n a_k y_{i_k} \right\| - \left\| \sum_{k=1}^n a_k y_{j_k} \right\| \right| < \varepsilon_n$$

Thus

$$\lim_{i_1 \to \infty} \cdots \lim_{i_n \to \infty} \left\| \sum_{k=1}^n a_k y_{i_k} \right\|$$

exists and as is easy to see defines a norm $\|\cdot\|$ on c_{00} which makes the unit vector basis (e_i) into a normalized basis. $((e_i), \|\cdot\|)$ or sometimes $E = \overline{(c_{00}, \|\cdot\|)}$ is called a *spreading model* of (y_i) or of X.

Also if (x_i) is weakly null then it can be shown that (e_i) is 2-unconditional and even without this assumption $(e_1 - e_2, e_3 - e_4, \ldots)$ is always unconditional. One proves, when (e_i) is weakly null, that (e_i) is *suppression-1 unconditional* ($\|\sum_E a_i e_i\| \leq \|\sum_1^n a_i e_i\|$ for all $n \in \mathbb{N}$,

$(a_i)_1^n \subseteq \mathbb{R}$ and $E \subseteq \{1, \ldots, n\}$) using Mazur's theorem that some convex block basis of (y_i) is norm null. The spreading model E is *finitely representable* in X: for all finite dimensional $F \subseteq E$ and $\varepsilon > 0$ there exists $G \subseteq X$ with $d(G, F) < 1 + \varepsilon$. Thus Ramsey theory tells us that asymptotically (in the sense of behavior at infinity) one can find an unconditional sequence in X.

Arguably the best Banach space theorem of the 70's was Rosenthal's ℓ_1 theorem. A sequence (x_i) is *weak Cauchy* if $\lim_{i \to \infty} x^*(x_i)$ exists for all $x^* \in X^*$. Basic sequences (x_i) and (y_i) are *K-equivalent* if there exist A, B with $AB \le K$ and for all scalars (a_i)

$$A^{-1}\| \sum a_i y_i\| \le \| \sum a_i x_i\| \le B\| \sum a_i y_i\| .$$

The condition just insures that the natural map $T : [(x_i)] \to [(y_i)]$ that sends x_i to y_i is an isomorphism.

2.6 Theorem. [R1] *Let (x_i) be a bounded sequence with no weak Cauchy subsequence. Then a subsequence of (x_i) is equivalent to the unit vector basis of ℓ_1.*

Rosenthal's original proof did not directly use Ramsey theory. However as observed by Farahat [Fa] one could simplify the argument by using the following Ramsey theorem. $[\mathbb{N}]^\omega$ is given the pointwise topology, i.e., the relative product topology on $[\mathbb{N}]^\omega \subseteq 2^{\mathbb{N}}$.

2.7 Theorem. [GP] *Let $\mathcal{A} \subseteq [\mathbb{N}]^\omega$ be a Borel set and let $M \in [\mathbb{N}]^\omega$. Then there exists $N \in [M]^\omega$ so that either $[N]^\omega \subseteq \mathcal{A}$ or $[N]^\omega \subseteq \sim \mathcal{A}$.*

Many people were involved in the development of infinitary Ramsey theory and stronger versions of this theorem exist (see [El], [Sil], [NW], [Mat] and the expository paper [O1] on applications of Ramsey theory to Banach space theory). For example $\mathcal{A}$ need only be analytic or co-analytic. Most generally the theorem remains valid if $\mathcal{A}$ has the Baire property for the topology generated by the family of basic open sets of the form

$$N(F, M) = \{L \in [\mathbb{N}]^\omega : L \cap \{1, \ldots, \max F\} = F$$
$$\text{and } L \cap \{\max F + 1, \max F + 2, \ldots\} \subseteq M\}$$

for all finite $F \subseteq \mathbb{N}$ and $M \in [\mathbb{N}]^\omega$.

Rosenthal's theorem provided part of the solution to the following famous companion problem to (P1).

(P6) Does every X contain Y which is either reflexive or isomorphic to c_0 or ℓ_1?

X is reflexive iff B_X is weakly compact. Thus by the Eberlein-Smulian theorem, X is reflexive iff every bounded sequence in X admits a weak Cauchy subsequence (x_n) and X is weakly sequentially complete (every weak Cauchy sequence is weakly convergent).

An affirmative answer to (P2) would yield an affirmative answer to (P6) [J2]. Indeed suppose that (x_i) is unconditional but $[(x_i)]$, the closed linear span of (x_i), is not reflexive. As shown by James one obtains that either some normalized block basis of (x_i) is not weakly null (i.e., (x_i) is not *shrinking*) which easily gives that some block basis is equivalent to the unit vector basis of ℓ_1) or that (x_i) is not *boundedly complete*. This means there is a block basis (y_i) satisfying $\sup_n \| \sum_1^n y_i \| < \infty$ yet $\sum_1^\infty y_i$ is divergent and one obtains c_0 from this.

In the late 70's the following problem was addressed. From Rosenthal's theorem we obtain that a space X must contain either ℓ_1 or a normalized weakly null sequence. Given a normalized weakly null sequence (x_i) is some subsequence unconditional? (It is easy to get a $1 + \varepsilon$-basic subsequence by an argument much like that given above to show that X contains a $1 + \varepsilon$-basic sequence.) This was solved in the negative by Maurey and Rosenthal [MR] in 1977.

2.8 Example. *There exists a normalized weakly null basic sequence (x_i) having the property that the summing basis (s_i) is uniformly block finitely representable in all subsequences.*

Let's explain what this means. The *summing basis* is the most famous of all *conditional* bases:

$$\left\| \sum_1^n a_i s_i \right\| = \sup_{1 \le m \le p \le n} \left| \sum_{i=m}^p a_i \right|$$

It is not unconditional because

$$\left\| \sum_1^n (-1)^i s_i \right\| = 1 \quad \text{yet} \quad \left\| \sum_1^n s_i \right\| = n \ .$$

The summing basis is equivalent to the basis $(e_1, e_1 + e_2, e_1 + e_2 + e_3, \ldots)$ for c_0. Maurey and Rosenthal constructed (x_i) so that for some K, in every subsequence one could find a block basis K-equivalent to (s_i). They also localized their example to obtain $(x_i) \subseteq C(\omega^{\omega^2})$ so that (s_i) was uniformly *block finitely representable* in all subsequences: for some

$K < \infty$ and all $n \in \mathbb{N}$ and all subsequences one could find a block basis $(y_i)_1^n$ which was K-equivalent to $(s_i)_1^n$. But this was a subsequence example and did not examine all block bases. However the trick of the construction would appear again 16 years later when the unconditional basic sequence problem was solved.

In spite of this negative solution, J. Elton [Elt] was able to use Ramsey theory (Theorem 2.7) to obtain that a subsequence of a normalized weakly null sequence did possess some unconditionality and of a different type than that given by the theory of spreading models. A sketch of the proof appears in [O1].

2.9 Theorem. *If (x_i) is a normalized weakly null sequence then some subsequence (y_i) is nearly unconditional. Precisely for all $\delta > 0$ there exists $K(\delta) < \infty$ so that if $n \in \mathbb{N}$ and $(a_i) \subseteq [-1,1]^n$ and $G \subseteq \{i \leq n : |a_i| \geq \delta\}$ then*

$$\left\| \sum_G a_i y_i \right\| \leq K(\delta) \left\| \sum_{i=1}^n a_i y_i \right\| .$$

Another remarkable theorem of the 70's was due to J.L. Krivine [K] (see also [Le], [R2] or [MS]).

2.10 Theorem. *Let (x_i) be a basic sequence. Then there exists $1 \leq p \leq \infty$ so that ℓ_p is block finitely representable in (x_i). Precisely, for all $n \in \mathbb{N}$ and $\varepsilon > 0$ there exists a normalized block basis $(y_i)_1^n$ of (x_i) which is $1 + \varepsilon$-equivalent to the unit vector basis of ℓ_p^n.*

Moreover the proof yields the following. The sequence $(y_i)_1^n$ can be chosen according to the following scheme. We say "$x < y$" w.r.t (x_i) if $x, y \in \langle x_i \rangle$, the linear span of (x_i) and $\operatorname{supp} x < \operatorname{supp} y$ (if $x = \sum_E a_i x_i$ and $a_i \neq 0$ for $i \in E$, $E = \operatorname{supp} x$ is the *support of* x).

"$\forall \ k_1 \ \exists \ y_1 \in \langle x_i \rangle$ with $y_1 > x_{k_1} \ \forall \ k_2 \ \exists \ y_2 \in \langle x_i \rangle$ with $y_2 > x_{k_2} \cdots \forall \ k_n \ \exists \ y_n \in \langle x_i \rangle$ with $y_n > x_{k_n}$ so that $(y_i)_1^n$ is $1 + \varepsilon$-equivalent to the unit vector basis of ℓ_p^n."

We will return to this scheme later.

Ramsey theory has become a standard tool in Banach space theory. But its limitations showed the need for a "block Ramsey theory" for Banach spaces. The search concluded only in the 90's. We discuss this in Chapter 4.

Some other famous problems from this era or even much earlier are

(P7) Is X isomorphic to its closed hyperplanes?

$Y \subseteq X$ is a *hyperplane* of X if $\dim X/Y = 1$. All closed hyperplanes of a space X are mutually isomorphic. Other problems related to the unconditional basic sequence problem were of the type,

Can you find nontrivial operators on a given space X or a subspace of X?

(P8) Does X contain a subspace Y which has a nontrivial (not finite dimensional nor finite codimensional) complemented subspace? i.e., is $Y = Z \oplus W$ (both infinite dimensional)?

And finally,

(P9) Does there exist a space X so that if T is a bounded linear operator on X then for some $\lambda \in \mathbb{R}$, $T = \lambda I + S$ where S is strictly singular (and $I =$ identity operator)?

An operator T is *strictly singular* if its restriction to any infinite dimensional subspace is not an (into) isomorphism.

3

The unconditional basic sequence problem and connections with distortion

We have seen that there is a strong connection between (P1) and the notion of distortion. In this chapter we address the solution of (P2), the unconditional basic sequence problem and also explore its connection with distortion.

The Maurey-Rosenthal example (Example 2.8) cited in the previous chapter gave a technique for spoiling unconditionality at least for block bases which were in fact subsequences of the basis. We sketch the example. A certain weakly compact subset $\mathcal{F}$ of B_{c_0} (the unit ball of c_0) is chosen. Then the space is defined to be the completion of c_{00} under

$$\|x\| = \|x\|_{\mathcal{F}} \equiv \sup\{|\langle x, f\rangle| : f \in \mathcal{F}\} . \tag{3.1}$$

The set $\mathcal{F}$ is constructed as follows. A certain lacunary subsequence $(m_n) \in [\mathbb{N}]^\omega$ is chosen. φ is taken to be any one-to-one function into $\{m_n\}$ whose domain consists of all $(E_1, \ldots, E_n)$ where $E_1 < \cdots < E_n$ are nonempty subsets of $\mathbb{N}$, $n \in \mathbb{N}$, so that $\varphi(E_1, \ldots, E_n) > |E_n|$ for all $E_1 < \cdots < E_n$. Place $f \in \mathcal{F}_0$ if $f = \sum_{i=1}^n \frac{1_{E_i}}{\sqrt{|E_i|}}$ for some $n \in \mathbb{N}$, $E_1 < \cdots < E_n$ where $|E_1| = 1$ and $|E_{i+1}| = \varphi(E_1, \ldots, E_i)$ for $i < n$.

Then take $\mathcal{F}$ to be the set

$$\{P_n f : \exists (f_i) \subseteq \mathcal{F}_0 \text{ with } f_i \to f \text{ pointwise}, n \in \mathbb{N} \cup \{\omega\}\}$$

where P_n is the natural projection (restriction) onto $\langle e_i \rangle_1^n$.

The point of the construction is that if $f = \sum_1^\infty \frac{1_{E_i}}{\sqrt{|E_i|}} \in \mathcal{F}$ then the sequence $x_i = \frac{1_{E_i}}{\sqrt{|E_i|}}$ is equivalent to the summing basis. And of course such a block basis (x_i) may be formed inside of any subsequence of (e_i).

To see this let $\mathcal{G}_n = \{1_E \sqrt{|E|} : |E| = m_n\}$ (we assume $(m_n) = $ range φ restricted to all such $(E_1, \ldots, E_i)$ as appear in the definition of $\mathcal{F}_0$). For $n \in \mathbb{N}$ define the seminorm $\|\cdot\|_n \equiv \|\cdot\|_{\mathcal{G}_n}$ on c_{00} (see 3.1).

208

Clearly $\|\cdot\|_n \leq 2\|\cdot\|$. For each subsequence (e_{k_i}) of (e_i) and every i_0 there exists $x = \frac{1_E}{\sqrt{|E|}} \in \mathcal{G}_{i_0} \cap \langle e_{k_i} \rangle$. Thus $\|x\|_{i_0} = 1$ and a simple calculation shows that for $i \neq i_0$, $\|x\|_i \leq \min(\frac{\sqrt{m_{i_0}}}{\sqrt{m_i}}, \frac{\sqrt{m_i}}{\sqrt{m_{i_0}}}) \leq \varepsilon_{\max(i,i_0)}$, where $\varepsilon_i = \frac{\sqrt{m_{i-1}}}{\sqrt{m_i}}$ for $i > 1$.

Suppose now that $f = \sum_1^n \frac{1_{E_i}}{\sqrt{|E_i|}} \in \mathcal{F}_0$ and set $x_i = \frac{1_{E_i}}{\sqrt{|E_i|}} \in \mathcal{G}_{n(i)}$. Then

$$\|x_i\| \leq \sum_{j=1}^\infty \|x_i\|_j \leq (n(i)-1)\varepsilon_{n(i)} + 1 + \sum_{j=n(i)+1}^\infty \varepsilon_j < 1 + \varepsilon$$

(here the lacunary condition enters). Furthermore if $(a_i)_1^n \subseteq [-1,1]^n$ and $k \leq n$ then

$$\left\| \sum_1^n a_i x_i \right\| \geq \left| \left\langle \sum_{j=1}^k \frac{1_{E_i}}{\sqrt{|E_i|}}, \sum_1^n a_i x_i \right\rangle \right| = \left| \sum_{i=1}^k a_i \right|.$$

Finally let $g = \sum_1^m \frac{1_{G_j}}{\sqrt{|G_j|}} \in \mathcal{F}_0$.

Either $G_1 \neq E_1$ or the two sequences $(G_i)_1^m$ and $(E_j)_1^n$ agree up to $j = k < \min(m,n)$ and then differ or they agree up to $k = \min(n,m)$. An easy calculation, similar to that used to estimate $\|x_i\|$ above yields for $x = \sum_1^n a_i x_i$,

$$|\langle x, g \rangle| \leq \left| \sum_{j=1}^k a_j \right| + \sum_{j=1}^\infty j\varepsilon_j < \left| \sum_{j=1}^k a_j \right| + \varepsilon .$$

It follows that $(x_i)_1^n$ is equivalent to $(s_i)_1^n$, uniformly in n.

What's the moral to this story? Suppose we could find a set $\mathcal{F} \subseteq B_{c_0}$, a corresponding norm $\|\cdot\|_\mathcal{F}$, subsets $\mathcal{G}_i \subseteq \mathcal{F}$ and $\varepsilon_i \downarrow 0$ so that in every block basis of (e_i) and for all i_0 there exists x with $\|x\|_{\mathcal{G}_{i_0}} = 1$ and $\|x\|_{\mathcal{G}_i} \leq \varepsilon_{\max(i,i_0)}$ for $i \neq i_0$. Then we could obtain a counterexample to (P2). That this is impossible was observed by Pei-Kee Lin [Lin]. However a slightly weaker condition does lead somewhere as we shall soon see.

It is perhaps first worth observing that the norm in T can naturally be described as $\|\cdot\|_\mathcal{F}$ for some $\mathcal{F}$. However $\mathcal{F}$ is not as simply described as the $\mathcal{F}$ in the Maurey-Rosenthal example. $\mathcal{F}$ is taken to be the smallest pointwise closed set of functions which contain $\pm e_n$ for all n and satisfy if $(f_i)_1^n \subseteq \mathcal{F}$ are 1-admissible then $\frac{1}{2}\sum_{i=1}^n f_i \in \mathcal{F}$. This was in fact Tsirelson's original description of the unit ball of T^*, the original

Tsirelson space. Equivalently we describe the norm in T is as follows. Let $\| \cdot \| \in \mathcal{N}$ be a norm that generates a space $X = \overline{(c_{00}, \| \cdot \|)}$. Set

$$A_n^*(X) = \left\{ \frac{1}{2} \sum_1^n f_i : (f_i)_1^n \subseteq B_{X^*} \text{is 1-admissible} \right\} .$$

Then one lets T be the space X whose norm satisfies for $x \in c_{00}$,

$$\|x\| = \|x\|_\infty \vee \sup \left\{ |f(x)| : f \in A_n^*(X) \text{ for some } n \in \mathbb{N} \right\} \qquad (3.2)$$

In 1993 W.T. Gowers and B. Maurey [GM1] gave a counterexample to (P2). The norm was defined by an equation in a form roughly resembling the one just given. Of course their norm had to be more complicated, in that (3.2) is inherently unconditional. They needed an equation that was highly conditional and this was obtained by adding a particular third term to the equation. Also the A_n^*'s were somewhat modified for reasons we are about to explain.

The conditions satisfied by the Maurey-Rosenthal seminorms $\| \cdot \|_n$ ($\forall M \in [\mathbb{N}]^\omega \; \forall i_0 \; \exists x \in \langle e_i \rangle_M$ so that $\|x\|_{i_0} = 1$ and $\|x\|_i < \varepsilon_{\max(i,i_0)}$ for $i \neq i_0$) are a weak form of distortion. Recall that X is λ-distortable if there exists an equivalent norm $| \cdot |$ on X so that for all $Y \subseteq X$

$$\sup \left\{ \frac{|x|}{|y|} : x, y \in (S_Y, \| \cdot \|) \right\} \geq \lambda .$$

X is *arbitrarily distortable* if X is λ-distortable for all $\lambda > 1$. Let's see why T is distortable.

The fact that ℓ_1 is not distortable localizes via a compactness argument to give the following. For all $n \in \mathbb{N}$, $\varepsilon > 0$, $K < \infty$ there exists m so that if $(y_i)_1^m$ is K-equivalent to the unit vector basis of ℓ_1^m then some normalized block basis $(x_i)_1^n$ is $1 + \varepsilon$-equivalent to the unit vector basis of ℓ_1^n. A direct proof of this can also be given. One can show that if a normalized sequence $(y_i)_1^{n^2}$ is K^2-equivalent to the unit vector basis of $\ell_1^{n^2}$ then a block basis of length n is K-equivalent to the unit vector basis of ℓ_1^n. One need only break $(y_i)_1^{n^2}$ into n successive groupings each of length n. If no grouping has the desired property, form a normalized block basis, each of whose elements has ℓ_1 norm (w.r.t. the coefficients of some grouping of (y_i)) exceeding K. Then this block basis must be K-equivalent to the unit vector basis of ℓ_1^n (by the last part of the argument that ℓ_1 is not distortable).

T is distortable via the equivalent norms

$$\|x\|_n \equiv \sup\left\{ \frac{1}{2}\sum_{i=1}^{n} \|E_i x\| : E_1 < \cdots < E_n \right\}.$$

To see this first note that if $(x_i)_1^m$ *is 1-admissible* in T (i.e., $(\operatorname{supp} x_i)_1^m$ is 1-admissible or equivalently $e_m \leq \operatorname{supp} x_1 < \cdots < \operatorname{supp} x_m$) then $\|\sum_1^m x_i\| \geq \frac{1}{2}\sum_1^m \|x_i\|$. Thus by the localized result just discussed of James' theorem that ℓ_1 is not distortable, for all $\varepsilon > 0$ and $m \in \mathbb{N}$ in any block subspace Y of T we can find a 1-admissible normalized block basis $(x_i)_1^m$ which is $1 + \varepsilon$ equivalent to the unit vector basis of ℓ_1^m. Consider $x = \frac{1}{m}\sum_1^m x_i$. $\|x\| \geq \frac{1}{1+\varepsilon}$ and $\|x\|_n = \frac{1}{2}\sum_1^n \|E_j x\|$ for some $E_1 < \cdots < E_n$. For $m \gg n$ we obtain $\|x\|_n \approx 1/2$. The idea is that most of the x_i's, of which there are relatively many compared to the few E_j's, cannot be affected by more than one E_j. Precisely, if $F \equiv \{i \leq m : |\{j : E_j x_i \neq 0\}| > 1\}$ then $|F| < n$ and so (think $m \gg n$)

$$\frac{1}{2}\sum_{j=1}^{n} \|E_j x\| \leq \frac{1}{2}\sum_{j=1}^{n}\left(\left\|E_j\left(\frac{1}{m}\sum_{i\in F} x_i\right)\right\| + \left\|E_j\left(\frac{1}{m}\sum_{i\notin F} x_i\right)\right\| \right)$$

$$\leq \frac{1}{m}\left\|\sum_{i\in F} x_i\right\| + \frac{1}{2}\frac{1}{m}\sum_{i\notin F}\|x_i\|$$

$$< \frac{n}{m} + \frac{1}{2}.$$

Since m can be as large as we wish it follows that for all $\varepsilon > 0$ there exists $y \in S_Y$ with $\|y\|_n < \frac{1}{2} + \varepsilon$.

We call this x an (ℓ_1^m, ε) *average*. Next consider an "average" of such vectors $y = \frac{2}{n}\sum_1^n x_i$ where each x_i is an $(\ell_1^{m_i}, \varepsilon)$ average where m_i is large depending on $\max \operatorname{supp} x_{i-1}$. In [GM1] this is called an "RIS" for rapidly increasing sequence. Suppose $\|y\| = \frac{1}{2}\sum_1^k \|E_i y\|$ where $(E_i)_1^k$ is 1-admissible. Let i_0 be minimal so that $E_i x_{i_0} \neq 0$ for some i. Then there are relatively few E_i's relative to the length of the average of x_j for $j > i_0$. Hence an argument like that above yields

$$\|y\| \leq \frac{2}{n}\left(\frac{1}{2}n + \varepsilon + 1\right) = 1 + \frac{1+\varepsilon}{n}.$$

Also $\|y\|_n \geq \frac{1}{1+\varepsilon}$. Thus $\|\cdot\|_n$ can be seen to $2/(1+\frac{2}{n})$ distort T.

If the constant $\frac{1}{2}$ in the definition of T (equation 2.1) is replaced by $\frac{1}{k}$ one can show that the ensuing space is $k - \varepsilon$ distortable for all $\varepsilon > 0$. This leads to the definition of S, the first arbitrarily distortable space constructed by Th. Schlumprecht in 1991 [Sch1, Sch2].

As we shall soon see the creation of S unlocked the key to the solution of the unconditional basic sequence problem. S is the completion of c_{00} under the implicit norm (taking $f(n) = \log_2(n+1)$)

$$\|x\| = \|x\|_\infty \vee \sup\left\{ \frac{1}{f(n)} \sum_{i=1}^{n} \|E_i x\| : 2 \leq n \in \mathbb{N}, \; E_1 < \cdots < E_n \right\}.$$

The admissibility condition in the definition of T is no longer needed because of the damping factor $\frac{1}{f(n)}$. The unit vector basis is not only 1-unconditional but also *subsymmetric*: $\|\sum_1^\infty a_i e_i\| = \|\sum_1^\infty a_i e_{n_i}\|$ if $n_1 < n_2 < \cdots$. S is still asymptotically close to ℓ_1 but not as close as T. If $(x_i)_1^n$ is a block basis of (e_i) then $\|\sum_1^n x_i\| \geq \frac{1}{f(n)} \sum_1^n \|x_i\|$. This is enough to insure either by Krivine's theorem or a mild extension of the localization of the proof that ℓ_1 is not distortable that ℓ_1 is block finitely representable in all block subspaces of S. Schlumprecht showed that his space is arbitrarily distortable via the sequence of equivalent norms

$$\|x\|_n = \|x\|_\infty \vee \sup\left\{ \frac{1}{f(n)} \sum_{i=1}^{n} \|E_i x\| : E_1 < \cdots < E_n \right\}.$$

He also showed that every block subspace of S contains a block basis $1 + \varepsilon$ equivalent to the unit vector basis of S. This proof used RIS's. One shows that if (x_i) is a normalized block basis of (e_i) where each (x_i) is an $(\ell_1^{m_i}, \varepsilon_i)$ average, m_{i+1} depending on $\max \operatorname{supp} x_i$ and $\varepsilon_i \downarrow 0$ suitably fast, then (x_i) is $1 + \varepsilon$-equivalent to (e_i). Since $\|\sum_1^n e_i\| = \frac{n}{f(n)}$ [Sch1, Sch2] this ultimately yields that $\|\cdot\|_n$ $f(n)$-distorts S. Further results on S appear in [AS1] and [AS2].

Gowers and Maurey began their paper solving the unconditional basic sequence problem by observing that S has a stronger distortion property, one much like that in the Maurey-Rosenthal example.

3.1 Definition. X *is* biorthogonally distortable *if there exist a sequence of subsets* $\mathcal{G}_n \subseteq B_{X^*}$, $n \geq 1$, *and* $\varepsilon_n \downarrow 0$ *so that setting* $\|\cdot\|_n = \sup\{|g(x)| : g \in \mathcal{G}_n\}$, *for all i_0 and all $Y \subseteq X$ there exists $y \in S_Y$ with $\|y\|_{i_0} > 1/2$ and $\|y\|_j \leq \varepsilon_{\min(i_0,j)}$ for $j \neq i_0$.*

A biorthogonally distortable space is easily seen to be arbitrarily distortable but the converse is unknown. It is not even known if an arbitrarily distortable space must contain a biorthogonally distortable subspace.

The equivalent norms $\|\cdot\|_n$ on S for n in some lacunary subsequence of $\mathbb{N}$ biorthogonally distort S. As observed by Gowers and Maurey a biorthogonally distortable space X has the following property.

3.2 Theorem. [GM1] *Let X be biorthogonally distortable. For all $\lambda > 1$*

there exists an equivalent norm $|\cdot|$ on X so that no sequence in X is λ-unconditional.

In fact using essentially the same Maurey-Rosenthal like argument of [GM1] one can prove that for all n and $\varepsilon > 0$ there exists an equivalent norm $|\cdot|$ on X so that assuming X has a basis, every monotone basis of length n (including the summing basis) is $1 + \varepsilon$-block represented in all block bases [OS2]. For example to renorm S to uniformly admit the summing basis $(s_i)_1^m$ in all block bases we take

$$A_n^*(S) = \left\{ \frac{1}{f(n)} \sum_1^n f_i : f_1 < \cdots < f_n \ , \ f_i \in B_{S^*} \text{ for } i \leq n \right\}.$$

Let (m_n) be a lacunary subsequence of $\mathbb{N}$. Choose an injection φ into $\{m_n\}$ from the set of all finite sequences $(f_1, \ldots, f_i)$ where $f_1 < \cdots < f_i$ are in B_{S^*} and each f_i has rational coordinates. Then let for $n \in \mathbb{N}$,

$$\mathcal{F}^n = \left\{ \sum_{i=1}^n f_i : f_1 < \cdots < f_n \ , \ f_1 \in A_{m_n}^*(S) \right.$$
$$\left. \text{and } f_{i+1} \in A_{\varphi(f_1,\ldots,f_i)}^*(S) \text{ for } 1 \leq i < n \right\}.$$

Define $|\cdot| = \|\cdot\|_{\mathcal{F}^m}$. As in the Maurey-Rosenthal example one obtains that if X is any block subspace of S then there exist $x_1 < \cdots < x_m$ in X which are 10-equivalent to the summing basis $(s_i)_1^n$. The difference between the Maurey-Rosenthal example and biorthogonal distortion, in addition to finding y in all block subspaces rather than just those spanned by subsequences, is that the condition $\varepsilon_{\max(i_0,j)}$ is replaced by $\varepsilon_{\min(i_0,j)}$. This prevents one from having an infinite block basis which is not unconditional but is sufficient to allow the local version.

To obtain the more general result cited above of [OS2] for a biorthogonally distortable X one first observes that it suffices to find such a norm that works for a fixed monotone normalized basis $(w_i)_1^n$. Indeed it is easy to construct a "universal" monotone basis $(w_i)_1^\infty$ for which every monotone $(z_i)_1^n$ is $1 + \varepsilon_n$ equivalent to some $(w_i)_{i=j+1}^{j+n}$. The equivalent norm that works for $(w_i)_1^n$ up to constant 15 (rather than $1 + \varepsilon$) is defined by $\|\cdot\| = \|\cdot\|_{\mathcal{F}}$ where (z_i^*) is an enumeration of all elements of c_{00} with rational coefficients and $\mathcal{F} = \{\sum_{i=1}^n b_i z_{k_j}^* : z_{k_1}^* < \cdots < z_{k_n}^*, n_1$ sufficiently large, $z_{k_i}^* \in \mathcal{G}_{n_1}$ and $z_{k_{i+1}}^* \in \mathcal{G}_{k_i}$ and $\|\sum_1^n b_i w_i^*\| \leq 1\}$. The condition on n_1 is to insure that $n^2 \varepsilon_{n_1}$ is small. The argument then proceeds as in [MR]. To get the norm that gives $1 + \varepsilon_n$ equivalence one need only spread things out a bit more using functionals of the type

$$\sum_{i=1}^{n} b_i \sum_{j=(i-1)n+1}^{in} z^*_{k_j}$$

to make the errors in the estimates become small. Since we deal only with finding $(w_i)_1^n$, $1+\varepsilon_n$-representable in all block bases we do not need the stronger $\varepsilon_{\max(j,i)}$ estimate of [MR].

So how do you take this and construct a space without an unconditional basic sequence? The norm in S is unconditional so even if you λ-spoil it, it will still be unconditional. Gowers and Maurey achieve this by constructing a new Tsirelson-Schlumprecht like space which has built into it the fact that all block bases are not created equally. The definition is (we again set $f(n) = \log_2(n+1)$)

$$\|x\| = \|x\|_\infty \quad \vee \quad \left\{ \frac{1}{f(n)} \sum_{i=1}^{n} \|E_i x\| : 2 \leq n , \ E_1 < \cdots < E_n \right.$$

$$\left. \text{are intervals of integers} \right\}$$

$$\vee \quad \{|\langle g, Ex\rangle| : g \text{ is a special functional,}$$

$$E \subset \mathbb{N} \text{ is an interval}\}$$

The term special functional needs to be explained.

Let Q be all elements of $B_{c_0} \cap c_{00}$ having only rational values. $J = (j_i)$ is a certain rapidly growing subsequence of $\mathbb{N}$ and $K = (j_{2i-1})$, $L = (j_{2i})$. σ is an injection from all block bases $(f_1, \ldots, f_n)$ in Q into L satisfying $\sigma(f_1, \ldots, f_n)$ is much larger than $\max \operatorname{supp} f_n$. If $(Y, \| \cdot \|)$ is a space resulting from $\| \cdot \| \in \mathcal{N}$ set

$$A^*_m(Y) = \left\{ \frac{1}{f(m)} \sum_{1}^{m} f_i : f_1 < \cdots < f_m , \ f_i \in B_{Y^*} \right\} .$$

For $k \in \mathbb{N}$ set

$$b^*_k(Y) = \{(g_1, \ldots, g_k) : g_1 < \cdots < g_k , \ g_i \in Q \text{ all } i , \ g_1 \in A^*_{j_{2k}}(Y)$$
$$\text{and} \quad g_{i+1} \in A^*_{\sigma(g_1, \ldots, g_i)}(Y) \text{ for } i \leq k-1\} .$$

Let

$$B^*_k(Y) = \left\{ \frac{1}{\sqrt{f(k)}} \sum_{i=1}^{k} g_i : (g_1, \ldots, g_k) \in b^*_k(Y) \right\} .$$

An element of $B^*_k(Y)$ is called a *special functional* (in Y). Then one proves that there exists a $\| \cdot \| \in \mathcal{N}$ for some space X so that

$$\|x\| = \|x\|_\infty \quad \vee \quad \sup\{|g(x)| : g \in A_m^*(X) , \; m \geq 2\}$$
$$\vee \quad \sup\{|g(Ex)| : g \in B_k^*(X) , \; k \geq 2 , \; E \text{ an interval}\} .$$

By virtue of the definition of the norm one can find, just as in S, in any block subspace for all $\varepsilon > 0$ and $n \in \mathbb{N}$ a $1 + \varepsilon$-ℓ_1^n block sequence. Hence one can construct an RIS of arbitrary length and thus a normalized RIS average. Using this and quite a few technical calculations, Gowers and Maurey produce for $k \in K$ a block basis $(x_i)_1^k$ of normalized RIS-averages so that (essentially)

$$\Big\| \sum_{i=1}^k x_i \Big\| \geq \frac{1}{2} \frac{k}{\sqrt{f(k)}} \quad \text{while} \quad \Big\| \sum_{i=1}^k (-1)^k x_i \Big\| \leq \frac{2k}{f(k)} .$$

In particular no block basis of (e_i) is unconditional.

The [GM1] space above actually has the following property. Let (e_i) be the basis for X. Given $\varepsilon > 0$ and block subspaces Y and Z of (e_i) there exists m so that

$$\forall \, n_1 \; \exists \, e_{n_1} < y_1 \in S_Y \; \forall \, n_2 \; \exists \, e_{n_2} < z_2 \in S_Z$$
$$\cdots \forall \, n_{2m-1} \; \exists \, e_{2m-1} < y_m \in S_Y \; \forall \, n_{2m} \; \exists \, e_{n_{2m}} < z_m \in S_Z$$

so that setting $\bar{y} = \sum_1^n y_i$ and $\bar{z} = \sum_1^m z_i$ we have $\|\bar{y} + \bar{z}\| \geq \frac{1}{2} \frac{k}{\sqrt{f(k)}}$ and $\|\bar{y} - \bar{z}\| \leq \frac{2k}{f(k)}$. It follows that $\inf\{\|y - z\| : y \in S_Y, z \in S_Z\} = 0$ for all $Y, Z \subseteq X$ and thus X is *H.I.* (*hereditarily indecomposable*), i.e., X contains no subspace having a nontrivial complemented subspace. Note also that no block basis of (e_i) is *asymptotically unconditional*. This means there does not exist $K < \infty$ and a block basis (b_i) of (e_i) so that if $(x_i)_1^n$ is 1-admissible w.r.t. (b_i), then $(x_i)_1^n$ is K-unconditional.

Gowers and Maurey also showed the remarkable results [GM1]:

· An H.I. space is not isomorphic to its hyperplane nor to any proper subspace.

· If T is a bounded linear operator on a complex H.I. space then $T = \lambda I + S$ for some $\lambda \in \mathbb{C}$ and some strictly singular (not an into isomorphism when restricted to any infinite dimensional subspace) operator S. The same property holds for the real space (with $\lambda \in \mathbb{R}$) of [GM1] described above, and now denoted by GM in the literature.

So in some sense GM has few operators on it. However in another sense it has many operators. Indeed in [AS3] G. Androulakis and Th. Schlumprecht have constructed many strictly singular non-compact operators on GM, enough to conclude that ℓ_∞ embeds into the Banach space of bounded linear operators on GM.

Thus problems (P2), (P7), (P8) and (P9) were solved in one fell swoop. Gowers [G2] subsequently solved (P6) in the negative. A weaker formulation of (P6) remains open.

(Q5) If ℓ_1 is not finitely representable in X does X contain Y with Y^* separable? Does X contain a subspace which is either reflexive or isomorphic to c_0 or ℓ_1?

R.C. James' famous space, the James tree space, JT [J3] has the property that it does not contain ℓ_1 yet JT^* is nonseparable. Another space with this property, the James function space, JF is constructed in [LS].

The technique of [GM1] has led to the discovery of a number of interesting spaces with varying lists of properties. For example P. Habala [Ha] has constructed a space X so that no subspace has the Gordon-Lewis property, a weaker property than having an unconditional basis. V. Ferenczi [F1] has constructed a uniformly convex H.I. space. Then Habala and Ferenczi [FH] constructed a uniformly convex space so that no subspace has the Gordon-Lewis property. V. Ferenczi [F2] has shown that the space of Gowers and Maurey [GM1] is *quotient H.I.* (every infinite dimensional quotient is H.I.). He also constructs a reflexive H.I. space whose dual space is not H.I. Other fascinating spaces are constructed in [GM2]. Some other discoveries are mentioned in later chapters.

The existence of H.I. Banach spaces was unknown until [GM1]. Once observed they began to appear in many forms and many places. Quite recently Argyros and Felouzis [AF] have shown that many spaces (such as L_p for $1 < p < \infty$, c_0 and T) are quotients of H.I. spaces. In addition they proved the following remarkable theorem.

3.3 Theorem. [AF] *Every space X contains either an isomorph of ℓ_1 or a subspace Y which is a quotient of an H.I. space.*

Any space having ℓ_1 as a quotient must contain ℓ_1 so ℓ_1 is not a quotient of an H.I. space. The proof of Theorem 3.3 uses interpolation (in the [DFJP] sense) and numerous results from [AMT], some of which are discussed in Chapter 6. S. Argyros [Ar2] has also proved that if a separable space contains isomorphs of all separable reflexive H.I. Banach spaces then it must contain an isomorph of $C[0,1]$.

4
Gowers' dichotomy: A block Ramsey Theorem

W.T. Gowers [G1] discovered the long sought after block Ramsey theorem for Banach spaces and used it to prove his famous

4.1 Theorem (Gowers' Dichotomy Theorem). *Every X contains a subspace Y which either has an unconditional basis or is H.I.*

Before stating the block Ramsey theorem we set some notation. Let (e_i) be a basis for a space X. Σ denotes the set of all finite normalized block bases (including the null sequence) of (e_i). If Y is a block subspace of X and $\sigma \subseteq \Sigma$ we say σ *covers* Y if there exists $(x_i)_1^n \in \sigma$ with $x_i \in Y$ for all i. σ is *large* if σ covers all block subspaces of X.

Suppose σ is large. We consider a game played by S who chooses block subspaces of X, and V who then chooses elements of the unit sphere of the block subspace chosen by S. The plays alternate $S, V, S, V, \ldots$. V *wins* if at some point the choices of V, $(x_1, \ldots, x_k) \in \sigma$. S wins if for all k, the plays of V, $(x_1, \ldots, x_k) \notin \sigma$.

Let $\Delta = (\delta_1, \delta_2, \ldots)$ be a sequence of positive numbers. We define

$$\sigma_\Delta = \{(y_1, \ldots, y_k) \in \Sigma : \exists\, (x_1, \ldots, x_k) \in \sigma$$
$$\text{with } \|x_i - y_i\| \le \delta_i \text{ for } 1 \le i \le k\}$$

4.2 Theorem. (Gowers' block Ramsey Theorem). *Let $\sigma \subseteq \Sigma$ be large for X. Let Δ be any sequence of positive numbers. Then there exists a block subspace Z of X so that V has a winning strategy inside Z for σ_Δ.*

The conclusion means that V can always win provided that S is only permitted to choose block subspaces of Z.

To deduce the dichotomy theorem from this we first quantify the notion H.I. For $\varepsilon > 0$, X is $H.I.(\varepsilon)$ if for all subspaces $Y, Z \subseteq X$ there exist

$y \in Y$, $z \in Z$ such that

$$\|y - z\| < \varepsilon\|y + z\| \ .$$

(Of course if X has a basis it suffices to only consider block subspaces Y and Z.) It can be easily seen that X is H.I. iff X is $H.I.(\varepsilon)$ for all $\varepsilon > 0$. The dichotomy theorem thus follows from a standard diagonal argument and

4.3 Proposition. *Let $C > 1$ and suppose that X contains no C unconditional basic sequence. Then for $\varepsilon > 1/C$, X contains a $H.I.(\varepsilon)$ subspace.*

Proof. We may assume X has a basis (e_i) of basis constant 2. Let σ be all normalized block bases $(x_i)_1^k$ of (e_i) so that

$$\left\|\sum_{i=1}^{k}(-1)^i\lambda_i x_i\right\| < \frac{1}{C}\left\|\sum_1^k \lambda_i x_i\right\|$$

for some sequence of scalars $(\lambda_i)_1^k$. σ is large since X contains no C unconditional basic sequence. Indeed, otherwise we could find a block subspace Y of X so that for all normalized block bases $(y_i)_1^k$ of (e_i) with $y_i \in Y$ for $i \leq k$, for all $(\lambda_i)_1^k$

$$\left\|\sum_1^k \lambda_i y_i\right\| \leq C\left\|\sum_{i=1}^k (-1)^i\lambda_i y_i\right\|$$

or equivalently

$$\left\|\sum_1^k (-1)^i\lambda_i y_i\right\| \leq C\left\|\sum_1^k \lambda_i y_i\right\| \ .$$

It follows that if $(x_i)_1^\infty \subseteq Y$ is a normalized block basis of (e_i) then for all (a_i), and all choices of $\pm$

$$\left\|\sum_1^k \pm a_i x_i\right\| \leq C\left\|\sum_1^k a_i x_i\right\| \ .$$

So (x_i) would be C-unconditional.

Choose $\Delta = (\delta_i)$ and $\varepsilon' > 0$ so that $\frac{C}{1+\varepsilon'} - 3\sum_1^\infty \delta_i > \frac{1}{\varepsilon}$ and so that if (x_i) is any normalized block basis of (e_i) then any Δ-perturbation of (x_i) is $1 + \varepsilon'$-equivalent to (x_i). By Theorem 4.2 there exists a block subspace $Z \subseteq X$ so that V has a winning strategy for σ_Δ inside Z. Let W and Y be block subspaces of Z. If S plays $W, Y, W, Y, \ldots$ then V can choose $(x_1, \ldots, x_k) \in \sigma_\Delta$ with $x_i \in W$ if i is odd and $x_i \in Y$ if i is even. Thus there exist $(x'_1, \ldots, x'_k)$ and $(\lambda_1, \ldots, \lambda_k)$ with $\|x'_i - x_i\| < \delta_i$ and

$$\left\| \sum_1^k \lambda_i x'_i \right\| > C \left\| \sum_1^k (-1)^i \lambda_i x'_i \right\|.$$

Hence by our choice of ε', Δ and the triangle inequality,

$$\left\| \sum_1^k \lambda_i x_i \right\| > \frac{1}{\varepsilon} \left\| \sum_1^k (-1)^i \lambda_i x_i \right\|.$$

Let $w = \sum_{i \, \text{odd}} \lambda_i x_i$, $y = \sum_{i \, \text{even}} \lambda_i x_i$. Then $w \in W$, $y \in Y$ and $\|w + y\| > \frac{1}{\varepsilon} \|w - y\|$. Hence Z is $H.I.(\varepsilon)$. $\qquad\square$

We shall not prove Theorem 4.2. However we shall present Maurey's proof of Proposition 4.3 [Ma1] (see also [TJ1]) which uses only the essential elements of the proof of Theorem 4.2 required to establish it. Those familiar with the Nash-Williams Ramsey theorem [NW] will note some similarities.

Maurey's proof of Proposition 4.3. We shall prove a slightly different variant. Given $\varepsilon > 0$, X contains either a $\frac{2}{\varepsilon}$-unconditional basic sequence or a subspace which is $H.I.(2\varepsilon)$.

We assume X has a 2-basis (e_i) and let X_0 be the $\mathbb{Q}$-linear subspace generated by (e_i). All $\mathbb{Q}$-subspaces we henceforth consider will be the $\mathbb{Q}$-linear subspaces generated by block bases in X_0. Thus our $\mathbb{Q}$-subspaces are countable sets. For $(x, y) \in X_0 \times X_0$ we say (x, y) *accepts* the $\mathbb{Q}$-subspace Z of X_0 if for all $\mathbb{Q}$-subspaces $U, V \subseteq Z$ there exists $(u, v) \in U \times V$ with

$$(x + u, y + v) \in A \equiv \{(w, z) \in X_0 \times X_0 : \|w - z\| < \varepsilon \|w + z\|\}.$$

(x, y) *rejects* Z if (x, y) does not accept any $\mathbb{Q}$-subspace of Z.

Let $X_0 \times X_0 = \{(x_n, y_n) : n \in \mathbb{N}\}$. Inductively we choose $\mathbb{Q}$-subspaces $X_1 \supseteq X_2 \supseteq \cdots$ so that for each n either (x_n, y_n) rejects X_n or (x_n, y_n) accepts X_n. Let Z_0 be a $\mathbb{Q}$-diagonal space of the X_i's. Thus for all n either (x_n, y_n) rejects Z_0 or it accepts Z_0. If $(0, 0)$ accepts Z_0 then $\bar{Z}_0$ is $H.I.(2\varepsilon)$. Otherwise we shall construct a block basis (z_k) with $1 \leq \|z_k\| \leq 2$ so that for all m and scalars $(a_i)_1^m$, all $\pm$,

$$\left\| \sum_{k=1}^m a_k z_k \right\| \leq \frac{1}{\varepsilon} \left\| \sum_{k=1}^m \pm a_k z_k \right\|,$$

provided that $a_k = j_k / N 2^k$ for some integer $|j_k| \leq N 2^k$ for all $k \leq m$, where $N > 16/\varepsilon$ is a fixed integer. This suffices to yield (as in the proof of Proposition 4.3 given above with an additional perturbation argument) that (z_k) is $\frac{2}{\varepsilon}$-unconditional. In other words, we shall construct the z_k's so that $\|x - y\| \geq \varepsilon \|x + y\|$ whenever $x = \sum_I a_k z_k$, $y = \sum_J a_k z_k$ where a_k's have the form above and I, J partition $\{1, \ldots, m\}$; i.e., we wish $(x, y) \notin A$ for such x and y. Thus it suffices to show such an (x, y) rejects Z_0. We shall call such an (x, y) formed from $(z_i)_1^m$ a *reasonable pair*.

Once more, our goal is to use that $(0, 0)$ rejects Z_0 to construct (z_k) so that every reasonable pair formed from (z_k) rejects. First we note that if (x, y) rejects Z_0 then for all $\mathbb{Q}$-subspaces $W \subseteq Z_0$ there exists a $\mathbb{Q}$-subspace $W' \subseteq W$ so that for all $w' \in W'$, $(x + w', y)$ rejects. Otherwise there exists such a W so that for all $\mathbb{Q}$-subspaces $U \subseteq W$ there exists $u_0 \in U$ so that $(x + u_0, y)$ accepts. Thus for all $V \subseteq W$ there exists $(u_1, v) \in U \times V$ so that $(x + u_0 + u_1, y + v) \in A$, hence (x, y) accepts W which contradicts (x, y) rejects Z_0.

Assume $(z_i)_1^n$ are chosen so that all reasonable pairs formed from $(z_i)_1^n$ reject. By our observation above (there are only finitely many such reasonable pairs) there exists $W \subseteq Z_0$ so that for all $w \in W$, and all reasonable pairs (x, y), $(x + w, y)$ rejects. Choose $1 < \|z_{n+1}\| < 2$ with $z_{n+1} \in W$. Let (x', y') be a reasonable pair formed from $(z_i)_1^{n+1}$. Then (x', y') is of the form $(x + a z_{n+1}, y)$ or $(x, y + a z_{n+1})$ for some reasonable pair (x, y) formed from $(z_i)_1^n$. Both forms reject. That the second case rejects follows from the symmetry of A and of reasonable pairs. $\qquad\qquad\qquad\qquad\square$

Some generalizations of Theorems 4.1 and 4.2 have been found. R. Wagner [W1] has produced a local version of Theorem 4.1: Every X contains either an asymptotically unconditional basic sequence or a basic sequence (x_i) satisfying the following. $\forall \varepsilon > 0 \ \exists \ n \forall$ block subspaces $Y, Z \ \exists \ y_1 < z_1 < \cdots < y_n < z_n$, $y_i \in Y$, $z_i \in Z$, with

$$\left\| \sum_1^n y_i - \sum_1^n z_i \right\| < \varepsilon \left\| \sum_1^n y_i + \sum_1^n z_i \right\|.$$

Gowers [G3] has found the following more general result. Let X have basis (e_i). $\Sigma = \Sigma(X)$ is the set of all infinite block bases (x_i) of (e_i) with $0 \leq \|x_i\| \leq 1$. V and S play the game as above. Given $\sigma \subseteq \Sigma$, V wins if no matter what block subspaces X_n are chosen by S, V can choose $x_n \in X_n$ so that $(x_1, x_2, \ldots) \in \sigma$. σ is *weakly Ramsey* if for all positive sequences Δ there exists a block subspace Y of X so that all block bases of Y, (y_i) with $0 \leq \|y_i\| \leq 1$, belong to $(\sigma^c)_\Delta$ or V has a winning strategy for σ_Δ inside Y.

Σ is given the relative topology of the product space $X^{\mathbb{N}}$.

4.4 Theorem. [G3] *Let $\sigma \subseteq \Sigma(X)$ be an analytic set. Then σ is weakly Ramsey.*

In the same paper Gowers obtains the following application refining Theorem 4.1. We first state the result and then define the relevant terms.

4.5 Theorem. [G3] *Every X contains a subspace Y satisfying one of the following four mutually exclusive properties.*

(1) Y *is H.I. (and every operator from a subspace Z of Y into Y is a strictly singular perturbation of a multiple of the inclusion map).*

(2) Y *has an unconditional basis and every isomorphism between block subspaces W and Z of Y is a strictly singular perturbation of the restriction of some invertible diagonal operator on Y.*

(3) Y *has an unconditional basis and is strictly quasi-minimal.*

(4) Y *has an unconditional basis and is minimal.*

Y is *minimal* if Y embeds isomorphically into every subspace of Y. For example, ℓ_p is minimal $(1 \leq p < \infty)$ as are c_0, T^* [CJT] and S [Sch2] but T is not [CO]. No H.I. space is minimal. X and Y are *totally incomparable* if no subspace of X is isomorphic to a subspace of Y (ℓ_p and ℓ_q are totally incomparable for $1 \leq p \neq q < \infty$). X is *quasi-minimal* if it does not contain a pair of totally incomparable subspaces. X is *strictly quasi-minimal* if X is quasi-minimal and X does not contain a minimal subspace. T is strictly quasi-minimal.

R. Wagner [W2] has considered a higher order extension of the Gowers game above. Let $\sigma \subseteq \Sigma$, the set of finite normalized block bases of (e_i). In the 1-game S chooses a block subspace Y and V chooses $y \in S_Y$. In the $\alpha+1$-game, S chooses a block subspace Y and V chooses $y \in S_Y$ and then they play the α-game. If α is a limit ordinal S chooses $\beta < \alpha$ and a block subspace Y. V chooses $y \in S_Y$ and an ordinal α' with $\alpha \leq \alpha' < \beta$. Then they play the α'-game. V has a winning strategy if V can make sure that the resulting sequence it chooses is in σ. The set σ naturally defines a tree and thus can be given an order (see Chapter 7).

4.6 Theorem. [W2] *Let $\alpha < \omega_1$ and $\sigma \subseteq \Sigma$. Suppose that σ and σ_Δ are well founded trees and that both have the same index α in every block subspace. Then there exists $Y \subseteq X$ so that V has a winning strategy for the α-game inside Y for σ_Δ.*

Wagner then uses this to obtain a result like that in [W1] concerning "β-asymptotically unconditional" sequences.

5

Distortion

The distortion problem, (P4), was solved in [OS2] (see also [OS3, OS4]). In fact ℓ_p is biorthogonally distortable for $1 < p < \infty$.

5.1 Theorem. *Let $1 < p < \infty$. There exist sequences of sets $A_n \subseteq S_{\ell_p}$, $A_n^* \subseteq B_{\ell_p^*}$ so that each A_n is asymptotic in ℓ_p (i.e., $A_n \cap Y \neq \emptyset$ for all $Y \subseteq \ell_p$) and $\varepsilon_n \downarrow 0$ so that*

(i) $\forall n \; \forall x \in A_n \exists x^* \in A_n^*$ with $x^*(x) > 1 - \varepsilon_n$
(ii) $\forall n \neq m \; \forall x \in A_n \; \forall x^* \in A_m^*$,

$$|x^*(x)| < \varepsilon_{\min(n,m)}$$

Moreover the sets A_n are spreading (if $(a_i) \in A_n$ then $(0, a_1, 0, 0, a_2, \ldots)$ $\in A_n$) and lattice ($x \in A_n \Leftrightarrow |x| \in A_n$).

Later [Ma4] it was shown that the sets A_n could also be taken to be *symmetric* ($(a_i) \in A_n \Leftrightarrow (a_{\pi(i)}) \in A_n \forall$ permutations π of $\mathbb{N}$).

How does one produce such sets? Let's consider an "easier" problem. Show that ℓ_2 is distortable. This is equivalent to producing sets $A, B \subseteq S_{\ell_2}$ so that A and B are *asymptotic* ($\forall X \subseteq \ell_2$, $A \cap X \neq \emptyset$ and $B \cap X \neq \emptyset$) and $D(A, B) \equiv \inf\{\|a - b\| : a \in A, b \in B\} > 0$. Roughly, if such sets A and B exist then one defines a norm that distorts ℓ_2 by taking the ball of this norm to be $\overline{co}(A \cup -A)$. And if ℓ_2 is distortable then there exists an equivalent norm $|\cdot|$ on ℓ_2 so that $|\cdot| : S_{\ell_2} \to [0, \infty)$ does not stabilize. Thus there exist $X \subseteq \ell_2$ and $a < b$ so that for all $Y \subseteq X$ there exist $x, y \in S_Y$ with $|x| > b$ and $|y| < a$. Now $X = \ell_2$ and we take $B = \{x \in S_X : |x| > b\}$, $A = \{x \in S_X : |x| < a\}$. These sets are asymptotic and separated. How would you pick the sets A and B? How do you distinguish between points in S_{ℓ_2}? The history here is that

223

given an explicit formula for a norm on ℓ_2 one can show, sometimes after much work, that it does not distort ℓ_2 (see (Q3)).

The approach of [OS2] is indirect. First we note that one could transfer distortion, should it exist, between one ℓ_p space and another through the Mazur map. This is defined as follows.

$$M_p : S_{\ell_1} \to S_{\ell_p} \text{ is given by } M_p(a_i)_{i=1}^{\infty} = (\text{sign } a_i |a_i|^{1/p})_{i=1}^{\infty} .$$

M_p is a uniform homeomorphism between the unit spheres [Rib]. For $1 < p < \infty$, ℓ_p is distortable iff there exist asymptotic sets $A, B \subseteq S_{\ell_p}$ with $D(A, B) > 0$. Since M_p preserves block bases of (e_i), M_p preserves asymptotic sets. It follows that ℓ_p is distortable iff S_{ℓ_1} contains a pair of separated asymptotic sets and this in turn can be moved to any ℓ_q via M_q.

Fine and dandy except the problem is no easier in any ℓ_p, nor is it easy to find separated asymptotic sets in ℓ_1. The approach of [OS2] was to use a generalized Mazur map to transfer distortion from a known Tsirelson type space to ℓ_1 and then out to ℓ_p. Of course the separated asymptotic sets do not lead to a distortion of ℓ_1 but they do show that not every Lipschitz function on S_{ℓ_1} (e.g., $f(x) = \text{dist}(A, x)$) is oscillation stable. A consequence of this approach is to reveal the significance of Tsirelson type spaces. They are not just some collection of pathological counterexamples. Analyzing the structure of these spaces leads to a new fundamental discovery concerning the structure of Hilbert space.

The generalized Mazur map that we needed was prepared for use in [Gil], [Lo]. Suppose that (e_i) is a normalized 1-unconditional basis for X. Define the *entropy* map

$$E_X : (\ell_1 \cap c_{00}) \times X \to [-\infty, \infty)$$

by

$$E_X(h, x) \equiv \sum_i |h_i| \log |x_i|$$

where $h = (h_i)$, $x = \sum x_i e_i$, under the convention $0 \log 0 = 0$. Then we define

$$F_X : S_{\ell_1} \cap c_{00} \to S_X$$

by

$$F_X(h) = \text{the unique } x \in S_X$$

so that if $x = \sum x_i e_i$ then

(i) $E(h,x) \geq E(h,y)$ for all $y \in S_X$

(ii) $\text{supp } h = \text{supp } x$

(iii) $\text{sign } x_i = \text{sign } h_i$ for all i.

Of course such an x must be shown to exist. If X is uniformly convex and uniformly smooth it can be shown [OS2] that F_X extends to a uniform homeomorphism between S_{ℓ_1} and S_X. More generally one can prove

5.2 Theorem. [OS2] *Let X have an unconditional basis. Then S_X and S_{ℓ_1} are uniformly homeomorphic if (and only if) c_0 is not finitely representable in X.*

The "only if" direction is due to Enflo [E2]. Theorem 5.2 has been extended to nondiscrete Banach lattices by F. Chaatit [C]. N.J. Kalton [Ka1] and M. Daher [Dah] have discovered proofs of these results using complex interpolation theory (see also [BeL, Chapter 9]).

The map F_X can be used to prove Theorem 5.1 by transferring, in a somewhat indirect way, the biorthogonal distortion of S to ℓ_1 (which does not yield distortion there but does yield some interesting subsets of S_{ℓ_1}) and then out to ℓ_p via M_p.

Let's return to (P3) and rephrase the problem as follows:

(P3)$'$ For what X is every Lipschitz $f : S_X \to \mathbb{R}$ oscillation stable?

If some such f is not oscillation stable then there exists $Y \subseteq X$ and $a < b$ so that for all $Z \subseteq Y \; \exists \; z, w \in S_Z$ with $f(z) < a$ and $f(w) > b$. Thus S_Y contains a pair A, B of separated asymptotic sets. Conversely given such sets A and B the function

$$f(x) = \text{dist}(x, A) \text{ is Lipschitz but not oscillation stable on } X \text{ .}$$

It follows that every Lipschitz $f : S_X \to \mathbb{R}$ is oscillation stable iff for all $Y \subseteq X$, S_Y satisfies the approximate Ramsey property described in the introduction. Namely if $C_1, \ldots, C_k$ is a finite coloring of S_Y and $\varepsilon > 0$ then there exists $Z \subseteq Y$ and i_0 with $S_Z \subseteq (C_{i_0})_\varepsilon$, the ε-neighborhood of C_{i_0}. From our above discussion ℓ_1 fails this property as does ℓ_p if $1 < p < \infty$. What about c_0? Gowers [G4] in a deep piece of work using ultrafilters proved that c_0 has this property. Combined with Milman's theorem [M1] we obtain X has the Lipschitz stabilization property of (P3)$'$ iff X is c_0-*saturated* (i.e., for all $Y \subseteq X$, Y contains c_0). As far as distortion goes, X is not distortable implies that every $Y \subseteq X$ must contain c_0 or ℓ_1.

The relations between the notions X is distortable, X is arbitrarily distortable and X is biorthogonally distortable remain unclear. It is not known if every distortable space contains an arbitrarily distortable subspace. Thus we make the following definition.

If $(X, \| \cdot \|)$ is given an equivalent norm $| \cdot |$,

$$d(X, | \cdot |) = \inf_{Y \subseteq X} \sup \left\{ \frac{|x|}{|y|} : x, y \in (S_Y, \| \cdot \|) \right\} .$$

X is of *D-bounded distortion* if for all equivalent norms $| \cdot |$ on X and for all $Y \subseteq X$, $d(Y, | \cdot |) \leq D$.

(Q6) Does there exist a distortable space of bounded distortion? Is T of bounded distortion?

The existence of a distortable space of bounded distortion would be quite fascinating. Such a space would exhibit a new type of geometry other than that which we presently know to exist. In terms of colorings induced by equivalent norms this would be a sort of weak Ramsey property. Certain bad colorings exist but not too bad.

As noted previously T is $2 - \varepsilon$ distortable for all $\varepsilon > 0$ but nothing more is known. There are other reasons why T might be such a space. In [MT1] it is proved that if X has bounded distortion then X contains a subspace which is asymptotic ℓ_p for some $1 \leq p < \infty$ or asymptotic c_0.

5.3 Definition. *Let X have a basis (e_i), $1 \leq p < \infty$. X is asymptotic ℓ_p if there exists $K < \infty$ so that for all n if $(x_i)_1^n$ is a normalized block basis of $(e_i)_n^\infty$ then $(x_i)_1^n$ is K equivalent to the unit vector basis of ℓ_p^n.*

Following this result the next theorem was obtained

5.4 Theorem. [TJ2] *Let X have bounded distortion. Then X contains an unconditional basic sequence.*

We sketch the proof. Actually we present, with apologies to the author, a proof which is less elegant but which allows us to introduce and discuss the generalized Schreier classes $(S_\alpha)_{\alpha < \omega_1}$ which we will need in the sequel. Furthermore the published proof [TJ2] does not require the result of [MT1] which we shall use. The classes $(S_\alpha)_{\alpha < \omega_1}$ appeared in [AO] for $\alpha < \omega$ and in [AA] for general $\alpha < \omega_1$. We have already defined S_0 and S_1 in Chapter 2. If S_α has been defined, let

$$S_{\alpha+1} = \left\{ \bigcup_1^n E_i : (E_i)_1^n \text{ is 1-admissible and } E_i \in S_\alpha \text{ for } i \leq n, \, n \in \mathbb{N} \right\} .$$

If α is a limit ordinal choose and fix $\alpha_n \uparrow \alpha$.

$$S_\alpha = \{E : \exists\, n \in \mathbb{N} \text{ with } E \in S_{\alpha_n} \text{ and } n \leq \min E\}\,.$$

One disturbing thing about the definition of S_α for $\alpha > \omega$ is that the classes depend upon one or more choices at limit ordinals. However no useful theorem developed thus far concerning the S_α's depend upon these choices. The utility of the S_α's is in their complexity. We digress and discuss some of their properties.

Each class S_α is *regular*. This means it is

hereditary: $F \subseteq G \in S_\alpha \Rightarrow F \in S_\alpha$

spreading: $F = (n_1, \ldots, n_k) \in S_\alpha$ with $n_1 < \cdots < n_k$ and $m_1 < \cdots < m_k$ satisfies $n_i \leq m_i$ for $i \leq k \Rightarrow (m_1, \ldots, m_k) \in S_\alpha$

pointwise closed: i.e., closed in $2^{\mathbb{N}}$.

Moreover we have the following proposition. If $\mathcal{F} \subseteq [\mathbb{N}]^{<\omega}$ and $\mathcal{G} \subseteq [\mathbb{N}]^{<\omega}$ are regular we define

$$\mathcal{F}[\mathcal{G}] \;=\; \left\{ \bigcup_1^n G_i : n \in \mathbb{N},\ G_1 < \cdots < G_n,\ G_i \in \mathcal{G} \text{ for } i \leq n \right.$$
$$\left. \text{and } (\min(G_i)_1^n) \in \mathcal{F} \right\}.$$

If $N = (n_i)$ is a subsequence of $\mathbb{N}$,

$$\mathcal{F}(N) = \{(n_i)_{i \in F} : F \in \mathcal{F}\}\,.$$

5.5 Proposition. [OTW]

(a) Let $\alpha < \beta < \omega_1$. Then there exists $n \in \mathbb{N}$ so that $n \leq F \in S_\alpha \Rightarrow F \in S_\beta$.

(b) For all $\alpha, \beta < \omega_1$ there exists $N \in [\mathbb{N}]^\omega$ so that $S_\alpha[S_\beta](N) \subseteq S_{\beta+\alpha}$

(c) For all $\alpha, \beta < \omega_1$ there exists $M \in [\mathbb{N}]^\omega$ so that $S_{\beta+\alpha}(M) \subseteq S_\alpha[S_\beta]$.

It is also easy to see that the Cantor-Bendixson index of the compact metric space S_α in $2^{\mathbb{N}}$ is ω^α. The S_α's are large enough to measure the complexity of any regular $\mathcal{F} \subseteq [\mathbb{N}]^\omega$. One has

5.6 Proposition. [OTW] *If $\mathcal{F}$ is regular with Cantor-Bendixson index less than or equal to ω^α then there exists $M \in [\mathbb{N}]^\omega$ with $\mathcal{F}(M) \subseteq S_\alpha$.*

R. Judd [Ju] via the notion of Schreier games obtained a pretty dichotomy theorem subsequently generalized as follows (see [Far1], [Kir] and for the simplest argument using Ramsey theory [Ga4]).

5.7 Theorem. *Let $\mathcal{F}$ be a hereditary family in $[\mathbb{N}]^{<\omega}$ and let $\alpha < \omega_1$. Given $N \in [\mathbb{N}]^{\omega}$ there exists $M \in [N]^{\omega}$ so that either $S_\alpha \cap [M]^{<\omega} \subseteq \mathcal{F}$ or $\mathcal{F} \cap [M]^{<\omega} \subseteq S_\alpha$. (In fact S_α can be replaced by any hereditary family.)*

Proof of Theorem 5.4. Let X have D-bounded distortion. By the result cited above [MT1] we may assume X is asymptotic ℓ_p for some $1 \leq p < \infty$ w.r.t. the basis (e_i) (the asymptotic c_0 case is similar). We shall prove that for $D' > D$ and for each $\alpha < \omega_1$ there is a block basis (x_i^α) of (e_i) which is $4D'$ S_α-*unconditional*. By this we mean that if $(y_i)_1^n$ is α-*admissible* w.r.t. (x_i^α) (i.e., $y_1 < \cdots < y_n$ and $(\min(\operatorname{supp} y_i))_1^n \in S_\alpha$ where "supp" is taken w.r.t. (x_i^α)) then $(y_i)_1^n$ is $4D'$-unconditional. This is sufficient to deduce the theorem. Indeed (see Chapter 7) the tree of all normalized $4D'$-unconditional sequences in X then has unbounded order and hence X contains an unconditional basic sequence.

We do this by producing (x_i^α) so that if $(E_i)_1^n$ are α-admissible intervals then for $x \in \langle x_i^\alpha \rangle$

$$\left\| \sum_1^n \pm E_i x \right\| \leq 4D' \|x\| .$$

This is easy for $\alpha = 0$. Assume (x_i^α) has been defined. We define an equivalent norm $|\cdot|$ on $\langle x_i^\alpha \rangle$ by $|x| = \sup\{\|\sum_1^n \pm E_i x\| : E_1 < \cdots < E_n$ are intervals and $(E_i)_1^n$ is $\alpha + 1$-admissible$\}$. In the definition of $|x|$, Ex is w.r.t. the basis (x_i^α). Clearly $\|x\| \leq |x|$. Also since (x_i^α) is S_α-unconditional and asymptotic ℓ_p, $|\cdot|$ is an equivalent norm. Indeed let $|x| = \|\sum_1^n \pm E_i x\|$ and form sets A_j so that $|x| = \|\sum_{j=1}^m \sum_{i \in A_j} \pm E_i x\|$ so that $(\sum_{i \in A_j} \pm E_i x)_{j=1}^m$ is 1-admissible and for each j, $(\pm E_i x)_{i \in A_j}$ is α-admissible. Thus

$$|x| \sim \left(\sum_{j=1}^m \left\| \sum_{i \in A_j} \pm E_i x \right\|^p \right)^{1/p} \leq \left(\sum_{j=1}^m K \left\| \sum_{i \in A_j} E_i x \right\|^p \right)^{1/p} \leq CK \|x\|$$

for appropriate C and K. We claim that if $(E_i)_1^n$ is $\alpha + 1$ admissible then $|\sum_1^n \pm E_i x| \leq 4|x|$ for $x \in \langle x_i^\alpha \rangle$. Assuming the claim and the fact that our space is of D bounded distortion there exists a block basis $(x_i^{\alpha+1})$ of (x_i^α) so that

$$\left\| \sum_1^n \pm E_i x \right\| \leq 4D' \|x\|$$

whenever $(E_i)_1^n$ is $\alpha + 1$-admissible. The latter holds w.r.t. (x_i^α) and hence also w.r.t. $(x_i^{\alpha+1})$. The limit ordinal case is easy.

The claim follows from the combinatorial

5.8 Lemma. [TJ2] *Let $\beta < \omega_1$ and let $(E_i)_1^n$ and $(F_j)_1^m$ be β-admissible families of intervals. Let $\mathcal{H}$ be the family of intervals obtained from these by considering all (nonempty) intervals of the forms $E_i \cap F_j$, $E_i \setminus \bigcup_1^m F_j$, $F_j \setminus \bigcup_1^n E_i$. Then there exist four families $(\mathcal{F}_i)_1^4$ of β-admissible families of intervals so that $\mathcal{H} = \bigcup_1^4 \mathcal{F}_i$.* □

Maurey extended Theorem 5.1 to a larger class of Banach spaces.

5.9 Theorem. [Ma2] *Let X be a Banach space with an unconditional basis. If ℓ_1 is not finitely representable in X then X contains an arbitrarily distortable subspace.*

Combining these results one obtains that the search for a space of bounded distortion can be confined to spaces X having an unconditional basis which are asymptotic ℓ_p for some $1 \leq p < \infty$ or asymptotic c_0 and if $p \in (1, \infty)$, ℓ_1 is finitely representable in X. This last proviso may be removable and leads to a special case of (Q6).

(Q7) Can an asymptotic ℓ_p space for $1 < p < \infty$ be of bounded distortion?

Following [GM1] and [OS2] in the early 90's some short lived problems arose concerning unconditional basic sequences and distortion. Suppose that (x_i) is asymptotically unconditional (i.e., S_1-unconditional). Must some block basis be unconditional? Suppose X is asymptotically ℓ_1. Can X be arbitrarily distortable? W.T. Gowers rapidly dispensed with the first problem [G5] producing a counterexample. A bit later Argyros and Deliyanni [AD] produced a) an asymptotic ℓ_1 space with an unconditional basis which was arbitrarily distortable and b) an asymptotic ℓ_1 space containing no unconditional basic sequence. The example for a) was a beautiful new type of *mixed Tsirelson space* which we now describe. Let $1 > \theta_1 \geq \theta_2 \geq \cdots \to 0$ and let $\alpha_1 < \alpha_2 < \cdots < \omega_1$. The space $T(\theta_i, S_{\alpha_i})_{\mathbb{N}}$ is the completion of c_{00} under the implicit norm

$$\|x\| = \|x\|_\infty \vee \sup_n \sup \left\{ \theta_n \sum_1^m \|E_i x\| : (E_i)_1^m \text{ is } \alpha_n\text{-admissible} \right\}.$$

Note that the definition includes that of T. Indeed $T = T(2^{-i}, S_i)_{\mathbb{N}}$. Argyros and Deliyanni proved that by suitably choosing θ_i and α_i one could produce an arbitrarily distortable space. The key was to keep the coefficients from being geometric in relation to the α_i's, i.e., unlike the situation for T. The arbitrary distortions are given by the norms

$$\| \cdot \|_n = \sup\left\{ \theta_n \sum_1^m \|E_i x\| : (E_i)_1^m \text{ is } \alpha_n\text{-admissible} \right\}.$$

For example one can prove that $T(\frac{1}{i+1}, S_i)_{\mathbb{N}}$ is arbitrarily distortable. More generally $T(\theta_n, S_n)_{\mathbb{N}}$ is arbitrarily distortable if $\frac{\theta_n}{\theta^n} \to 0$ where $\theta \equiv \lim \theta_n^{1/n}$ (which may be assumed to exist) [AnO] (see also [Ga3]). The example in b) is a complicated conditionalization of this example along the lines of [GM1].

Another property not shared by all Tsirelson type spaces X is the property that c_0 may be finitely representable in X. If X has an unconditional basis then if c_0 is finitely represented in X one has that for all $\varepsilon > 0$, $n \in \mathbb{N}$ there exist disjointly supported $(x_i)_1^n$ in X which are $1 + \varepsilon$-equivalent to the unit vector basis of ℓ_∞^n [Jo2].

This cannot happen in T since [CO] T is naturally isomorphic to T_m, *modified Tsirelson space*. The norm in T_m is given by

$$\|x\| = \|x\|_\infty \quad \vee \quad \sup\left\{ \frac{1}{2} \sum_1^n \|E_i x\| : n \in \mathbb{N} \text{ and } (E_i)_1^n \right.$$

$$\left. \text{are disjoint and } n \le \min E_i \text{ for all } i \right\}.$$

P.-K. Lin and Denka Kutzarova [LK] proved that c_0 is finitely representable in Schlumprecht's space S (which is not quite asymptotic ℓ_1). Roughly one forms the appropriate vectors to look like the markings on a meter stick. All marks of the same height form one vector. The number of markings and their heights are carefully controlled. In [ADKM] modified mixed Tsirelson spaces are constructed and studied. It is proved that if $\theta_n^{1/n} \to 1$ then c_0 is finitely represented in every block subspace of $T(\theta_i, S_i)_{\mathbb{N}}$. Hence the spaces $T(\theta_i, S_i)_{\mathbb{N}}$ and their modified versions $T_M(\theta_i, S_i)_{\mathbb{N}}$ are *totally incomparable* (no subspace of one is isomorphic to a subspace of the other). They also produce a *boundedly modified mixed Tsirelson space* which is arbitrarily distortable. In this definition one defines the norm using "disjoint S_1 families" but only S_n-admissible families for $n > 1$. Thus c_0 is not finitely representable in this space. Finally they produce a variation of the last example which is H.I..

The classes S_α lead to a higher order study of asymptotic ℓ_1 spaces.

Let (x_i) be a basis for X.

$$\dot{\delta}_\alpha(x_i) = \sup\left\{\delta \geq 0 : \left\|\sum_1^n y_i\right\| \geq \delta \sum_1^n \|y_i\| \text{ whenever } (y_i)_1^n \right.$$
$$\left. \text{is } \alpha\text{-admissible w.r.t. } (x_i)\right\}.$$

These constants were defined and studied in [OTW]. They measure the complexity of the asymptotic ℓ_1 structure of (x_i). In addition, the following definitions were made:

$$\dot{\delta}_\alpha(X) = \dot{\delta}_\alpha(x_i) = \sup\{\dot{\delta}_\alpha(y_i) : (y_i) \text{ is a block basis of } (x_i)\}$$
$$\ddot{\delta}_\alpha(X) = \ddot{\delta}_\alpha(x_i) = \sup\{\dot{\delta}_\alpha(X, |\cdot|) : |\cdot| \text{ is an equivalent norm on } X\}.$$

Using the discussion in Chapter 7 it is easy to see that X contains ℓ_1 iff $\dot{\delta}_\alpha(X) > 0$ for all $\alpha < \omega_1$.

By passing to block bases one can stabilize these constants. We say that the basic sequence (x_i) Δ-*stabilizes* $\gamma = (\gamma_\alpha)_{\alpha<\omega_1} \subseteq \mathbb{R}$ if there exist $\varepsilon_n \downarrow 0$ so that for all $\alpha < \omega_1$ there exists $m \in \mathbb{N}$ so that for all $n \geq m$, if (y_i) is a block basis of $(x_i)_n^\infty$ then

$$|\dot{\delta}_\alpha(y_i) - \gamma_\alpha| < \varepsilon_n.$$

The Δ-*spectrum* of X, $\Delta(X)$, is defined to be all γ so that some block basis of X stabilizes γ. $\ddot{\Delta}(X)$ is all such γ's obtained from equivalent norms on X. The *spectral index* of X, $I_\Delta(X) = \inf\{\alpha < \omega_1 : \dot{\delta}_\alpha(X) < 1\}$ (provided X does not contain ℓ_1).

5.10 Theorem. [OTW] *Let X be an asymptotic ℓ_1 space not containing ℓ_1. Then $I_\Delta(X) = \omega^\alpha$ for some $\alpha < \omega_1$. If $I_\Delta(X) = \alpha_0$ and $\ddot{\delta}_{\alpha_0}(X) = \theta$ then $\ddot{\delta}_{\alpha_0 \cdot n + \beta} = \theta^n$ for all n and $\beta < \alpha_0$. $\ddot{\delta}_\beta(X) = 0$ if $\alpha_0 \cdot \omega \leq \beta < \omega_1$.*

For Tsirelson's space T we have

5.11 Theorem. [OTW] *Let (x_i) be a block basis of (e_i) in T.*

(a) $\delta_n(x_i) = 2^{-n}$ for all $n \in \mathbb{N}$
(b) For all $\gamma \in \Delta(T)$, $\gamma_n = 2^{-n}$
(c) For all $\gamma \in \ddot{\Delta}(T)$, $\gamma_n \leq 2^{-n}$
(d) $I_\Delta(X) = 1$ for all block subspaces $X \subseteq T$.

Other results appear in [OTW] which study the parameters above in terms of bounded distortion. As a whole these give weight to the study of T as the natural candidate, if such a space exists, for a distortable space of bounded distortion.

We also note the following result from [OTW]. If $(x_i) \subseteq T$ stabilizes γ for some equivalent norm $|\cdot|$ and $\gamma_1 = 1/2$ then $|\cdot|$ cannot D-distort X (for some absolute constant D). Moreover no such norm $|\cdot|$ is known for which $\gamma_1 < 1/2$ for some $\gamma \in \Delta(T, |\cdot|)$. Further results on certain norms that cannot arbitrarily distort T are given in [OT].

Finally we note that the noncommutative analogue of the ℓ_p spaces, the Schatten classes C_p of operators on Hilbert space, have been shown to be arbitrarily distortable for $1 < p < \infty$ [TJ3].

6

Asymptotic structure

There is more than one notion of asymptotic structure; we shall consider two distinct aspects. The first, that of spreading models, was discussed briefly in Chapter 2. The asymptotic behavior of a normalized basic sequence (x_i) is stabilized resulting in a relatively well behaved spreading model (e_i). Information about the finite dimensional subspaces of $[(e_i)]$ passes back to $[(x_i)]$ but infinite dimensional information need not. For example while T is reflexive all of its spreading models are isomorphic to ℓ_1. In fact (see Chapter 5) T is asymptotic ℓ_1. Also H. Rosenthal [R3] proved that every X admits a 1-unconditional spreading model yet as we have seen by [GM1] X need not even contain an unconditional basic sequence. A replacement problem for (P1) arose following Tsirelson's example.

(P10) Does every X have a spreading model isomorphic to ℓ_p for some $1 \leq p < \infty$ or c_0?

The answer is no by the following example [OS5]. Let $f(n) = \log_2(n+1)$ and choose $(n_k) \in [\mathbb{N}]^\omega$ so that $\sum_{k=1}^\infty \frac{1}{f(n_k)} < \frac{1}{10}$. X is the completion of c_{00} under the implicit norm

$$\|x\| = \|x\|_\infty \vee \left(\sum_{k=1}^\infty \|x\|_{n_k}^2 \right)^{1/2} \quad \text{where for } k \geq 2 ,$$

$$\|x\|_k = \max\left\{ \frac{1}{f(k)} \sum_{i=1}^k \|E_i x\| : E_1 < \cdots < E_k \right\} .$$

X is reflexive and (e_i) is a normalized 1-unconditional subsymmetric bases for X. Clearly, since one need only check spreading models generated by block bases of (e_i), no spreading model of X can be ℓ_p for $1 < p < \infty$ or c_0. This is because of the lower S-norm estimate. The only way normalized vectors in ℓ_2 can add in an ℓ_1 fashion is if they are essentially the same or at least share a common component (after passing to a subsequence). But the nature of $\| \cdot \|_{n_k}$ forbids having $\| \sum_{j=1}^{r} x_i \|_{n_k} \approx \sum_{j=1}^{r} \|x_i\|_{n_k}$ for large r when $x_1 < \cdots < x_r$. In fact one obtains that no spreading model of X contains ℓ_p $(1 \le p < \infty)$ or c_0. This raises the question posed to us by V.D. Milman.

(Q8) Does every X admit a spreading model which is either reflexive or isomorphic to ℓ_1 or c_0?

One can also ask the following question. We write $X \to E$ if E is a spreading model of (some sequence in) X.

(Q9) For every X is there a finite chain $X \to E_1 \to \cdots \to E_n$ with E_n isomorphic to some ℓ_p or c_0?

The notion of a spreading model involves stabilization of norms at infinity. One could hope for a stronger type of stabilization. One result of this kind is the following. Let (x_i) generate a spreading model (e_i). By passing to a subsequence of (x_i) using Ramsey's theorem we may assume for all $x \in X$, $(a_i)_1^k \subseteq \mathbb{R}$

$$\lim_{n_1 \to \infty} \cdots \lim_{n_k \to \infty} \left\| x + \sum_{i=1}^{k} a_i x_{n_i} \right\| \equiv \left\| x + \sum_1^k a_i e_i \right\|$$

exists. Thus we get a norm on $X \oplus E$. If it happens that $\|x+e\| = \|x+e'\|$ for all $x \in X$ and all $e, e' \in \langle e_i \rangle$ with $\|e\| = \|e'\|$ then it can be shown that X contains almost isometric copies of some ℓ_p or c_0. Thus such stabilization is too much to hope for in general.

Another generalization of spreading models is given in [ORS]. Given a sequence of finite dimensional spaces $F_n \subseteq X$ with $\dim F_n \to \infty$ there exist finite dimensional spaces $G_n \subseteq F_{k_n}$ for some $k_1 < k_2 < \cdots$ with $\dim G_n \to \infty$ and a spreading model (e_i) so that if $x_i \in S_{G_i}$ for all i then (x_i) generates the spreading model (e_i). Higher order spreading models have also been considered in several contexts. One does not have the stability of the ordinary spreading model but rather must consider a larger class of bases. If C is a class of normalized bases of finite length and $\alpha < \omega_1$ we say that a normalized basis (x_i) has an $\alpha - C$ *spreading model* if for all $F \in S_\alpha$, $(x_i)_{i \in F} \in C$. This notion has been used in particular for C being either the class of all finite bases K-equivalent to the unit vector basis of ℓ_p^n for some n or those which are K-unconditional (e.g., [OTW], [AMT], [AG], [W2], [Far2], [Far3] and numerous other papers).

So how do we get infinite dimensional information about X from knowledge about its spreading models? There are two results of this type. The first involves isometric information about all spreading models. The framework of the theorem we are about to state is the following. Let X have a basis (e_i) and let $\mathcal{S}(X)$ denote the set of all spreading models of normalized block bases of (e_i). Suppose $|\mathcal{S}(X)| = 1$, i.e., there is only one such spreading model. Then by the proof of Krivine's theorem, Theorem 2.10, one obtains that this unique spreading model must be 1-equivalent to the unit vector basis of ℓ_p for some $1 \leq p < \infty$ or c_0. Indeed Krivine's proof yields that the ℓ_p^n's one finds can be taken to be given by an identically distributed block basis of length n. Thus if (x_i) has a spreading model then some block basis of the form $(\sum_{i=1}^k a_i x_{nk+i})_{n=1}^\infty$ has a spreading model (z_i) where $(z_i)_1^n$ is $1+\varepsilon$-equivalent to the unit vector basis of ℓ_p^n. This yields that if $|\mathcal{S}(X)| = 1$ then the unique spreading model is some ℓ_p (or c_0). Theorem 6.1 says that if this spreading model is ℓ_1 (or c_0) then X contains ℓ_1 (or c_0). The case where the spreading model is ℓ_p $(1 < p < \infty)$ remains open, (Q10) below.

6.1 Theorem. [OS6] *Let (e_i) be a basis for X. If all spreading models of normalized block bases of (e_i) are 1-equivalent to the unit vector basis of ℓ_1 (respectively, c_0) then X contains an isomorph of ℓ_1 (respectively, c_0).*

(Q10) Let $1 < p < \infty$ and let (e_i) be a basis for X so that all spreading models of any normalized block basis are 1-equivalent to the unit vector basis of ℓ_p. Does X contain ℓ_p?

The proof of 6.1 actually only requires that $\|y_1 + y_2\| = 2$ (respectively, 1) if (y_i) is such a spreading model. The proof of the ℓ_1 case makes use of some beautiful work on the complexity of weakly null sequences by Argyros, Merkourakis and Tsarpalias [AMT], which we describe.

First recall that by Mazur's theorem if (e_i) is normalized weakly null then some *convex* (i.e., $x_i \in \operatorname{co}(e_j)$ for all i) block basis (x_i) is norm null. Ptak proved

6.2 Theorem. [P] *Let $\mathcal{F}$ be an hereditary collection of finite subsets of $\mathbb{N}$ and let $\varepsilon > 0$. Suppose that for all $a_i \geq 0$ with $\sum a_i = 1$ there exists $F \in \mathcal{F}$ with $\sum_F a_i > \varepsilon$. Then there exists $M \in [\mathbb{N}]^\omega$ with $[M]^{<\omega} \subseteq \mathcal{F}$.*

Both Mazur's and Ptak's theorems were localized in [AMT] in terms of the Schreier classes S_α (see Chapter 5). For (e_i), the unit vector basis of c_{00}, and $\alpha < \omega_1$, $M \in [\mathbb{N}]^\omega$ we define a convex block basis (α_n^M) of

α-*averages* of $(e_i)_M$ as follows. If $M = (m_i)$ we set (the zero in this formula is really a zero)

$$0_n^M = e_{m_n} \quad \text{and} \quad 1_1^M = \frac{1}{m_1} \sum_{i=1}^{m_1} 0_i^M \, .$$

Thus 1_1^M is the longest average of $(e_{m_i})_1^\infty$ whose support is in S_1. We continue taking 1_2^M to be the longest average of the next vectors in (e_{m_i}), $(e_{m_{m_1+1}}, e_{m_{m_1+2}}, \ldots)$ whose support is in S_1 And so on. In general

$$(\alpha+1)_1^M = \frac{1}{m_1} \sum_{i=1}^{m_1} \alpha_i^M \quad \text{and}$$

$$(\alpha+1)_{n+1}^M = \frac{1}{m_{k_n}} \sum_{i=1}^{m_{k_n}} \alpha_i^{M'} \quad \text{where}$$

$$M' = \{m \in M : m > \mathrm{supp}(\alpha+1)_n^M\} \text{ and } m_{k_n} = \min M' \, .$$

If α is a limit ordinal and S_α was defined via the sequence $\alpha_n \uparrow \alpha$ we set

$$\alpha_1^M = (\alpha_{m_1})_1^M \text{ and for } k > 1 \, , \ \alpha_k^M = (\alpha_{n_k})_1^{M_k} \text{ where}$$
$$M_k = \{m \in M : m > \mathrm{supp}\, \alpha_{k-1}^M\} \text{ and } n_k = \min M_k \, .$$

Let $\mathcal{F} \subseteq [\mathbb{N}]^{<\omega}$ be hereditary and let $\alpha < \omega_1$ and $\varepsilon > 0$. $\mathcal{F}$ is said to be (M, α, ε) *large* if for all $N \in [M]^\omega$ and $n \in \mathbb{N}$,

$$\sup_{F \in \mathcal{F}} \langle \alpha_n^N, 1_F \rangle > \varepsilon \, .$$

6.3 Theorem. [AMT] *If $\mathcal{F}$ is (M, α, ε) large then there exists $N \in [M]^\omega$ with $\mathcal{F} \supseteq S_\alpha(N)$.*

Recall $S_\alpha(N) = \{(n_i)_{i \in F} : F \in S_\alpha\}$ where $N = (n_i)$. The generalization of Mazur's theorem becomes

6.4 Theorem. [AMT] *Let (e_i) be normalized weakly null. Then there exists $\alpha < \omega_1$ so that for all $M \in [\mathbb{N}]^\omega$, $\lim_n \|\alpha_n^M\| = 0$.*

These results are used in [AMT] to deduce certain information about higher order ℓ_1 spreading models in a given space. It is also worth mentioning that the following dual form of Elton's theorem (Chapter 2) is obtained in [AMT] by using Ramsey theory in a manner similar to that of Elton. A normalized sequence (e_i) is *convexly unconditional* if for every $\delta > 0$ there exists $C(\delta) > 0$ so that if $x = \sum a_i x_i$ with $\sum |a_i| = 1$ and $\|x\| > \delta$ then $\|\sum \varepsilon_i a_i x_i\| > C(\delta)$ for all choice of $\varepsilon_i = \pm 1$.

6.5 Theorem. [AMT] *If (e_n) is normalized weakly null then some subsequence is convexly unconditional.*

What if we replace the hypothesis of Theorem 6.1 by the weaker statement that in every block basis some spreading model is 1-equivalent to the unit vector basis of ℓ_1? Is this enough to insure that X contains ℓ_1? The answer is no. T can be shown to have this property [OS6]. But there is a theorem if one allows for renormings. First we recall a problem raised by V.D. Milman in 1971 [M1].

(P11) Let X be a reflexive Banach space. Does there exist an equivalent norm $\|\cdot\|$ on X so that for all $(x_n) \subseteq X$ satisfying

$$\lim_m \lim_n \|x_m + x_n\| = 2\lim_n \|x_n\| \, ,$$

(x_n) must be convergent?

By James' theorem on the sup of linear functionals [J4], if a norm $\|\cdot\|$ satisfies

$$\begin{cases} \text{for all } (x_n) \subseteq X \text{ if} \\ \lim_m \lim_n \|x_n + x_m\| = 2\lim_n \|x_n\| \\ \text{then } (x_n) \text{ converges} \end{cases} \qquad (6.1)$$

then X must be reflexive. Indeed given $x^* \in X^*$ with $\|x^*\| = 1$, choose (x_n), $\|x_n\| = 1$ with $\lim x^*(x_n) = 1$. Then by (6.1) (x_n) must converge to some x, $\|x\| = 1$. Thus $x^*(x) = \|x^*\|$. So every x^* achieves its norm and hence, by James, X is reflexive.

Condition (6.1) is a search for a geometric characterization of reflexivity. One result of this type was due independently to D.P. Milman [Mil] and B.J. Pettis [Pe] who proved that a uniformly convex space must be reflexive. However as observed first by Day [Da] a reflexive space need not be isomorphic to a uniformly convex space. T is another example of such a space. The solution to (P11) yielded a bonus: further characterizations of when a space X contains ℓ_1 or c_0.

The solution to (P11) follows as a corollary to the next theorem.

6.6 Theorem. [OS7] *Every X admits an equivalent strictly convex norm $\|\cdot\|$ with the following properties:*

(a) *If $(x_m) \subseteq X$ is relatively weakly compact and if $\lim_m \lim_n \|x_m + x_n\| = 2\lim_n \|x_n\|$ then (x_n) is norm convergent.*

(b) *If $(x_n) \subseteq X$ satisfies*

$$\lim_m \lim_n \|x_m \pm x_n\| = 2\lim_n \|x_n\| > 0$$

> then some subsequence of (x_n) is equivalent to the unit vector
> basis of ℓ_1.

(c) *If (x_n) is weakly null and satisfies*

$$\lim_m \lim_n \|x_m + x_n\| = \lim_n \|x_n\| > 0$$

> then some subsequence of (x_n) is equivalent to the unit vector
> basis of c_0.

Another corollary of this theorem is

6.7 Corollary. *The following are equivalent for a given space X.*

(a) *X contains an isomorph of ℓ_1 (respectively c_0).*

(b) *For all equivalent norms $\|\cdot\|$ on X there exists a normalized se-
quence in X having a spreading model which is 1-equivalent to
the unit vector basis of ℓ_1 (respectively, c_0).*

(c) *For all equivalent norms $\|\cdot\|$ on X there exists a normalized (and
respectively weakly null) sequence in X having a spreading model
(e_i) satisfying $\|e_1 \pm e_2\| = 2$ (respectively, $\|e_1 + e_2\| = 1$).*

The new content of this theorem are the implications b) $\Rightarrow$ a) and c)
$\Rightarrow$ a). The reverse implications follow from James' proof that ℓ_1 and c_0
are not distortable.

The theorem extends results of Maurey on ℓ_1 and ℓ_∞-types [Ma3] (see
also [R4]) to the setting of renormings.

Another notion of asymptotic structure was presented in [MT1] and
[MMT] continuing ideas introduced in [M1] and [M4]. This can be done
in several settings and we present here the simplest one (for more general
versions see [MMT] and [MT2]). Let (e_i) be a basis for X and let $n \in \mathbb{N}$.
A finite normalized basic sequence $(b_i)_1^n \in \{X, (e_i)\}_n$ if

$$\forall \, \varepsilon > 0 \; \forall \, k_1 \; \exists \, x_1 \in S_{\langle e_i \rangle_{k_1}^\infty} \; \forall \, k_2 \exists \, x_2 \in S_{\langle e_i \rangle_{k_2}^\infty} \ldots \forall \, k_n \; \exists \, x_n \in S_{\langle e_i \rangle_{k_n}^\infty}$$

so that $(x_i)_1^n$ is $1 + \varepsilon$ equivalent to $(b_i)_1^n$. It is easy to see that if $(b_i)_1^n$
is a normalized block basis of some spreading model of a block basis of
(e_i), then $(b_i)_1^n \in \{X, (e_i)\}_n$. In particular the latter is nonempty.

We say X is *Asymptotic* ℓ_p (note the capital A) if there exists $K < \infty$
so that $(b_i)_1^n$ is K-equivalent to the unit vector basis of ℓ_p^n for all $(b_i)_1^n \in
\{X, (e_i)\}_n$ and for all n. Actually it is shown in [MMT] that for $p > 1$
one only need assume that $d(\langle b_i \rangle_1^n, \ell_p^n) \leq K$ for some K. It remains open
if this is sufficient for $p = 1$. How does this notion compare with saying
(e_i) is asymptotic ℓ_p (Chapter 5)? This is answered by the following
theorem.

First recall that a sequence (E_n) of finite dimensional spaces is an FDD (*finite dimensional decomposition*) for X if $X = \overline{\langle E_n \rangle}$ and for some $C < \infty$ every nonzero sequence (z_n) with $z_n \in E_n$ is C-basic. Equivalently for all $x \in X$ there exist unique vectors $w_n \in E_n$ so that $x = \sum w_n$. A *blocking* $(E_n)_{n=1}^{\infty}$ of (e_i) is a collection of finite dimensional subspaces of X of the form

$$E_n = \langle e_i \rangle_{i=p_{n-1}+1}^{p_n} \quad \text{for some} \quad 0 = p_0 < p_1 < \cdots .$$

(H_n) is a *skipped blocking* of (E_n) if $H_n = \langle E_i \rangle_{i=r_n}^{q_n}$ for some $1 \leq r_1 \leq q_1 < q_1 + 1 < r_2 \leq q_2 < q_2 + 1 < r_3 + \leq q_3 < \ldots..$

6.8 Theorem. [KOS] *Let (e_i) be a basis for X and let $\varepsilon_i \downarrow 0$. Then there exists a blocking $(E_n)_{n=1}^{\infty}$ of (e_i) with the following property. If $(H_k)_{k=1}^n$ is any finite skipped blocking of $(E_i)_{i=n}^{\infty}$ and $x_i \in S_{H_i}$ for $i \leq n$ then $(x_i)_1^n$ is $1 + \varepsilon_n$ equivalent to some $(b_i)_{i=1}^n \in \{X, (e_i)\}_n$.*

Using this the following can be proved.

6.9 Theorem. *Let (e_i) be a boundedly complete basis for X. Suppose that X is Asymptotic ℓ_p for some $1 \leq p < \infty$ (or Asymptotic c_0) w.r.t. (e_i). Then there exists a blocking (H_j) of (e_i) and $K < \infty$ so that if $n \in \mathbb{N}$, $(G_j)_1^n$ is any blocking of $(H_j)_n^{\infty}$ and $b_j \in S_{G_j}$ for $i \leq n$, then $(b_j)_1^n$ is K-equivalent to the unit vector basis of ℓ_p^n (or ℓ_{∞}^n).*

In particular any block basis of (H_j) is asymptotic ℓ_p (or c_0) with a fixed constant. We say (H_j) is an asymptotic ℓ_p FDD for X.

Krivine's theorem (Chapter 2) yields that there exists $1 \leq p \leq \infty$ so that for all n the unit vector basis of ℓ_p^n belongs to $\{X, (e_i)\}_n$. Thus this notion of asymptotic structure corrects the "ℓ_p deficiency" exhibited by spreading models (the negative solution to (P10)). Furthermore the problem partially addressed concerning small asymptotic structure in (P11) has a complete solution in this setting. Suppose that $|\{X, (e_i)\}_2| = 1$. Then the only element of $\{X, (e_i)\}_2$ must be the unit vector basis of ℓ_p^2 for some $1 \leq p \leq \infty$. It is not hard to show in such a case [MMT] that X contains almost isometric copies of ℓ_p (or c_0 if $p = \infty$). In fact if X is reflexive it can be shown [OS8] that in this case X can be embedded into $(\sum F_i)_{\ell_p}$, an ℓ_p sum of finite dimensional spaces.

If (y_i) is a block basis of (e_i) then $\{Y, (y_i)\}_n \subseteq \{X, (e_i)\}_n$ and both sets are closed subsets of the compact space of all bases of length n

having basis constant not exceeding that of (e_i) under the metric

$$d_b((b_i)_1^n, (c_i)_1^n) \;=\; \inf\left\{ AB : A^{-1}\left\|\sum_1^n a_i b_i\right\| \le \left\|\sum_1^n a_i c_i\right\|\right.$$

$$\left. \le B\left\|\sum_1^n a_i b_i\right\| \text{ for all } (a_i)_1^n \subseteq \mathbb{R}\right\}.$$

(Actually $\log d_b(\cdot, \cdot)$ is the metric.)

Furthermore if (z_i) is almost a block basis of (y_i) (i.e., for some n, $(z_i)_n^\infty$ is a block basis of (y_i)) then $\{Z, (z_i)\}_m \subseteq \{Y, (y_i)\}_m$ for all m. Thus [MMT] (see also [MT2]) we may stabilize the asymptotic structure: There exists a block basis (y_i) of (e_i) so that for all block bases (z_i) of (y_i) and all n, $\{Z, (z_i)\}_n = \{Y, (y_i)\}_n$. It is natural to ask how small this stabilized set can be. Must it contain the unit vector basis of ℓ_p^n for only one value of p?

The answer is no. The set can be quite large.

6.10 Theorem. [OS1] *There exists a reflexive space X with a basis (e_i) so that for all block bases (y_i) of (e_i), for all n and for all monotone normalized bases $(b_i)_1^n$, $(b_i)_1^n \in \{Y, (y_i)\}_n$.*

Of course such a space cannot contain an unconditional basic sequence. Moreover this space has the property which [MR] suggested might be possible: the summing basis is block finitely representable in all block bases. The norm is a cousin of the norm in [GM1] but is too technical to bother to define here.

The following problem does remain open. Assume $\{Y, (y_i)\}_n$ is stabilized for all n. $p \in [1, \infty]$ is a *Krivine number* for Y if the unit vector basis of ℓ_p^n belongs to $\{Y, (y_i)\}_n$ for all n.

(Q11) Let Y have stabilized [MMT] asymptotic structure. Is the set of Krivine numbers for Y a (closed) interval?

It is easy to see that $\{Y, (y_i)\}_n$ is always a connected set under d_b. In [HT] it is shown that there exists a "block quotient" Z of Y with a (natural) basis (z_i) with the property that if $p < q$ are Krivine numbers for Y then ℓ_r is finitely block representable in (z_i) for all $p \le r \le q$.

The notion of asymptotic structure (in the sense of [MMT]) has been extended to operators [MW].

As noted in Theorem 6.8 there is a direct connection between asymptotic structures $\{X, (e_i)\}_n$ and the skipped asymptotic block sequences w.r.t. a certain blocking of (e_i) into an FDD. It is natural to inquire about infinite results. Given a class C of infinite normalized bases and a space X with a basis (e_i) when is it possible to find a blocking (E_n) of (e_i) so that every skipped normalized block basis of (E_n) nearly belongs to C? The next result answers this question. First we need some definitions. The use of trees to study asymptotic structure appears in [MiS], [MMT], [KOS] and other places as well.

A class C of infinite normalized bases is *closed* if whenever $(x_i)_1^\infty$ is a normalized basis satisfying for all $k \in \mathbb{N}$ there exists $(x_i^k)_{i=1}^\infty \in C$ with $x_i^k = x_i$ for $i \le k$, then $(x_i) \in C$.

$T_\omega = \{(n_1, \ldots, n_j) : j \in \mathbb{N}, \, n_i \in \mathbb{N} \text{ for } i \le j\}$ is a countably branching tree ordered by $(n_1, \ldots, n_j) \le (m_1, \ldots, m_k)$ if $j \le k$ and $n_i = m_i$ for $i \le j$. We say $\mathcal{T} \in T_\omega(X)$ if

$$\mathcal{T} = \{x(n_1, \ldots, n_j) : (n_1, \ldots, n_j) \in T_\omega\} \subseteq S_X \ .$$

$\mathcal{T}$ is a *block tree* w.r.t. the basis (e_i) for X if $(x(n_1, \ldots, n_j, n))_{n=1}^\infty$ are block bases of (e_i) for all $(n_1, \ldots, n_j) \in T_\omega \cup \varphi$. Similarly $\mathcal{T} \in T_\omega(X)$ is *weakly null* if the successors of every node (including φ) form a weakly null sequence.

6.11 Theorem. [OS8] *Let X have a basis (e_i) and let C be a closed class of infinite normalized bases. Suppose that whenever $\mathcal{T} \in T_\omega(X)$ is a block tree w.r.t. (e_i) then some branch of $\mathcal{T}$ lies in C. Then for all $\varepsilon > 0$ there exists a blocking (E_n) of (e_i) so that if (H_i) is any skipped blocking of $(E_n)_{n=2}^\infty$ and $x_i \in S_{H_i}$ for $i \in \mathbb{N}$ then $d_b((x_i), C) < 1 + \varepsilon$.*

The proof can be deduced from Martin's theorem that Borel games are determined [Mar]. Indeed consider the game played by players (I) and (II). (I) chooses $n_1 \in \mathbb{N}$. Then (II) chooses $x_1 \in S_{\langle e_i \rangle_{n_1}^\infty}$. (I) chooses $n_2 > \max \operatorname{supp}(x_1)$, (II) chooses $x_2 \in S_{\langle e_i \rangle_{n_2}^\infty}$ and so on. (I) wins if $(x_i)_1^\infty \in C$. Otherwise (II) wins. One can easily check that Martin's theorem applies and thus either (I) or (II) has a winning strategy. The hypothesis on C rules out a winning strategy for (II): if (II) wins then "$\forall \, n_1 \, \exists \, x_1 \in S_{\langle e_i \rangle_{n_1}^\infty} \, \forall \, n_2 \, \exists \, x_2 \in S_{\langle e_i \rangle_{n_2}^\infty} \cdots$ with $(x_i) \notin C$." One can use this to obtain a block tree $\mathcal{T} \in T_\omega(X)$ having no branches in C. Thus we are left with "$\exists \, n_1 \, \forall \, x_1 \in S_{\langle e_i \rangle_{n_1}^\infty} \, \exists \, n_2 \, \forall \, x_2 \in S_{\langle e_i \rangle_{n_2}^\infty} \cdots$ with $(x_i)_1^\infty \in C$." This plus a standard approximation argument yields the theorem.

Something can also be said even if X does not have a basis. If $C \subseteq S_X$ we let $\overline{C}$ denote the closure of C w.r.t. the product topology $X^\mathbb{N}$ where

X is given the discrete topology. For $\varepsilon > 0$,

$$C_\varepsilon = \{(y_i)_1^\infty \subseteq S_X : \exists\, (x_i) \in C \text{ with} \\ \|x_i - y_i\| < \varepsilon/2^i \text{ for } i \in \mathbb{N}\}.$$

We then consider, given $C \subseteq S_X^{\mathbb{N}}$, a 2 player game [MMT] where player
(I) chooses $Y_1 \in \mathrm{cof}(X) \equiv$ the set of finite codimensional subspaces of X
and player (II) chooses $y_1 \in S_{Y_1}$. Then (I) chooses $Y_2 \in \mathrm{cof}(X)$ and (II)
chooses $y_2 \in S_{Y_2}$ and so on. Player (I) wins if $(y_i) \in C$. By $(W_I(C))$ we
shall mean that player (I) has a winning strategy.

6.12 Theorem. [OS8] *Let $\mathcal{B}$ be a countable collection of subsets of S_X^ω.
Then there exists an isometric embedding of X into a space Z having an
FDD (E_i), so that for $A \in \mathcal{B}$ the following are equivalent.*

a) *$\forall \varepsilon > 0$ $(W_I(\overline{A_\varepsilon}))$.*
b) *For every $\varepsilon > 0$ there is a blocking (G_i) of (E_i) and a sequence $\delta_i \searrow 0$,
so that for every sequence $(x_n) \subset S_X$, satisfying for some sequences
$1 = k_0 < k_1 < \cdots$ in $\mathbb{N}$*

$$\|(\mathrm{Id} - P_{[G_j]_{j=k_{n-1}+1}^{k_n-1}})(x_n)\| < \delta_n \text{ for all } n,$$

$(x_n) \in \overline{A_\varepsilon}$.
c) *For every $\varepsilon > 0$ there is a blocking (G_i) of (E_i), so that for every
sequence $(x_n) \subset S_X$ which is a skipped block w.r.t. (G_i), $(x_n) \in \overline{A_\varepsilon}$.*

*If X has a separable dual (E_i) can be chosen to be shrinking and
independent from $\mathcal{B}$, and, furthermore, if X is reflexive, Z can be chosen
to be reflexive. In these cases (a) is equivalent to*

d) *For every $\varepsilon > 0$ every weakly null tree $\mathcal{T} \in T_\omega(X)$ has branch in $\overline{A_\varepsilon}$.*

6.13 Remark. Note that Theorem 6.12 means the following. Assume
for all $\varepsilon > 0$ Player I has a winning strategy for the $\overline{A_\varepsilon}$-game. Then
given $\overline{A_\varepsilon}$ Player I can embed X into a space with an appropriate FDD
(F_i), and use the following strategy:

Take $Y_1 = \oplus_{i=2}^\infty F_i \cap X$.
If Player II has chosen the vector x_{n-1} in the $n-1$st round,
choose $N \in \mathbb{N}$ so that $\|P_{\oplus_{i=N}^\infty F_i}(x_{n-1})\| < \delta_n$ and put
$Y_n = \oplus_{i=N+1}^\infty F_i \cap X$.

Using this theorem one can prove the following

6.14 Theorem. [OS8] *Let X be a separable reflexive Banach space. Let $1 < p < \infty$ and assume that for some $K < \infty$ whenever $T \in T_\omega(X)$ is weakly null then some branch of T is K-equivalent to the unit vector basis of ℓ_p. Then for all $\varepsilon > 0$ there exists a finite codimensional subspace X_0 of X which $K^2 + \varepsilon$-embeds into the ℓ_p sum of finite dimensional spaces.*

An analogous theorem for c_0 has been proved by N.J. Kalton [Ka2]. The reflexive hypothesis is replaced by "X does not contain an isomorph of ℓ_1" and the conclusion is that X embeds into c_0. Also Theorem 6.14 generalizes a result in [KW]: Assume X does not contain ℓ_1. Suppose there exists $1 < p < \infty$ so that for all normalized weakly null $(x_n) \subseteq X$ there exists a subsequence (x'_n) so that for all $x \in S_X$ and $t > 0$, $\lim_n \|x + t x_n\| = (1 + t^p)^{1/p}$. Then for all $\varepsilon > 0$, X $1 + \varepsilon$-embeds into the ℓ_p sum of finite dimensional spaces.

7

Ordinal Indices

Certain ordinal indices for a space X have proven quite useful in infinite dimensional theory. One has a property of a space such as X does not contain an isomorph of ℓ_p or X^* is separable. A countable ordinal index is defined which measures the complexity of the property with the limiting case of index equal to ω_1 being equivalent to failure of the property. We have encountered this idea previously when we discussed S_α-unconditional basic sequences in the proof of Theorem 5.4. The first such ordinal index in Banach space theory was constructed by Szlenk for the property, X^* is separable. [Szl].

Let X^* be separable. Thus B_{X^*} is a compact metric space in its ω^* topology. We define for $\varepsilon > 0$

$$
\begin{aligned}
K_0(X, \varepsilon) &= B_{X^*} \\
K_{\alpha+1}(X, \varepsilon) &= \{x^* \in K_\alpha(X, \varepsilon) : \exists\ (x_n^*) \subseteq K_\alpha(X, \varepsilon) \\
&\qquad \text{with } x_n^* \xrightarrow{\omega^*} x^* \text{ and } \underline{\lim} \|x_n^* - x\| \geq \varepsilon\}
\end{aligned}
$$

If α is a limit ordinal

$$
K_\alpha(X, \varepsilon) = \bigcap_{\beta < \alpha} K_\alpha(X, \varepsilon) .
$$

We then define

$$
\begin{aligned}
S(X, \varepsilon) &= \sup\{\alpha : K_\alpha(X, \varepsilon) \neq \emptyset\} \text{ and} \\
S(X) &= \sup\{S(X, \varepsilon) : \varepsilon > 0\} .
\end{aligned}
$$

The index does make sense. Each $K_\alpha(X, \varepsilon)$ is a ω^* closed subset of B_{X^*} and because X^* is separable one obtains that $K_\alpha(X, \varepsilon) \not\supseteq K_\beta(X, \varepsilon)$ if $\alpha < \beta$ and $K_\alpha(X, \varepsilon) \neq \emptyset$. Thus $S(X) < \omega_1$ since a compact metric space cannot admit a strictly decreasing family of closed sets $(K_\alpha)_{\alpha < \omega_1}$.

Also $S(X)$ is an isomorphic invariant, $S(X) = S(Y)$ if X and Y are isomorphic.

Actually Szlenk's index was defined in a somewhat more complicated manner. However it can be seen using Rosenthal's ℓ_1 theorem (Theorem 2.6) that the above definition is equivalent to Szlenk's for spaces X not containing ℓ_1. One could try to define $S(X)$ as above for general X but it turns out that $S(X) < \omega_1$ iff X^* is separable.

Szlenk used his index to prove that there is no universal space in the class of all spaces with separable dual. X is *universal* for a class C of spaces if each $Y \in C$ is isomorphic to a subspace of X and $X \in C$.

There is a connection between spaces X with small Szlenk index and the ω^*-UKK (weak* uniform Kadec-Klee) property. X has the ω^*-*UKK* if for all $\varepsilon > 0$ there exists $\delta = \delta(\varepsilon) > 0$ so that if $\|x_n^*\| = 1$, $x_n^* \xrightarrow{\omega^*} x^*$ and $\|x_n^* - x^*\| \geq \varepsilon$ for all n then $\|x^*\| \leq 1 - \delta$. Clearly if X has the ω^*-UKK then $S(X, \varepsilon) < \omega$ for all $\varepsilon > 0$.

Huff [Hu] asked if the converse was true (actually he asked this in the reflexive setting). If $S(X, \varepsilon) < \omega$ for all $\varepsilon > 0$ can X be given an equivalent ω^*-UKK norm? The answer is yes [KOS] and the proof actually involves asymptotic structure. In the setting where X has a shrinking basis (e_i), which just means that the biorthogonal functionals (e_i^*) are a basis for X^*, one can prove that there exists $1 \leq p < \infty$ and $c > 0$ so that if $(b_i)_1^n \in \{X^*, (e_i^*)\}_n$ then $\|\sum_1^n a_i b_i\| \geq c(\sum_1^n |a_i|^p)^{1/p}$. This is used to solve the renorming problem.

7.1 Theorem. *Let X be such that $S(X, \varepsilon) < \omega$ for all $\varepsilon > 0$. Then there exists an equivalent norm $|\cdot|$ on X so that X has the ω^*-UKK. Moreover the modulus $\delta(\varepsilon)$ is of power type, i.e., there exists $C > 0$ and $1 \leq p < \infty$ so that for all $\varepsilon > 0$, $\delta(\varepsilon) \geq C\varepsilon^p$.*

More precise quantitative results have been obtained by N.J. Kalton, G. Godefroy and G. Lancien [GKL].

A more general version of Huff's problem remains unsolved. It comes from certain moduli created by V.D. Milman ([M1], [MP]) and was communicated to us by W.B. Johnson. If $x \in S_X$ and $t > 0$ we set

$$\delta_X(t, x) = \delta_X^{(1)}(t, x) = \liminf_{\substack{x_1 \to \infty \\ x_1 \in S_X}} (\|x + tx_1\| - 1) \text{ by which we mean}$$

$$\equiv \sup\left\{ \inf_{x_1 \in S_Y} (\|x + tx_1\| - 1) : Y \subseteq X, \ \dim \frac{X}{Y} < \infty \right\},$$

$$\delta_X(t) = \delta_X^{(1)}(t) = \inf\{\delta_X(t, x) : x \in S_X\}.$$

X is called *asymptotically uniformly convex* [JLPS] if $\delta_X(t) > 0$ for all $t > 0$. For $n \in \mathbb{N}$ define

$$\delta_X^{(n)}(t,x) = \liminf_{\substack{x_1 \to \infty \\ x_1 \in S_X}} \ldots \liminf_{\substack{x_n \to \infty \\ x_n \in S_X}} \left(\left\| x + \sum_{i=1}^n t x_i \right\| - 1 \right)$$

and

$$\delta_X^{(n)}(t) = \inf\{\delta^{(n)}(t,x) : x \in S_X\} \ .$$

(Q12) Suppose X is such that for all $t > 0$ there exists $n_0 \in \mathbb{N}$ with $\delta_X^{(n_0)}(t) > 0$. Does there exist an equivalent norm $|\cdot|$ on X so that $(X, |\cdot|)$ is asymptotically uniformly convex?

From [KOS] it follows that the answer is yes if X is reflexive. A very nice discussion of asymptotic uniform convexity (and the dual notion of asymptotic uniform smoothness) along with a reinterpretation of the results of [KOS] and [GKL] is given in [JLPS].

Bourgain considered several ordinal indices for spaces X in the 70's. His ℓ_1 index [Bo1] is defined as follows. A *tree T on X* is a tree whose nodes are finite sequences $(x_i)_{i=1}^n$ where $x_i \in S_X$ for all i. Furthermore if $(x_i)_1^n \le (y_i)_1^m$ then $n \le m$ and $x_i = y_i$ for $i \le n$. T is *closed* if it is closed in the product topology: If $(x_i^j)_{i=1}^n \in T$ and $\lim_j x_i^j = x_i$ for $i \le n$ then $(x_i)_1^n \in T$.

For $K < \infty$ an *ℓ_1-K tree on X* is a tree on X such that all nodes are K equivalent to the unit vector basis of ℓ_1^n for some n. If T is a closed tree on X possessing no infinite branches (i.e., $\nexists\ (x_i)_1^\infty \subseteq S_X$ so that $(x_i)_1^n \in T$ for all n) then by the Kunen-Martin boundedness principle (see [Del]) T has order $o(T) < \omega_1$. The order of T is defined as follows.

$$
\begin{aligned}
T^0 &= T \\
T^{\alpha+1} &= \{(x_i)_1^n \in T : (x_i)_1^{n+1} \in T \text{ for some } x_{n+1}\} \\
T^\alpha &= \bigcap_{\beta<\alpha} T^\beta \text{ if } \alpha \text{ is a limit ordinal} \\
o(T) &= \inf\{\alpha : T^\alpha = \emptyset\} \ .
\end{aligned}
$$

Bourgain's ℓ_1 index is

$$I(X) = \sup\{I(X,K) : K > 1\}$$

where

$$I(X,K) = \sup\{o(T) : T \text{ is an } \ell_1\text{-}K \text{ closed tree on } X\}$$

Thus $I(X) < \omega_1 \Leftrightarrow X$ does not contain ℓ_1.

Certain calculations about $I(X)$ appear in [JO]. For example it is shown that $I(X)$ is of the form ω^α for some α, but not all such ordinals can be the index of a space X. The index for Tsirelson's space is $I(T) = \omega^\omega$. Also a distortion type result is obtained: For all $\alpha < \omega_1$, $K < \infty$ and $\varepsilon > 0$ there exists $\beta < \omega_1$ so that if $I(X, K) \geq \beta$ then $I(X, 1 + \varepsilon) \geq \alpha$. $I(X)$ is one way to measure the complexity of the ℓ_1^n's that might live inside a space not containing ℓ_1. Other measures were given in Chapter 5.

ℓ_1 could of course be replaced by other structures. This idea was used by Bourgain, Rosenthal and Schechtman [BRS] to construct uncountably many mutually non-isomorphic complemented subspaces of L_p. Consequently L_p is the only space universal for the class of complemented subspaces of L_p. Bu [Bu] used this sort of index to give an alternate proof of the Krivine-Maurey theorem on types.

One could also define an unconditional index. Given K, one can consider the tree of all normalized K-unconditional sequences in X and obtain the index $U(X) = \sup\{U(X, K) : K \geq 1\}$. As in the ℓ_1-theory it follows that X contains an unconditional basic sequence iff $U(X) = \omega_1$. The proof of Theorem 5.4 used this idea. The fact that (x_i^α) was S_α-unconditional with constant $4D'$ yields that $U(X, 4D') \geq \omega^\alpha$, the Cantor-Bendixson index of S_α.

A theorem of R.C. James [R5] and D.P. and V.D. Milman [MM] can be rephrased in terms of indices. A normalized basic sequence (x_i) is K-ℓ_1^+ if (x_i) is K basic and $\|\sum a_i x_i\| \geq K^{-1}\sum a_i$ whenever $(a_i) \subseteq [0, \infty)$. They observed that X is reflexive iff X contains no ℓ_1^+ sequence. X is thus *superreflexive* (if Y is finitely representable in X then Y is reflexive) iff there exists K and $n \in \mathbb{N}$ so that X does not contain $(x_i)_1^n$ which is K-ℓ_1^+. X is superreflexive iff it can be given an equivalent uniform convex norm [E3] (see also [Pi1]).

We can define the ℓ_1^+-index, $I^+(X)$ just like the ℓ_1-index using ℓ_1^+ trees. Thus we have

7.2 Theorem. *(a)* X *is reflexive iff* $I^+(X) < \omega_1$.
(b) X *is superreflexive iff* $I^+(X) = \omega$.

Also there are connections between $I^+(X)$ and the Szlenk index, $S(X)$ [AJO]. One can define $I_\omega^+(X)$ much like $I^+(X)$ except that the ℓ_1^+-trees in consideration have the additional property that the successors of any nonmaximal element $(x_i)_1^n$ are of the form $(x_1, \ldots, x_n, y_m)$ where (y_m) is a weakly null sequence. One obtains that $I_\omega^+(X) = S(X) = \omega^\alpha$ for some $\alpha < \omega_1$ when X^* is separable. Furthermore a wide variety of indices of this sort, including the I^+ index, are shown to be equal to ω^α for some $\alpha < \omega_1$ if they are countable.

Another ordinal index for Banach spaces is the Baire-1 index. This was defined independently by Kechris and Louveau [KL] and also in [HOR]. In the latter a connection was made with spreading models.

Let K be compact metric, $\varepsilon > 0$ and let $F : K \to \mathbb{R}$ be a bounded function.

$$
\begin{aligned}
H_0(F,\varepsilon) &= K \\
H_{\alpha+1}(F,\varepsilon) &= \{k \in K : \exists\, (k_n) \subseteq H_\alpha(F,\varepsilon)\,,\; k_n \to k \\
&\qquad \text{and } \varliminf |F(k_n) - F(k)| \geq \varepsilon\} \\
H_\alpha(F,\varepsilon) &= \bigcap_{\beta<\alpha} H_\beta(F,\varepsilon)\,,\; \text{if } \alpha \text{ is a limit ordinal.}
\end{aligned}
$$

We set

$$
H(F,\varepsilon) = \inf\{\alpha < \omega_1 : H_\alpha(F,\varepsilon) = \emptyset\}
$$

provided the set is nonempty and $H(F,\varepsilon) = \omega_1$ otherwise. Then F is Baire-1 iff $H(F) \equiv \sup\{H(F,\varepsilon) : \varepsilon > 0\} < \omega_1$. Furthermore [CMR] $\sup_{\varepsilon>0} H(F,\varepsilon) < \omega$ iff $F = G - H$ for two (upper) bounded semicontinuous functions G and H. Such a function is called DBSC, the difference of bounded upper semicontinuous functions. F is called *Baire-1/2* if $H(F,\varepsilon) < \omega$ for all $\varepsilon > 0$. F is Baire-1/2 iff F is the uniform limit of $(F_n) \subseteq DBSC$ [HOR].

The focal point of [HOR] was as follows. X may be naturally regarded as a subspace of $C(K)$ where K is the compact metric space B_{X^*} given the ω^* topology. For $x^{**} \in X^{**}$ we let $F_{x^{**}} = x^{**}|_K$. Of course $F_{x^{**}} \in C(K)$ iff $x^{**} \in X$. If $x^{**} \in X^{**} \setminus X$ then $F_{x^{**}}$ is discontinuous and the character of its discontinuities lead one to information about the subspaces of X. For example X does not contain ℓ_1 iff every function $F_{x^{**}}$ is Baire-1 [OR]. In this case $\sup\{H(F_{x^{**}}) : x^{**} \in X^{**}\} < \omega_1$. Also if $F_{x^{**}}$ is Baire-1 then there exists $(x_n) \subseteq X$ with $x_n \xrightarrow{\omega^*} x^{**}$ in X^{**}. In the case where $F_{x^{**}}$ is Baire-1 we have the following theorem.

7.3 Theorem. [HOR]

1) *Let $x^{**} \in X^{**}$ and assume $F_{x^{**}}$ is Baire-1 but not Baire-1/2. Let $(x_n) \subseteq X$ with $x_n \xrightarrow{\omega^*} x^{**}$ in X^{**}. Then some subsequence of (x_n) has spreading model equivalent to the unit vector basis of ℓ_1.*

2) Let $x^{**} \in X^{**}$ and assume $F_{x^{**}}$ is Baire-1/4 but $x^{**} \notin X$. Then if $(x_n) \subseteq X$, $x_n \xrightarrow{\omega^*} x^{**}$ in X^{**}, some convex block basis of (x_n) has spreading model equivalent to the summing basis for c_0.

Baire-1/4 is defined as follows. We say F is Baire-1/4 if there exists $K < \infty$ so that F is the uniform limit of functions $(F_n) \subseteq DBSC$ satisfying $\sup_n |F_n|_D < \infty$. Here the DBSC norm $|\cdot|_D$ is given by

$$|F|_D = \inf \left\{ \sup_{k \in K} \sum_{j=1}^{\infty} |\varphi_j(k)| : (\varphi_j) \subseteq C(K) \text{ and } f = \sum_1^{\infty} \varphi_j \text{ pointwise on } K \right\}$$

The space $(DBSC(K), |\cdot|_D)$ is a Banach space. Further characterizations of DBSC, Baire 1/2, functions Baire-1/4 functions and other classes of Baire-1 functions can be found in [R7] and [Far4], [Far5], [Far6] where other indices are also introduced.

Recently D.H. Leung and W.-K. Tang [LeT] have shown the compatibility of the oscillation index given above and another index, the "convergence" index.

Earlier we cited Rosenthal's characterization of ℓ_1. Characterizations of c_0 are also known and we recall some of them.

Let (x_i) be a normalized basic sequence. It is easy to see that (x_i) is equivalent to the unit vector basis of c_0 iff

$$\sup \left\{ \left\| \sum_F x_i \right\| : F \in [\mathbb{N}]^{<\omega} \right\} < \infty .$$

Using this and Ramsey's theorem W.B. Johnson proved that a normalized weakly null sequence admits either a c_0 subsequence (a subsequence equivalent to the unit vector basis of c_0) or a subsequence so that for all further subsequences (x_i), $\lim_{n \to \infty} \| \sum_1^n x_i \| = \infty$. Elton obtained from his theorem (Theorem 2.9) the following extension of this (see also [O1]).

7.4 Theorem. [Elt] *Let (x_n) be a normalized weakly null sequence. Then if (x_n) admits no c_0 subsequence, there exists a subsequence (y_n) satisfying*

$$\sup_k \left\| \sum_1^k a_n y_n \right\| = \infty$$

whenever $(a_n) \notin c_0$.

Recently S. Argyros and I. Gasparis have extended this as follows.

7.5 Theorem. *Let (x_n) be a normalized weakly null sequence. Then either a subsequence is boundedly convexly complete or some convex block basis is equivalent to the unit vector basis of c_0.*

(x_n) is *boundedly convexly complete* if whenever $\sup_n \| \sum_1^n a_j x_j \| < \infty$ then

$$\limsup_n \left\{ \left\| \sum_{j \in F} a_j x_j \right\| : n \leq \min F, \ \sum_{j \in F} |a_j| \leq 1 \right\} = 0 \ .$$

Pełczyński's characterization of spaces X which contain c_0 isomorphically is the following. We also include for comparison the [OR] result cited above.

7.6 Theorem. [Pel], [OR] *Let $K = (B_{X^*}, \omega^*)$.*

(a) X contains an isomorph of c_0 iff there exists $x^{**} \in X^{**} \setminus X$ with $F_{x^{**}} \in DBSC(K)$.

(b) X contains an isomorph of ℓ_1 iff there exists $x^{**} \in X^{**} \setminus X$ with $F_{x^{**}} \notin Baire\text{-}1(K)$.

A weak Cauchy sequence is *non-trivial* if it does not converge weakly. Such a sequence can be seen to have a basic subsequence (x_n) which dominates the summing basis (s_i), i.e., for some C,

$$\| \sum a_i x_i \| \geq C \| \sum a_i s_i \| \ .$$

A sequence (x_i) is *strongly summing* if it is weak Cauchy and basic and satisfies whenever $\sup_n \| \sum_1^n c_j x_j \| < \infty$ then $\sum_1^\infty c_j$ converges. Rosenthal recently obtained, in the spirit of his ℓ_1 theorem, the following characterization of c_0.

7.7 Theorem. [R5] *Every non-trivial weak Cauchy sequence in a Banach space has either a strongly summing subsequence or a convex block basis equivalent to the summing basis.*

The proof involves a variant of an ordinal index defined by Kechris and Louveau [KL] used to characterize DBSC. We recall this ordinal index. Let K be compact metric and let $f : K \to \mathbb{R}$ be a bounded function. We define for $x \in K$,

$$\hat{f}(x) = \overline{\lim}_{y \to x} f(y) \quad \text{and} \quad \check{f}(x) = \underline{\lim}_{y \to x} f(y) \ .$$

$\hat{f}$ is upper semicontinuous and $\check{f}$ is lower semicontinuous. Set $f^0 = \hat{f}$,

$f^{\alpha+1} = ((f^{\alpha} - f)^{\wedge} + f)^{\wedge}$ if f^{α} is defined and real valued and for limit ordinals β if f^{α} is real valued for $\alpha < \beta$ set

$$f^{\beta} = \left(\sup_{\alpha < \beta} f^{\alpha} \right)^{\wedge}.$$

The index $r_{ND}(f)$ is defined to be that $\alpha < \omega_1$ such that f^{α} is not real valued if such an α exists and ω_1 otherwise. Then $f \notin DBSC(K)$ iff $r_{ND}(f) < \omega_1$ ([KL], [R5]). Recently S. Argyros and V. Kanellopoulos [AK] have extended partial results of Rosenthal and Bossard [Bos] by proving that if

$$\sup\left\{ r_{ND}(F_{x^{**}}) : x^{**} \in X^{**} \setminus X \right\} = \omega_1$$

then X contains an isomorph of c_0.

8

The homogeneous Banach space problem

We cannot end before at least briefly discussing one other spectacular result of the 90's.

We recall that the homogeneous Banach space problem (P5) is: If X is isomorphic to all $Y \subseteq X$, is X isomorphic to ℓ_2? This was solved by combining two beautiful pieces of work, Gowers' dichotomy theorem (Theorem 3.1) and the following theorem of Komorowski and Tomczak-Jaegermann [KT1, KT2]. A nice exposition somewhat simplifying the argument appears in [TJ1].

8.1 Theorem. *If X is homogeneous and not isomorphic to ℓ_2 then X has a subspace without an unconditional basis.*

It then follows that no subspace of X has an unconditional basis and so by the dichotomy theorem plus the fact that X is homogeneous we have that X must be H.I. But in view of the result of [GM1] that an H.I. space is not isomorphic to any proper subspace, this is impossible. Thus the solution of the homogeneous Banach space problem is achieved.

Komorowski and Tomczak-Jaegermann actually prove something stronger. They show

8.2 Theorem. *Let X be a Banach space not containing a subspace isomorphic to ℓ_2. Then X contains a subspace without an unconditional basis.*

Even more recently Komorowski and Tomczak-Jaegermann have made substantial progress on (Q4): if all subspaces of X have an unconditional basis is X isomorphic to ℓ_2? They proved [KT3].

8.3 Theorem. *If every subspace of $(X \oplus X \oplus \cdots)_{\ell_2}$ has an unconditional basis then X is isomorphic to ℓ_2.*

The proof of Theorem 8.2 is well exposed in [TJ1] and we shall not repeat it here. It is interesting to note that spreading models enter at one part of the argument.

The finite dimensional version of (P5) was solved in 1989 [MTJ1]

8.4 Theorem. [MTJ1] *There exists a function* $f : (0,1) \times (1,\infty) \to (1,\infty)$ *with the following property. If* $0 < \varepsilon < 1$, $K > 1$ *and* $\dim X = n$ *are such that for* $m = [\![\varepsilon n]\!]$, $d(E, \ell_2^m) \leq K$ *for all* m-*dimensional subspaces* $E \subseteq X$ *then* $d(X, \ell_2^n) \leq f(\varepsilon, K)$.

A weaker version (for ε sufficiently small) was obtained earlier by J. Bourgain [Bo3].

9

Concluding remarks

We have not addressed certain important problems that remain unsolved after many years concerning the classical Banach spaces themselves.

(Q13) Let K be a compact metric space. Is every complemented subspace of $C(K)$ isomorphic to $C(L)$ for some compact metric space L?

It is known that if K is uncountable then $C(K)$ is isomorphic to $C[0,1]$. If K is countable then $C(K)$ is isomorphic to $C(\omega^{\omega^\alpha})$ for some $\alpha < \omega_1$. Every complemented subspace of c_0 (isomorphic to $C(\omega)$) is either finite dimensional or isomorphic to c_0 ([Pel]). If X is complemented in $C[0,1]$ and X^* is nonseparable then X is isomorphic to $C[0,1]$ [R6]. Every quotient of c_0 embeds isomorphically into c_0 but this does not hold in general for $C(\omega^{\omega^\alpha})$. A discussion of these and related results may be found in [A1, A2, A3, A4], [Ga1, Ga2], [Bo2].

The isomorphism types of the complemented subspaces of $L_1[0,1]$ remain unclassified.

(Q14) Let X be a complemented (infinite dimensional) subspace of $L_1[0,1]$. Is X isomorphic to L_1 or ℓ_1?

Every X which is complemented in ℓ_p ($1 \le p < \infty$) or c_0 is isomorphic to ℓ_p or c_0. There are known to be uncountably many mutually nonisomorphic complemented subspaces of $L_p[0,1]$ ($1 < p < \infty$, $p \ne 2$) [BRS] and all are known to have a basis [JRZ]. These spaces have been classified as $\mathcal{L}_p$ spaces ([LP], [LR]), provided they are not Hilbert spaces. X is $\mathcal{L}_p$-λ if $X = \overline{\bigcup E_n}$ where $E_1 \subseteq E_2 \subseteq \cdots$ are finite dimensional spaces having the property that $d(E_n, \ell_p^{\dim E_n}) \le \lambda$ for all n.

(**Q15**) Does every $\mathcal{L}_p$ space for $1 < p < \infty$ have an unconditional basis?

More generally we have

(**Q16**) If X is complemented in a space with an unconditional basis does X have an unconditional basis?

A weaker notion than having an unconditional basis is L.u.s.t.. X has *L.u.s.t.* if $X = \overline{\cup E_n}$ where $E_1 \subseteq E_2 \subseteq \cdots$ are finite dimensional and for some $K < \infty$ each E_n has a K-unconditional basis.

(**Q17**) If X has L.u.s.t. does X contain an unconditional basic sequence?

The answer is yes if c_0 is not finitely representable in X [FJT].

Also we have largely ignored nonseparable problems.

It is known for some time that there does exist a nonseparable space not containing ℓ_p or c_0 or even a subsymmetric basic sequence [O2]. Also, using for example the subsymmetric basis of Schlumprecht's space, one can construct Banach spaces of arbitrary size which do not contain c_0 or ℓ_p.

However, as this paper was being given its final edits we were informed of some new spectacular results obtained by S. Argyros [Ar1] in the nonseparable domain.

9.1 Theorem. [Ar1] *There exists a nonseparable H.I. space X. Moreover X can be constructed so that*

i) $X^* = W \oplus \ell_1(\Gamma)$ *with W separable.*
ii) *Every operator $T : X \to X$ has the form $T = \lambda I + S$ where S is a weakly compact operator with separable range.*

Lindenstrauss proved that every nonseparable reflexive space contains a nontrivial complemented subspace [L]. Also it is proved in [Ket] that if B is a Banach space of cardinality equal to a Ramsey cardinal $\mathcal{K}$ then B contains an unconditional basic sequence. In fact B contains a symmetric sequence of length $\mathcal{K}$. The biorthogonal distortion of ℓ_p $(1 < p < \infty)$ given in Theorem 4.1 involved certain subsymmetric (or even symmetric) classes of dual functionals. These can be used to show that Hilbert space of any infinite dimension is arbitrarily distortable. We could go on listing problems and results. For example we have only mentioned in passing a few results from local finite dimensional theory. Or we could discuss nonlinear problems in which great strides have been made [BeL] but that would be another survey. Readers interested in the problems we have stated (and the ones we have not) are urged to consult, when it appears, the Handbook of the Geometry of Banach Spaces [JL] currently under preparation.

Added in proof: (Q8) (see page 234) has been solved in the negative by the author, G. Androulakis, N. Tomczak-Jaegermann and Th. Schlumprecht.

References

[Ald] D.J. Aldous, *Subspaces of L^1 via random measures*, Trans. Amer. Math. Soc. **267** (1981), 445–453.

[A1] D. Alspach, *Quotients of $C[0,1]$ with separable dual*, Israel J. Math. **29** (1978), 361–384.

[A2] D. Alspach, *$C(K)$ norming subsets of $C[0,1]^*$*, Studia Math. **70** (1981), 27–67.

[A3] D. Alspach, *Operators on $C(\omega^\alpha)$ which do not preserve $C(\omega^\alpha)$*, Fund. Math. **153** (1997), 81–98.

[A4] D. Alspach, *Quotients of c_0 are almost isometric to subspaces of c_0*, Proc. A.M.S. **76** (1979), 285–288.

[AA] D. Alspach and S. Argyros, *Complexity of weakly null sequences*, Diss. Math. **321** (1992), 1–44.

[AO] D. Alspach and E. Odell, *Averaging weakly null sequences*, Functional Analysis, Springer-Verlag LNM 1332 (1988), 126–144.

[AJO] D. Alspach, R. Judd, E. Odell, *The Szlenk index and local ℓ_1 indices of a Banach space*, to appear in Positivity.

[AnO] G. Androulakis and E. Odell, *Distorting mixed Tsirelson spaces*, Israel J. Math. **109** (1999), 125–149.

[AS1] G. Androulakis and Th. Schlumprecht, *Block sequences in S*, preprint.

[AS2] G. Androulakis and Th. Schlumprecht, *On the subsymmetric sequences in S*, preprint.

[AS3] G. Androulakis and Th. Schlumprecht, *Strictly singular non-compact operators exist on the Gowers-Maurey space*, preprint.

[Ar1] S. Argyros, *Non-separable hereditarily indecomposable Banach spaces*, preprint.

[Ar2] S. Argyros, *A universal property of reflexive hereditarily indecomposable Banach spaces*, preprint.

[AD] S. Argyros and I. Deliyanni, *Examples of asymptotically ℓ^1 Banach spaces*, Trans. A.M.S. **349** (1997), 973–995.

[AF] S. Argyros and V. Felouzis, *Interpolating hereditarily indecomposable Banach spaces*, J. Amer. Math. Soc. **13** (2000), 243–294.

[AG] S. Argyros and I. Gasparis, *Unconditional structure of weakly null sequences*, preprint.

[ADKM] S. Argyros, I. Deliyanni, D.N. Kutzarova, A. Manoussakis, *Modified mixed Tsirelson spaces*, J. Funct. Anal. **159** (1998), 43–109.

[AK] S. Argyros and V. Kanellopoulos, *Optimal sequences of continuous functions converging to a Baire-1 function*, preprint.

[AMT] S. Argyros, S. Merkourakis and A. Tsarpalias, *Convex unconditionality and summability of weakly null sequences*, Israel J. Math. **107** (1998), 157–193.

[ANZ] S. Argyros, S. Negrepontis and Th. Zachariades, *Weakly stable Banach spaces*, Israel J. Math. **57** (1987), 68–88

[Ba] S. Banach, *Theorie des operations lineaires*, Warszawa, 1932.

[BL] B. Beauzamy and J.-T. Lapresté, *Modèles étalés des espace de Banach* Travaux en Cours, Herman, Paris, 1984.

[BeL] Y. Benyamini and J. Lindenstrauss, *Geometric nonlinear functional analysis*, AMS Colloq. Pub. vol.48, 2000.

[Bos] B. Bossard, *On a certain problem of H.P. Rosenthal*, preprint.

[Bo1] J. Bourgain, *On convergent sequences of continuous functions*, Bull. Soc. Math. Bel. **32** (1980), 235–249.

[Bo2] J. Bourgain, *The Szlenk index and operators on $C(K)$-spaces*, Bull. Soc. Math. de Belgique, Ser.B. **31** (1979), 87–117.

[Bo3] J. Bourgain, *On finite-dimensional homogeneous Banach spaces*, GAFA Israel Seminar 1986-87, ed. J. Lindenstrauss and V. Milman, LNM 1317 (1988), 232–239.

[BRS] J. Bourgain, H. Rosenthal, G. Schechtman, *An ordinal L^p-index for Banach spaces, with application to complemented subspaces of L^p*, Annals of Math. **114** (1981), 193–228.

[BS] A. Brunel and L. Sucheston, *B-convex Banach spaces*, Math. Systems Theory **7** (1974), 294–299.

[Bu] S.Q. Bu, *Deux remarques sur les espaces de Banach stables*, Compositio Math. **69** (3) (1989), 341–355.

[CS] T. Carlson and S. Simpson, *A dual form of Ramsey's theorem*, Advances in Math. **53** (1984), 265–287.

[CJT] P. Casazza, W.B. Johnson and L. Tzafriri, *On Tsirelson's space*, Israel J. Math. **47** (1984), 81–98.

[CO] P.G. Casazza and E. Odell, *Tsirelson's space and minimal subspaces*, Longhorn Notes: Texas Functional Analysis Seminar 1982-83, University of Texas, Austin, 61–72.

[CaS] P.G. Casazza and T.J. Shura, *Tsirelson's space*, Lectures Notes in Math., vol. 1363, Springer-Verlag, Berlin and New York, 1989.

[C] F. Chaatit, *On the uniform homeomorphisms of the unit spheres of certain Banach lattices*, Pacific J. Math. **168** (1995), 11–31.

[CMR] F. Chaatit, V. Mascioni and H. Rosenthal, *On functions of finite Baire index*, J. Funct. Anal. **42** (1996), 277–295.

[DFJP] W.J. Davis, T. Figiel, W.B. Johnson and A. Pełczyński, *Factoring weakly compact operators*, J. Funct. Anal. **17** (1974), 311–327.

[Dah] M. Daher, *Homéomorphismes uniformes entre les sphères unites des éspaces d'interpolation*, Thesis, Université Paris 7.

[Da] M.M. Day, *Reflexive spaces not isomorphic to uniformly convex Banach spaces*, Bull. Amer. Math. Soc. **47** (1941), 313–317.

[Del] C. Dellacherie, *Les derivations en theorie descriptive des ensembles et le theoreme de la borne*, Seminare de Prob. XI, Universite de Strasbourg, Springer-Verlag LNM 581 (1977), 34–46.

[De] R. Deville, *Geometrical implications of the existence of very smooth bump functions in Banach spaces*, Israel J. Math. **6** (1989), 1–22.

[D] J. Diestel, *Sequences and Series in Banach spaces*, Graduate Texts in Mathematics, Springer-Verlag (1984).

[DJLT] P.N. Dowling, W.B. Johnson, C.J. Lennard and B. Turett, *The optimaility of James' distortion theorems*, Proc. A.M.S. **125** (1997), 167–174.

[Dv]   A. Dvoretzky, *Some results on convex bodies and Banach spaces*, Proc. Sympos. Linear Spaces, Jerusalem, 1961, pp. 123–160.

[El]   E.E. Ellentuck, *A new proof that analytic sets are Ramsey*, J. Symbolic Logic **39** (1974), 163–165.

[Elt]   J. Elton, *Weakly null normalized sequences in Banach spaces*, Ph.D. Thesis, Yale University, New Haven, Ct. (1978).

[E1]   P. Enflo, *A counterexample to the approximation property in Banach spaces*, Acta Math. **130** (1973), 309–317.

[E2]   P. Enflo, *On a problem of Smirnov*, Ark. Mat. **8** (1969), 107–109.

[E3]   P. Enflo, *Banach spaces which can be given an equivalent uniformly convex norm*, Israel J. Math. **13** (1972), 281–288.

[Fa]   J. Farahat, *Espaces de Banach contenant ℓ_1 d'apres H.P. Rosenthal*, Seminaire Maurey-Schwartz, Ecole Polytechnique, 1973–74.

[Far1]   V. Farmaki, *Ramsey dichotomies with ordinal index*, preprint.

[Far2]   V. Farmaki, *Ordinal indices and Ramsey dichotomies measuring co-content and semi bounded completeness*, preprint.

[Far3]   V. Farmaki, *The uniform convergence ordinal index and the ℓ^1-behavior of a sequence of functions*, preprint.

[Far4]   V. Farmaki, *On Baire-1/4 functions and spreading models*, Mathematika **41** (1994), 251–265.

[Far5]   V. Farmaki, *On Baire-1/4 functions*, Trans. A.M.S. **348** (1996), 4023–4041.

[Far6]   V. Farmaki, *Calssifications of Baire-1 functions and c_0 spreading models*, Trans. A.M.S. **345** (1994), 819–831.

[F1]   V. Ferenczi, *A uniformly convex hereditarily indecomposable Banach space*, Israel J. Math. **102** (1997), 199–225.

[F2]   V. Ferenczi, *Quotient hereditarily indecomposable Banach spaces*, Canad. J. Math. **51** (1999), 566–584.

[FH]   V. Ferenczi and P. Habala, *A uniformly convex hereditarily indecomposable space whose subspaces fail Gordon-Lewis property*, Arch. Math. (Basel) **71** (1998), 481–492.

[FJ]   T. Figiel and W.B. Johnson, *A uniformly convex Banach space which contains no ℓ_p*, Compositio Math. **29** (1974), 179–190.

[FJT]   T. Figiel, W.B. Johnson and L. Tzafriri, *On Banach lattices and spaces having local unconditional structure with applications to Lorentz sequence spaces*, J. Approx. Theory **13** (1975), 297–312.

[FLM]   T. Figiel, J. Lindenstrauss and V.D. Milman, *The dimension of almost spherical sections of convex bodies*, Acta. Math. **139** (1977), 53–94.

[GP]   F. Galvin and K. Prikry, *Borel sets and Ramsey's theorem*, J. Symbolic Logic **38** (1973), 193–198.

[Ga1]   I. Gasparis, *Operators that do not preserve $C(\alpha)$-spaces*, preprint.

[Ga2]   I. Gasparis, *A class of ℓ_1-preduals which are isomorphic to quotients of $C(\omega^\omega)$*, Studia Math. **133** (1999), 131–143.

[Ga3]   I. Gasparis, *On the distortion of mixed Tsirelson spaces*, preprint.

[Ga4]   I. Gasparis, *A dichotomy theorem for subsets of the power set of the natural numbers*, Proc. A.M.S., (to appear)

[Gil]   T.A. Gillespie, *Factorization in Banach function spaces*, Indag, Math. **43** (1981), 287–300.

[Gl] E.D. Gluskin, *Finite-dimensional analogues of spaces without basis*, Doklady Acad. Nauk SSSR **216** (5) (1981), 146–150.

[GKL] G. Godefroy, N.J. Kalton and G. Lancien, *Szlenk indices and uniform homeomorphisms*, preprint.

[G1] W.T. Gowers, *A new dichotomy for Banach spaces*, Geom. Funct. Anal. **6** (1996), 1083–1093.

[G2] W.T. Gowers, *A space not containing c_0, ℓ_1 or a reflexive subspace*, Trans. Amer. Math. Soc. **344** (1994), 407–420.

[G3] W.T. Gowers, *Analytic sets and games in Banach spaces*, preprint IHES M/94/42.

[G4] W.T. Gowers, *Lipschitz functions on classical spaces*, European J. Combin. **13** (1992), 141–151.

[G5] W.T. Gowers, *A hereditarily indecomposable space with an asymptotic unconditional basis*, Oper. Theory: Adv. Appl. **77** (1995), 111–120.

[GM1] W.T. Gowers and B. Maurey, *The unconditional basic sequence problem*, J. Amer. Math. Soc. **6** (1993), 851–874.

[GM2] W.T. Gowers and B. Maurey, *Banach spaces with small spaces of operators*, Math. Ann. **307** (1997), 543–568.

[Gu] S. Guerre-Delabrière, *Classical Sequences in Banach Spaces*, Marcel Dekker, New York, 1992.

[Ha] P. Habala, *A Banach space whose subspaces fail Gordon-Lewis property*, Math. Ann. **310** (1998), 197–219.

[HT] P. Habala and N. Tomczak-Jaegermann, *Finite representability of ℓ_p in quotients of Banach spaces*, Positivity, to appear.

[HHZ] P. Habala, P. Hajek and V. Zizler, *Introduction to Banach spaces I and II*, Matfyzpress, Charles University, Prague, 1996.

[HM] R. Haydon and B. Maurey, *On Banach spaces with strongly separable types*, J. London Math. Soc. (2) **33** (1986), 484–498.

[HOR] R. Haydon, E. Odell and H. Rosenthal, *On certain classes of Baire-1 functions with applications to Banach space theory*, Functional Analysis, Springer-Verlag LNM 1470 (1991), 1–35.

[HORS] R. Haydon, E. Odell, H. Rosenthal and Th. Schlumprecht, *On distorted norms in Banach spaces and the existence of ℓ_p types*, unpublished manuscript.

[H] N. Hindman, *Finite sums from sequences within cells of a partition of N*, J. Combinatorial Theory (A) **17** (1974), 1–11.

[Hu] R. Huff, *Banach spaces which are nearly uniformly convex*, Rocky Mountain J. Math. **10** (1980), 743–749.

[J1] R.C. James, *Uniformly nonsquare Banach spaces*, Ann. of Math. (2)**80** (1964), 542–550.

[J2] R.C. James, *Bases and reflexivity of Banach spaces*, Ann. of Math. **52** (1950), 518–527.

[J3] R.C. James, *A separable somewhat reflexive Banach space with non-separable dual*, Bull. Amer. Math. Soc. **18** (1974), 738–743.

[J4] R.C. James, *Reflexivity and the sup of linear functionals*, Israel J. Math. **13** (1972), 289–300.

[J5] R.C. James, *Super-reflexive Banach spaces*, Can. J. Math. **24** (1972), 896–904.

[Jo1] W.B. Johnson, *Banach spaces all of whose subspaces have the Approximation Property*, in Special Topics in Applied Math., Proceedings Bonn 1979, North Holland, 1980, 15–26.

[Jo2] W.B. Johnson, *A reflexive Banach space which is not sufficiently Euclidean*, Studia Math. **55** (1976), 201–205.

[JL] W.B. Johnson and J. Lindenstrauss (editors), *Handbook of the geometry of Banach spaces*, in preparation.

[JLPS] W.B. Johnson, J. Lindenstrauss, D. Preiss and G. Schechtman, *Almost Fréchet differentiability of Lipschitz mappings between infinite dimensional Banach spaces*, preprint.

[JR] W.B. Johnson and H. Rosenthal, *On ω^*-basic sequences and their applications to the study of Banach spaces*, Studia Math. **43** (1972), 77–92

[JRZ] W.B. Johnson, H. Rosenthal and M. Zippin, *On bases, finite dimensional decompositions and weaker structures in Banach spaces*, Israel J. Math. **9** (1971), 488–508.

[Ju] R. Judd, *A dichotomy for Schreier sets*, Studia Math. **132** (1999), 245–256.

[JO] R. Judd and E. Odell, *Concerning the Bourgain ℓ_1 index of a Banach space*, Israel J. Math. **108** (1998), 145–171.

[Ka1] N.J. Kalton, private communciation.

[Ka2] N.J. Kalton, *On subspaces of c_0 and extensions of operators into $C(K)$-spaces*, preprint.

[KW] N.J. Kalton and D. Werner, *Property (M), M-ideals and almost isometric structure of Banach spaces*, J. Reine und Agnew. Math. **461** (1995), 137–178.

[KL] A. Kechris and A. Louveau, *A classification of Baire class 1 functions*, Trans. A.M.S. **318** (1990), 209–236.

[Ket] J. Ketonen, *Banach spaces and large cardinals*, Fund. Math. **81** (1974), 291–302.

[KOS] H. Knaust, E. Odell and Th. Schlumprecht, *On asymptotic structure, the Szlenk index and UKK properties in Banach space*, Positivity **3** (1999), 173–199.

[KT1] R. Komorowski and N. Tomczak-Jaegermann, *Banach spaces without local unconditional structure*, Israel J. Math. **89** (1995), 205–226.

[KT2] R. Komorowski and N. Tomczak-Jaegermann, *Erratum to "Banach spaces without local unconditional structure*, Israel J. Math. **105** (1998), 85–92.

[KT3] R. Komorowski and N. Tomczak-Jaegermann, *Subspaces of $\ell_p(X)$ and $\mathrm{Rad}(X)$ without local unconditional structure*, preprint.

[K] J.L. Krivine, *Sous espaces de dimension finie des espaces de Banach réticulés*, Ann. of Math. (2) **104** (1976), 1–29.

[KM] J.L. Krivine and B. Maurey, *Espaces de Banach stables*, Israel J. Math. **39** (1981), 273–295.

[Kir] P. Kyriakouli, *On hereditary families of finiite subsets of positive integers*,

[Le] H. Lemberg, *Nouvelle démonstration d'un théorème de J.L. Krivine sur la finie représentation de ℓ_p dans un espaces de Banach*, Israel J. Math. **39** (1981), 341–348.

[LeT] D.H. Leung and W.-K. Tang, *Functions of Baire class one*, preprint.

[Lin] Pei-Kee Lin, private communication.

[LK] Pei-Kee Lin and D. Kutzarova, *Remarks about Schlumprecht's space*, Proc. A.M.S., **128** (2000), 2059–2068.

[L] J. Lindenstrauss, *On non-separable reflexive Banach spaces*, Bull. Amer.

Math. Soc. **72** (1966), 967–970.

[LP] J. Lindenstrauss and A. Pełczyński, *Absolutely summing operators in* $\mathcal{L}_p$ *spaces and their applications*, Studia Math. **29** (1968), 275–326.

[LR] J. Lindenstrauss and H. Rosenthal, *The* $\mathcal{L}_p$ *spaces*, Israel J. Math. **7** (1969), 325–349.

[LS] J. Lindenstrauss and C. Stegall, *Examples of separable spaces which do not contain* ℓ_1 *and whose duals are non-separable*, Studia Math. **54** (1975), 81–105.

[LT1] J. Lindenstrauss and L. Tzafriri, *Classical Banach Spaces I*, Springer-Verlag, New York, 1977.

[LT2] J. Lindenstrauss and L. Tzafriri, *On Orlicz sequence spaces*, Israel J. Math. (1971), 379–390.

[LT3] J. Lindenstrauss and L. Tzafriri, *On complemented subspaces problem*, Israel J. Math. **9** (1971), 263–269.

[Lo] G. Ya. Lozanovskii, *On some Banach lattices*, Siberian Math. J. **10** (1969), 584–599.

[MTJ1] P. Mankiewicz and N. Tomczak-Jaegermann, *The solution of finite-dimensional homogeneous Banach space problem*, Israel J. Math. **75** (1991), 129–159.

[MTJ2] P. Mankiewicz and N. Tomczak-Jaegermann, *Schauder bases in quotients of subspaces of* $\ell_2(X)$, Amer. J. Math. **116** (1994), 1341–1363.

[Mar] D.A. Martin, *Borel determinancy*, Annals of Math. **102** (1975), 363–371.

[Mat] A.R.D. Mathias, *Happy families*, Annals of Math. Logic **12** (1977), 59–111.

[Ma1] B. Maurey, *Quelques progrés dans la compréhension de la dimension infinie*, in Espaces de Banach classiques et quantiques, Journée Annuelle, Soc. Math. de France, 1994, 1–29.

[Ma2] B. Maurey, *A remark about distortion*, Oper. Theory: Adv. Appl. **77** (1995), 131–142.

[Ma3] B. Maurey, *Types and* ℓ_1*-subspaces*, Longhorn Notes: Texas Functional Analysis Seminar 1982-83, University of Texas, Austin, 123–137.

[Ma4] B. Maurey, *Symmetric distortion in* ℓ_2, Oper. Theory: Adv. Appl. **77** (1995), 143–147.

[MMT] B. Maurey, V.D. Milman and N. Tomczak-Jaegermann, *Asymptotic infinite-dimensional theory of Banach spaces*, Oper. Theory: Adv. Appl. **77** (1994), 149–175.

[MR] B. Maurey and H. Rosenthal, *Normalized weakly null sequences with no unconditional subsequences*, Studia Math. **61** (1977), 77–98.

[Mil] D.P. Milman, *ON some criteria for the regularity of spaces of the type* (*B*), Dokl. Akad. nauk SSSR **20** (1938), 243–246, (in Russian).

[MM] D.P. Milman and V.D. Milman, *The geometry of nested families with empty intersection, the structure of the unit sphere in a non-reflexive space*, AMS Transl. 2 **85** (1969), 233–243.

[M1] V.D. Milman, *Geometric theory of Banach spaces II, geometry of the unit sphere*, Russian Math. Survey **26** (1971), 79–163, (trans. from Russian).

[M2] V.D. Milman, *The infinite dimensional geometry of the unit sphere of a Banach space*, Soviet Math. Dokl. **8** (1967), 1440–1444, (trans. from Russian).

[M3] V.D. Milman, *A new proof of the theorem of A. Dvoretsky on sections of convex bodies*, Functional Anal. Appl. **5** (1971), 28–37.

[M4] V.D. Milman, *The spectrum of bounded continuous functions which are given on the unit sphere of a B-space*, Funktsional Anal. i Prilozhen **3** (1969), 67–79, M.R. 40 4740.

[M5] V.D. Milman, *Dvoretsky's theorem — thirty years later*, GAFA **2** (1992), 455–479.

[MP] V.D. Milman and A. Perelson, *Infinite dimensional geometric moduli and type-cotype theory*, Geometric aspects of Banach spaces, 11–38, London Math. Soc. LNS 140, Cambridge Univ. Press (1989).

[MS] V.D. Milman and G. Schechtman, *Asymptotic theory of finite dimensional normed spaces*, Lecture Notes in Math., vol. 1200, Springer-Verlag, Berlin and New York, 1986, 156 pp.

[MiS] V.D. Milman and M. Sharir, *Shrinking minimal systems and complementation of ℓ_p^n-spaces in reflexive Banach spaces*, Proc. London Math. Soc. **39** (1979), 1–29.

[MT1] V.D. Milman and N. Tomczak-Jaegermann, *Asymptotic ℓ_p spaces and bounded distortions*, (Bor-Luh Lin and W.B. Johnson, eds.), Contemp. Math. **144** (1993), 173–195.

[MT2] V.D. Milman and N. Tomczak-Jaegermann, *Stabilized asymptotic structures and envelopes in Banach spaces*, preprint.

[MW] V.D. Milman and R. Wagner, *Asymptotic versions of operators and operator ideals*, Convex geometric analysis (Berkeley, CA, 1996), 165–179, Cambridge Univ. Press (1999).

[NW] C. St. J.A. Nash-Williams, *On well quasi-ordering transfinite sequences*, Proc. Camb. Phil. Soc. **61** (1965), 33–39.

[O1] E. Odell, *Applications of Ramsey theorems to Banach space theory*, Notes in Banach spaces, (H.E. Lacey, ed.), Univ. Texas Press, Austin, TX, (1980) pp. 379–404.

[O2] E. Odell, *On the types in Tsirelson's space*, Longhorn Notes, The University of Texas at Austin (1986), 61–72.

[OR] E. Odell and H.P. Rosenthal, *A double-dual characterization of separable Banach spaces containing ℓ^1*, Israel J. Math. **20** (1975), 375–384.

[ORS] E. Odell, H. Rosenthal and Th. Schlumprecht, *On weakly null FDD's in Banach spaces*, Israel J. Math. **84** (1993), 333–351.

[OS1] E. Odell and Th. Schlumprecht, *A Banach space block finitely universal for monotone bases*, Trans. Amer. Math. Soc. **352** No. 4 (1999), 1859–1888.

[OS2] E. Odell and Th. Schlumprecht, *The distortion problem*, Acta Math. **173** (1994), 259–281.

[OS3] E. Odell and Th. Schlumprecht, *The distortion of Hilbert space*, Geom. Functional Anal. **3** (1993), 201–207.

[OS4] E. Odell and Th. Schlumprecht, *Distortion and Stabilized Structure in Banach Spaces; New Geometric Phenomena for Banach and Hilbert Spaces*, Proc. International Congress of Mathematicians, vol.1,2, (Zürich, Switzerland 1994), Birkhäuser Verlag, Basel, Switzerland 1995, 955–965.

[OS5] E. Odell and Th. Schlumprecht, *On the richness of the set of p's in Krivine's theorem*, Oper. Theory: Adv. Appl. **77** (1995), 177–198.

[OS6] E. Odell and Th. Schlumprecht, *A problem on spreading models*, J. Funct. Anal. **153** (1998), 249–261.

[OS7] E. Odell and Th. Schlumprecht, *Asymptotic properties of Banach spaces under renormings*, J. Amer. Math. Soc. **11** (1998), 175–188.

[OS8] E. Odell and Th. Schlumprecht, *Trees and branches in Banach spaces*, preprint.

[OS9] E. Odell and Th. Schlumprecht, *Distortion and Aymptotic Structure*, preprint.

[OT] E. Odell and N. Tomczak-Jaegermann, *On certain equivalent norms on Tsirelson's space*, Illinois J. Math. **44** (2000), 51–71.

[OTW] E. Odell, N. Tomczak-Jaegermann and R. Wagner, *Proximity to ℓ_1 and distortion in asymptotic ℓ_1 spaces* J. Funct. Anal. **150** (1997), 101–145.

[Pel] A. Pełczyński, *Projections in certain Banach spaces*, Studia Math. **19** (1960), 209–228.

[Pe] B.J. Pettis, *A proof that every uniformly convex space is reflexive*, Duke Math. J. **5** (1939), 249–253.

[Pi1] G. Pisier, *Martingales with values in uniformly convex spaces*, Israel J. Math. **13** (1972), 361–378.

[Pi2] G. Pisier, *Weak Hilbert spaces*, Proc. London Math. Soc. **56** (1988), 547–579.

[P] V. Pitak, *A combinatorial theorem on systems of inequalities and its applications to analysis*, Czech. Math. J. **84** (1959), 629–630.

[Ra] F.P. Ramsey, *On a problem of formal logic*, Proc. London Math. Soc. (2) **30** (1929), 264–286.

[Rib] M. Ribe, *Existence of separable uniformly homeomorphic non isomorphic Banach spaces*, Israel J. Math. **48** (1984), 139–147.

[R1] H. Rosenthal, *A characterization of Banach spaces containing ℓ_1*, Proc. Nat. Acad. Sci. U.S.A. **71** (1974), 2411–2413.

[R2] H. Rosenthal, *On a theorem of Krivine concerning block finite representability of ℓ_p in general Banach spaces*, J. Funct. Anal. **28** (1978), 197–225.

[R3] H. Rosenthal, *Some remarks concerning unconditional basic sequences*, Longhorn Notes: Texas Functional Analysis Seminar 1982-83, University of Texas, Austin, 15–48.

[R4] H. Rosenthal, *Double dual types and the Maurey characterization of Banach spaces containing ℓ^1*, Longhorn Notes: Texas Functional Analysis Seminar 1983-84, University of Texas, Austin, 1–37.

[R5] H. Rosenthal, *A characterization of Banach spaces containing c_0*, J. Amer. Math. Soc. **7** (1994), 707–748.

[R6] H. Rosenthal, *On factors of $C[0,1]$ with non-separable dual*, Israel J. Math. **13** (1972), 361–378.

[R7] H. Rosenthal, *Differences of bounded semicontinuous functions, I*, preprint.

[Sch1] Th. Schlumprecht, *An arbitrarily distortable Banach space*, Israel J. Math. **76** (1991), 81–95.

[Sch2] Th. Schlumprecht, *A complementably minimal Banach space not containing c_0 or ℓ_p*, Seminar Notes in Functional Analysis and PDE's, LSU, 1991-92, pp. 169–181.

[Sc] J. Schreier, *Ein Gegenbeispiel zur Theorie der schwachen Konvergenz*, Studia Math. **2** (1930), 58–62.

[Sil] J. Silver, *Every analytic set is Ramsey*, J. Symbolic Logic **35** (1970), 60–64.

[S] A. Szankowski, *Subspaces without approximation property*, Israel J. Math. **24** (1978), 123–129.

[Sza1] S.J. Szarek, *A Banach space without a basis which has the bounded approximation property*, Acta Math. **159** (1987), 81–98.

[Sza2] S.J. Szarek, *The finite-dimensional basis problem with an appendix on nets of Grassman manifold*, Acta. Math. **151** (1983), 153–179.

[Szl] W. Szlenk, *The non-existence of a separable reflexive Banach space universal for all separable reflexive Banach spaces*, Studia Math. **30** (1968), 53–61.

[TJ1] N. Tomczak-Jaegermann, *A solution of the homogeneous Banach space problem*, Canadian Math. Soc. 1945-1995, Vol.3, J. Carrell and R. Murty (editors), CMS, Ottawa (1995), 267–286.

[TJ2] N. Tomczak-Jaegermann, *Banach spaces of type p have arbitrarily distortable subspaces*, GAFA **6** (1996), 1074–1082.

[TJ3] N. Tomczak-Jaegermann, *Distortions on Schatten classes C_p*, Operator Theory: Advances and Applications **77** (1995), 327–334.

[T] B.S. Tsirelson, *Not every Banach space contains ℓ_p or c_0*, Functional Anal. Appl. **8** (1974), 138–141.

[W1] R. Wagner, *Gowers dichotomy for asymptotic structure*, Proc. A.M.S. **124** (1996), 3089–3095.

[W2] R. Wagner, *Finite higher-order games and an inductive approach towards Gowers' dichotomy*, preprint.

Index

admissible
  1-admissible sets, 199
  1-admissible vectors, 211
  α-admissible, 228, 229
  α-admissible vectors, 231
almost isometric copies of ℓ_p, 198, 201, 239
approximation property, 195
asymptotic structure, 233, 238
  Asymptotic ℓ_p, 238, 239
  stabilized, 240
averages
  α-averages, 236
  (ℓ_1^m, ε) averages, 211, 215
  RIS averages, 211, 212, 215

Baire functions
  Baire class 1, 248, 250
  Baire class 1/2, 248
  Baire class 1/4, 249
  DBSC, 248, 250
  DBSC norm, 249
Banach lattice, 225
Banach Mazur distance, 198
Banach space
  arbitrarily distortable, 210–212
  complemented subspace, 192, 207, 255
  geometric characterization of reflexivity, 237
  H.I., 215, 216, 221
  H.I. (ε), 218, 219
  minimal, 221
  nonseparable, 195
  quasi-minimal, 221
  reflexive, 205, 237
  stable, 198
  strictly quasi-minimal, 221
  superreflexive, 247

  totally incomparable, 221, 230
  uniformly convex, 225, 237, 247
  uniformly smooth, 225
  weakly sequentially complete, 205
  weakly stable, 198
basis, 194, 195
  asymptotic ℓ_p, 226, 229, 231, 233
  asymptotically unconditional, 215, 220, 229
  basic, 195
  β-asymptotically unconditional, 222
  biorthogonal functional, 195
  block, 194, 201, 203
  block finitely representable, 205
  boundedly complete, 205
  conditional, 205
  constant, 195
  convexly unconditional, 236
  K-basic, 195
  K-equivalent, 204
  K-unconditional, 196, 213, 218
  monotone, 196
  nearly unconditional, 206
  projection, 194
  S_α-unconditional, 228
  shrinking, 205
  strongly summing, 250
  subsymmetric, 212
  summing, 205, 208, 213, 240, 250
  suppression 1-unconditional, 203
  unconditional, 252, 255
  unconditional basic sequence, 196, 197, 226, 247
  unconditional basic sequence problem, 194, 196, 201, 206, 208, 212, 215
  unconditional basis constant, 196
  universal monotone, 213

265

Borel games, 241

c_0-saturated, 225
c_{00}, 208
$C(K)$
    complemented subspaces, 254
Cantor-Bendixson index, 227
closed linear span $[\cdot]$, 195
cotype-q, 195, 252

Δ-spectrum, 231
Δ-stabilizes, 231
distortion, 225
    arbitrarily distortable, 226, 230
    biorthogonal, 212, 213, 223, 225,
        229, 255
    D-bounded, 226, 228, 229, 231
    distortable subspace, 197, 198
    distortion problem, 194
    James' c_0 and ℓ_1 theorem, 201,
        210, 211
    ℓ_p, 201, 223, 224
    λ-distortable, 200
    local result, 247
Dvoretsky's theorem, 199

entropy map E_X, 224
equivalent norms, 197

finite dimensional decomposition FDD,
    239
    blocking, 239
    skipped blocking, 239
finitely representable, 204, 216, 225, 229

Gordon-Lewis property, 216
Gowers' block Ramsey theorem, 217
Gowers' dichotomy theorem, 217, 252
Gowers-Maurey example, 210, 212–215

homogeneous Banach space problem,
    202, 252
hyperplane problem, 206, 215

isomorphism, 192

Krivine number, 240
Krivine's Theorem, 206, 212, 235, 239
Kunen-Martin boundedness principle,
    246

L.u.s.t., 255
large subsets, 217
linear span $\langle \cdot \rangle$, 196
L_p
    complemented subspaces, 247, 254

$\mathcal{L}_p$, 254
subspaces of L_1, 198
unconditional basis, 196

Maurey's proof, 219
Maurey-Rosenthal example, 205,
    208–210, 212, 213
Mazur map, 224
    generalized, 224

norm
    explicit, 200, 201, 224
    implicit, 200
norming, 196

ordinal index
    Baire-1 index, 248
    Bourgain's ℓ_1 index, 246
    DBSC index, 250
    $I^+(X)$, 247
    $I_\omega^+(X)$, 247
    Szlenk index, 244, 247
    unconditional index, 247
Orlicz space, 197

projection, 192

quotient, 192

Ramsey theory, 203, 204, 206, 217, 219,
    225, 234, 249
    weakly Ramsey, 220, 221
reasonable pair, 220
Rosenthal's ℓ_1 theorem, 204, 245

Schatten classes, 232
Schlumprecht's space S, 195, 211, 212,
    221
Schreier classes S_α, 199, 226, 227, 231,
    235
Schreier's space, 199
semicontinuous function, 248
separable quotient problem, 195
sets
    analytic, 204
    asymptotic, 223, 225
    co-analytic, 204
    hereditary, 227, 235
    lattice, 223
    pointwise closed, 227
    regular, 227
    separated, 223, 225
    spreading, 227
    symmetric, 223
special functional, 214
spectral index, 231
spreading model, 203, 233, 235

$\alpha - C$, 234
ℓ_p, 233, 235, 248
stabilization
    f stabilizes, 197, 224
    oscillation stable, 197, 225
strictly singular operator, 207, 215, 221
subspace, 194
    block, 194, 203
    complemented, 203
support of a vector, 206

tree
    block, 241
    closed tree, 246
    $\ell_1 - K$ tree, 246
    on X, 246
    well founded, 221
Tsirelson's space T, 194, 210, 226, 231, 237
    boundedly modified mixed, 230
    convexified Tsirelson space, 202
    distortion, 199, 211
    mixed Tsirelson space, 229
    modified, 230
    T^*, 221
    Tsirelson type norms, 200
type-p, 195
types, 198
    ℓ_∞, 238
    ℓ_1, 238
    trivial, 198

uniform Kadec-Klee property, 245
uniformly homeomorphic, 225
universal space, 245, 247

weak Cauchy sequence, 204, 250
weak Hilbert space, 202
winning strategy, 217, 218, 221

Printed in the United States
By Bookmasters